Rethinking Stevin, Stevin Rethinking

Nuncius Series

Studies and Sources in the Material and Visual History of Science

VOLUME 6

The titles published in this series are listed at *brill.com/nuns*

Rethinking Stevin, Stevin Rethinking

Constructions of a Dutch Polymath

Edited by

Karel Davids, Fokko Jan Dijksterhuis,
Rienk Vermij and Ida Stamhuis

BRILL

LEIDEN | BOSTON

Cover illustration: Statue of Simon Stevin on the Simon Stevinplein, Bruges, Belgium. The bronze statue is after a design from 1846 by sculptor Eugène Simonis (Brussels). Courtesy of the photographer Dennis Jarvis, Halifax, Canada.

The Library of Congress Cataloging-in-Publication Data is available online at http://catalog.loc.gov
LC record available at http://lccn.loc.gov/2020030406

Typeface for the Latin, Greek, and Cyrillic scripts: "Brill". See and download: brill.com/brill-typeface.

ISSN 2405-5077
ISBN 978-90-04-43290-1 (hardback)
ISBN 978-90-04-43291-8 (e-book)

This book is printed on acid-free paper and produced in a sustainable manner.

Contents

Preface

In December 2016, the Stevin Centre for History of Science and Humanities at the Vrije Universiteit Amsterdam organized a conference with the title *Stevin Inside Out. New perspectives on Simon Stevin*. At this meeting, held at the premises of the Stichting ter Bevordering van de Notariële Wetenschap in Amsterdam, scholars from Flanders, Germany, Italy, France, Sweden, Britain, the Netherlands and the United States presented papers which analyzed Stevin's methodology and made comparisons between the life and work of Stevin and that of other polymaths in Europe in the sixteenth and seventeenth centuries. The conference was generously supported by the Royal Netherlands Academy of Arts of Sciences, the Dr. C. Louise Thijssen-Schoute Stichting and the CLUE+ Research Institute of the Vrije Universiteit.

This volume is the outcome of that conference. As editors, we are grateful to Marco Beretta and Sven Dupré for their willingness to include this book in the NUNCIUS series and to an anonymous reviewer whose thoughtful comments were very helpful in improving the texts. Finally, we would like to thank Dr Vivienne Collingwood for her excellent editing of the manuscript.

The editors

Illustrations

Notes on Contributors

Marius Buning
is Research Fellow at Dahlem Humanities Center, Freie Universität Berlin.

H. Floris Cohen
is Professor Emeritus in Comparative History of Science at Utrecht University and former Editor of the History of Science Society, responsible for editing its flagship quarterly journal *Isis*.

Karel Davids
is Professor Emeritus of Economic and Social History at the Vrije Universiteit Amsterdam and associate of the Stevin Centre for History of Science and Humanities at the Vrije Universiteit Amsterdam.

David Dunér
is Professor in History of Science and Ideas at Lund University.

Maarten Van Dyck
is Associate Professor in Philosophy at Ghent University.

Fokko Jan Dijksterhuis
is Associate Professor of History of Science and Technology at the University of Twente, Louise Thijssen-Schoute Professor of Early Modern History of Knowledge at the Vrije Universiteit Amsterdam and board member of the Stevin Centre for History of Science and Humanities at the Vrije Universiteit Amsterdam.

Charles van den Heuvel
is Head of Department of History of Science and Scholarship at the ING Huygens Institute of the Royal Netherlands Academy of Arts and Sciences and Professor of Digital Methods and History, especially the History of Knowledge and Information Science, at the University of Amsterdam.

Julia Kursell
is Professor of Musicology at the University of Amsterdam.

Jan Lazardzig
is Professor of Theater Studies at the Freie Universität Berlin.

Pietro Daniel Omodeo
is Associate Professor at the Department of Philosophy and Cultural Heritage of the Ca' Foscari University of Venice.

Catherine Secretan
is Directrice de Recherche Emeritus at the Centre National de la Recherche Scientifique, Paris.

Ida Stamhuis
is Associate Professor Emeritus at the Vrije Universiteit Amsterdam, associate of the Stevin Centre for History of Science and Humanities at the Vrije Universiteit Amsterdam and former Honorary Professor at Aarhus University.

Rienk Vermij
is Professor at the Department of History of Science at the University of Oklahoma.

INTRODUCTION

Simon Stevin, Polymaths and Polymathy in the Early Modern Period

Karel Davids, Fokko Jan Dijksterhuis, Rienk Vermij and Ida Stamhuis

Simon Stevin (1548–1620) was a slightly older contemporary of Galileo, and can in many respects be considered his Dutch counterpart: a scholar who opened up new, mathematical, ways of studying the world. As such, he has rightly taken his place in the annals of early modern science. But Stevin was more than a clever theoretician; he was also an exponent of a new approach to knowledge in the early modern period, when established ways of understanding the world had been thrown into doubt. Ignoring these broader aspects, as has commonly been done, risks failing to grasp his full significance. Stevin remains badly understood, in many respects, and this volume is thus an attempt to understand him better by placing his work in this broader context.

In his approach to a wide range of topics, Stevin presents a case – and a very interesting one – of a more general trend in early modern learning, when knowledge was often produced by 'polymaths' rather than specialists. Analyzing his situation is therefore not only important as a means of assessing his aims and motives, but also leads to a better understanding of European intellectual history in general. Such a study will shed light on changes in mathematics and natural philosophy in the early modern period, and it will also contribute to a better understanding of the transformation of European learning.

Starting with Stevin, this collective volume aims to analyze the transformation of the European ideal of knowledge in the early modern period by comparing Stevin with similar multi-talented individuals ('polymaths') in the Low Countries and other European countries, and by examining the relations between the multiple strands of Stevin's thinking, writing and other activities.

1 Simon Stevin: a Brief Life

Most of what we know about Stevin has been derived from his many published works; there are hardly any letters or other sources of a more personal nature. His career can be reconstructed from official documents, but even here there

 | DOI:10.1163/9789004432918_002

are considerable gaps.[1] His life unfolded against a background of political turmoil in the sixteenth-century Netherlands, a conglomerate of disparate territories united under the House of Habsburg. The absolutism and religious policies of Philip II, the Habsburg ruler and king of Spain, who never returned to the Netherlands after he left in 1559, provoked revolts in the 1560s and 1570s, leading to a prolonged civil war. This eventually resulted in the separation between the northern provinces – which gained their independence as the Dutch Republic – and those in the south, roughly equivalent to modern-day Belgium, which continued under Habsburg rule for another two centuries.

Stevin was born in the south in or around 1548, in Bruges in the county of Flanders. He was the illegitimate son of Anthuenis Stevin and Catelyne vander Poort, who both came from respectable, well-to-do families. Stevin clearly received a good education (he later proved to know Latin and French) and became an office clerk for the 'Vrije van Brugge', one of the four constituent parts of the county. Bruges initially sided with the Habsburgs, but in 1578 the city was taken by the rebels, who imposed their Calvinist religion. Among the local leaders of the Revolt were members of the Sayon family, to whom Stevin was related by marriage; after Stevin's birth, his mother had married a certain Joost Sayon. Although there is no evidence that Stevin shared the ideas propagated by his stepfather's family, the political turmoil may well have contributed to his decision to leave his native Flanders.

In 1581, Stevin turned up in Leiden in the province of Holland, in the north of the Netherlands. There are some indications that he may have travelled in Europe before settling there, but these cannot be confirmed. Holland had been a major battlefield in the initial phase of the revolt, but by 1581 the hostilities had moved further south. Holland was now establishing itself as an independent territory and the nucleus of a new nation. Its economy was booming. Stevin settled there as a kind of civil engineer specializing in designing and constructing windmills, which were important for water drainage in Holland's polders. From 1584 onward he obtained several patents, and in 1588 he entered into a business venture with the Delft regent Jan Cornets de Groot, the father of Hugo Grotius.

At the same time, he had clearly set his sights on a more scholarly career. Immediately upon his arrival in 1581, he started publishing an impressive array

1 The standard biography is Dijksterhuis, *Simon Stevin* (1943) (in Dutch; Dijksterhuis, *Simon Stevin* (1970) is an abbreviated English version), but this is mostly a discussion of his works. Details about his life are dispersed through the literature. The more recent popular biography by Devreese and Vanden Berghe (2003) also presented some new information.

of books, most of which were on mathematical subjects. His books initially dealt with practical matters, but from the outset one can perceive an attempt to address these issues from a more general perspective (see the section on the *consten* below). The very first book, published under the name Simon Stephanus, proposes a general method for calculating each person's share in a joint venture.[2] The second, published the following year, is a book of interest tables. Stevin's stated motivation was to publicize the rules by which interest was calculated, for the common good. This rhetoric of serving the common good would also appear in his later works.

In the works he wrote in the following years, the theoretical approach became increasingly dominant. One of his best-known works, his propagation of decimal fractions (1585), was still mostly practical. On the other hand, topics that were typical of mathematical practice at that time, such as gnomonics, calendar-making, gauging and surveying, were left aside completely. Instead, Stevin wrote on problems in pure geometry, arithmetic, and the principles of statics and hydrostatics. This is not to say that he skipped topics that were potentially useful; on the contrary, he repeatedly stated that theory was justified because it could be put into practice. Such topics were addressed, however, on a theoretical level. This approach would earn him a reputation among historians of science as an early proponent of mathematical science. In particular, his work on statics and hydrostatics is generally considered a milestone in early modern science.

Without denying that these works were important, however, we should not forget that Stevin also wrote on subjects that lay entirely outside the mathematical disciplines: dialectics, politics and language, to name but a few. In his early years in Leiden, Stevin lived at the house of the rector of the Latin school, Nicolaus Stochius. In 1583, he enrolled in the newly-founded university that had opened in Leiden in 1575, to allow the rebel province to train its own elites henceforth. Leiden was still a small place, and Stevin rubbed shoulders with important scholars such as Lipsius and, later, Scaliger; something that left a clear impression on his work. On the other hand, it seems that these humanistically-oriented scholars did not take him very seriously.[3] Modern historians have tended to focus on Stevin's contributions to mathematics and engineering, and have paid less attention to these other fields. Nevertheless, these works are an important key to understanding Stevin's place in the world of learning as a polymath, rather than solely as a mathematician.

2 This work was only identified as a work by Stevin in 1996; see Oomes, 'Onbekend werk'.

3 For Stevin and humanism, see in particular De Smet, 'Simon Stevin'.

It was probably at Leiden, too, that he first met the man who would have a major impact on his career: Count Maurits of Nassau, the son of William, Prince of Orange, the leader of the Dutch Revolt. Born in 1567, Maurits had a passion for mathematics, and he enrolled at Leiden University in 1583. After the assassination of his father in 1584, he became stadholder of Holland and several of the other northern provinces. Moreover, as Captain-General of the army of the United Provinces, he was the young Republic's military leader in the ongoing war with Spain and something akin to a head of state. His position fell far short of that of a prince, however, in that he had no say in financial matters. The provincial states jealously kept the task of collecting taxes and making budgets to themselves.

Starting sometime in the 1590s, Stevin and Maurits began to work together closely. This marked a turn in Stevin's career. He became a main advisor to the stadholder on military and engineering matters, and he also re-organized the book-keeping of Maurits' domains. Consequently, his attention shifted back to practical issues, now mostly of a military nature. In the later part of his career, he published on fortification (one work appearing in 1617 claims to present the stadholder's 'order and use' in these matters) and on navigation, among other things. In 1600, Stevin famously drew up the curriculum of the engineering school that Maurits established at Leiden University, where land surveyors, military engineers and other practitioners received mathematical training in the vernacular.[4] In 1604, he obtained a formal position in the army of the United Provinces as quartermaster general. Until that time, he appears to have drawn a stipend as a member of Maurits' household.

During these years, the pace of Stevin's publications abated somewhat, but his scholarly pursuits were certainly not put to rest. Nor did he completely abandon his more theoretical interests. Stevin served as a private teacher of mathematics to the stadholder, and in 1606–1608 he published his *Wisconstighe ghedachtenissen* [Mathematical Memoirs], a compendium of mathematical works. As the title reads, it contained 'the things wherein his Excellency, Maurits, Prince of Orange, has exercised himself'. Indeed, many (though definitely not all) sections include comments made by Maurits, testifying to the interaction between Stevin the teacher and Maurits the pupil.

The *Wisconstighe ghedachtenissen* is a very mixed bag. It contains highly theoretical treatises, for example on the constitution of the universe (defending the Copernican system), on the tides, on perspective and so forth, as well as purely practical treatises on book-keeping, horse bridling, and a design for

4 See the chapter by Dijksterhuis; also Wiesenfeld, 'Duytsche Mathematique'.

a folding scoop for military use. In some sections, notably on mechanics, he re-used some of his earlier work. It also contains Stevin's exposition of the 'Wijsentijt' ('Age of the Sages'), on the nature and origin of wisdom. This rather notorious meditation is put in a new historical perspective by several authors of this volume (see below).

At the end of his life, Stevin appears to have been reasonably well off. He bought a house in The Hague and married a young woman, who had already borne him four children. He died in 1620 on an unknown date; his widow later married a local bailiff. The fate of his manuscripts is not altogether clear. Some remained with his widow and were studied by Isaac Beeckman and other scholars. A number of them were published by Stevin's son Hendrick; the originals have since been lost.[5] In 1636, however, Hendrick complained that the papers in his possession were incomplete, and surmised that several of them were still in the hands of the stadholder.[6] In the nineteenth century, some manuscripts were found among the papers of the seventeenth-century virtuoso Constantijn Huygens. They included two versions of Stevin's discourse on music, a work that was not then known in printed form.

Stevin's legacy largely remains an open issue. As of yet, there has been no systematic study of the reception of his works in the early modern period. Stevin's 'influence' at this time appears to have been largely indirect, in particular through the adoption of some of his ideas by Beeckman and the subsequent exchange with Descartes. Certainly, no 'Stevin school of thought' emerged.[7] When, in the nineteenth century, Dutch politicians and school teachers wanted to bolster Dutch national feeling and trawled history for paragons of national virtue, Stevin became one of their heroes. He has formed part of the public consciousness ever since. He was presented in a very narrow way, however; apart from his close collaboration with Prince Maurits in the struggle against the Spanish, just a few of his accomplishments were highlighted, and not always the most important ones – such as the construction of the sailing chariot. His propagation of the Dutch language also drew attention, especially in his native Flanders, where the struggle between Dutch and French speakers was still continuing. In the twentieth century, historians started to study Stevin's intellectual legacy more seriously. In 1943, the eminent Dutch historian of science E.J. Dijksterhuis, who had already included a section on Stevin in his book *The Mechanization of the World Picture*, published the first real scholarly biography, in which he surveyed and summarized all of Stevin's published works. Although

5 Hendrick Stevin, *Wisconstich filosofisch bedrijf* (1667) and *Burgherlicke stoffen* (1649).

6 Schouteet, 'Brief'.

7 Van Berkel, *Beeckman*; Schuster, *Descartes-Agonistes*.

he himself did not do any archival research, he obtained some information by mail from the Bruges archives, which he also included in his work. Dijksterhuis did discuss Stevin's theories of language and the 'Wijsentijt', but his main interest was in Stevin's contribution to the mathematical sciences. This biography undoubtedly contributed to Stevin's image as the 'Dutch Galileo'.

The renewed interest led to the publication, under the auspices of the Royal Netherlands Academy of Arts and Sciences, of the *Principal Works of Simon Stevin*, edited in five volumes between 1955 and 1966. The volumes contained a selection of Stevin's works, published both in facsimile and in an English translation. The various chapters were all edited by specialists in the respective fields, Dijksterhuis among them. This was a major event, but its impact was not entirely positive. The *Principal Works* had an effect on the Stevin historiography similar to that of the earlier *Oeuvres Complètes* in the case of Christiaan Huygens. Although the edition made many of his writings easily accessible, it also approached them from a rather limited perspective. The *Principal Works* divided up his works in accordance with modern disciplinary boundaries – sometimes even splitting individual publications – and portrayed him first and foremost as a 'mathematician' and engineer in the modern sense, focusing on traditional questions of discovery and theory development. It thereby emphasized Stevin's role in the history of mathematics and mechanics, particularly in the context of the sixteenth-century 'Archimedean turn' (a field in which he certainly was important).

Indeed, recent work on Stevin has largely focused on his accomplishments in the mathematical sciences. This has tended to take the form of case studies on specific subjects. For example, there have been thorough studies of Stevin's theory of perspective (Andersen), his music theory (Cohen, Rasch), and his ideas on architecture (Van den Heuvel). Other aspects have been covered in wider overviews, such as his ideas on geometry (Bos) and astronomy (Vermij). Due to his propagation of the Dutch language, there has long been a certain interest in Flanders and Holland in his linguistic work, too. De Mare analyzed Stevin's system of architectural knowledge as part of the general development of rules and conventions on images in the Dutch Golden Age. However, it has proven rather hard to paint a picture of the man himself: what moved him and what gave coherence to his work. The volume edited by Elkadhem and Vanden Berghe presented some archival material coupled with a few enlightening pieces of innovative historical interpretation, but offered no fresh perspective on Stevin.[8]

8 Elkhadem and Vanden Berghe, *Simon Stevin*. This work was published simultaneously in a French and a Dutch edition. The new information from the archives had earlier been published in Devreese and Vanden Berghe, '*Magic is No Magic*'.

Most scholars would nowadays agree that Stevin's relationship with humanist scholars in Leiden, and with humanist scholarship in general, holds one of the keys to his personality. Since Stevin's attitude appears to have been ambiguous in many respects, however, it has proven hard to pin down. Some bold new interpretations have been advanced. The historian of architecture Ed Taverne has claimed that in his *Castrametatio,* Stevin followed a geometrical order based on the Bible. More recently, in an unpublished dissertation, Kees Schilt has argued that an important Hermetic influence pervaded much of Stevin's work. These suggestions are still awaiting further substantiation.[9]

If we want to understand Stevin in his own context, we need to move beyond the narrow horizon of mathematics and engineering and avoid simple monocausal explanations. Stevin was active in many other fields. In that respect, he was similar to many early modern thinkers, who often pursued multiple interests. We argue that studying Stevin as a 'polymath' and acknowledging his place in fields such as the history of stoicism, Ramism and alchemy is more fruitful than defining him a priori as a mathematician.

This volume thus intends to move beyond the standard image of Stevin enshrined in the *Principal Works* by placing him in a broader European context, and widening the perspective by considering other aspects and connections of his work. Hence the title: 'Rethinking Stevin, Stevin rethinking'. We offer new insights into the phenomenon of 'polymaths' and 'polymathy' in general, and in the case of Simon Stevin in particular. The volume contains ten essays, grouped in two parts: 'Polymaths and polymathy in comparative perspective' and 'Stevin thinking: searching for keys in the *consten*'. Before discussing the individual contributions, let us first take a closer look at the terms 'polymath', 'polymathy' and related concepts.

2 Polymaths and Polymathy: a Conceptual Survey

Stevin is an example of a well-known type of early modern intellectual: the polymath. Although the terms 'polymath' and 'polymathy' did not become commonplace until the first half of the seventeenth century, the figure of the polymath had existed long before that time. The Hamburg philologist Johannes Wower, the first to publish a book with 'polymathy' in its title (*De polymathia tractatio,* 1603), considered Eratosthenes of Cyrene, who flourished in the third century BC, to be the quintessential polymath. Robert Burton, who with *The*

9 Taverne, *In 't land van belofte*; Schilt, *Simon Stevin.*

Anatomy of Melancholy introduced the word 'polymath' into written English in 1621, saw the perfect polymath as embodied in the pre-Socratic philosopher Democritus (c. 460–370 BC).[10] Present-day historians have identified polymaths in the Roman Republic and the Islamic world, including Marcus Terentius Varro (116–24 BC) and Majd al-Dīn al-Fīrūzābādī (1329–1415), as well as in Renaissance Italy, such as Leon Battista Alberti (1404–1472), Leonardo da Vinci (1452–1519), Cosimo Bartoli (1503–1572) and Girolamo Cardano (1501–1576).[11]

The Renaissance undoubtedly produced a congenial environment for polymaths. Wower's definition of 'polymathy' as the 'knowledge of diverse things, drawing on every kind of discipline and ranging very widely', chimes perfectly with the ideal of the 'universal man' that was articulated in fifteenth-century Italy and subsequently spread across Europe.[12] By the seventeenth century, however, the idea of general learning gradually became qualified and even criticized. 'Knowing everything' was no longer regarded as an aim worth pursuing per se; general learning was not necessarily always a good thing. Wower himself warned of the danger of polymathy turning into pedantry. Burton, in turn, was at pains to separate polymathy from polypragmatism; that is, to distinguish open-minded hunger for knowledge from pretentious claims to omniscience. In the eighteenth century, Diderot's *Encyclopédie* made a distinction between 'true' and 'fake' polymathy. 'True' polymathy, in the eyes of the Encyclopedists, was broad, thorough erudition regarding a large number of subjects, employed in a functional way, whereas 'fake' polymathy often consisted of little more than a confused mass of useless tidbits of knowledge, which were dished up for show regardless of their relevance. Polymathy in the true sense of the word was said to be exemplified by scholars such as Lipsius, Scaliger, Salmasius, Grotius, Kircher and Leibniz.[13]

The rise and decline of the polymath, and the question of the similarities or differences between polymathy and other concepts denoting general learning such as *polyhistorie* or encyclopedism, have been the subject of some historical discussion in recent years. Historians agree that polymaths multiplied from the Renaissance onwards, but that they had become a much rarer species by the end of the seventeenth century. Opinions still differ regarding who deserves

10 Dietz, 'Ioannes Wower', 142–149; Murphy, 'Robert Burton', 282–286.

11 *Varro Varius*; Strotmann, *Majd al-Dīn al-Fīrūzābādī*; Grafton, *Leon Battista Alberti*; Isaacson, *Leonardo da Vinci*; Bryce, *Cosimo Bartoli*; Maclean, 'Girolamo Cardano'.

12 Burke, *Social History of Knowledge. From Gutenberg to Diderot*, 84–85. Omodeo and Friedrich, *Duncan Liddel*.

13 Dietz, 'Ioannes Wower', 149; Murphy, 'Robert Burton', 282–286; Jaumann, 'Was ist ein Polyhistor?', 77, n. 2; 'Polymathie', in: Diderot and d'Alembert, eds., *Encyclopédie*, XII, 944.

the title of 'the last man who knew everything'. In historical accounts, Athanasius Kircher (1602–1680) and Wilhelm Leibniz (1646–1716) compete for the honor with Thomas Young (1773–1829), Alexander von Humboldt (1769–1859), Joseph Leidy (1823–1891) and Enrico Fermi (1901–1954). In the nineteenth century, the 'intellectual climate' was doubtless 'becoming more hostile to polymaths', as Peter Burke has observed. Thomas Young, for example, even published some of his articles anonymously to avoid being treated as a quack by his fellow physicians.[14]

As for the matter of conceptual kinship, historians view polymathy in the early modern period as being related, but not quite identical, to *polyhistorie* or 'encyclopedism'. The term 'polyhistor' became common after the publication of Daniel Georg Morhof's *Polyhistor* in 1688 (with expanded and revised editions appearing until 1747), which was both a guidebook to a vast body of literature and a manual on how to read and study texts. Polyhistors, according to Anthony Grafton's definition, sought both 'eloquence' and 'erudition'. They aspired to study and practice the art of rhetoric, and to acquire and organize as much knowledge as possible. As far as this latter ambition was concerned, *polyhistorie* was somewhat akin to 'encyclopedism', a term that originally referred to a drive to achieve comprehensive education, before it became associated in the eighteenth century with the compilation of a particular kind of book.[15] However, consensus about the exact meaning of *polyhistorie* remains a distant goal, as Hole Rössler has explained in a recent survey. Apart from the broadly shared assumption that *polyhistorie* formed part of the culture of scholars, interpretations of the phenomenon differed among early modern writers, and continue to do so among historians today. One of the reasons for this diversity of interpretations, as Rössler points out, lies in the fact that what German scholars call 'polyhistorie' is described as 'polymathie' in other European languages, even if the concepts are not quite identical. Another is that notwithstanding the frequent appearance of the title 'polyhistor' before the end of the seventeenth century, a discourse on 'polyhistorie' was notably absent.[16]

Moreover, studies on polymaths show that not all of them were learned in the same fields. Even 'all-knowing' people had their specialties. Some of them were eminently learned in languages, medicine or natural philosophy, others

14 *Athanasius Kircher*; Hockney, *The Last Man Who Knew Everything*; Robertson, *The Last Man Who Knew Everything*; Warren, *Joseph Leidy*; Schwartz, *The Last Man Who Knew Everything*; Burke, *Social History of Knowledge*, vol. II, 161–162.

15 Blair, *Too much to know*, 260–262; Grafton, 'The World of the Polyhistors', 33–37; Murphy, 'Robert Burton', 286–290; Jaumann, 'Was ist ein Polyhistor?', 76–84.

16 Rössler, 'Polyhistorie und Polymathie', 635, 644–645, 659–660, 673.

were highly accomplished in mathematics, architecture or engineering. Cardano, Lipsius and Scaliger boasted different kinds of strengths from Alberti, Da Vinci and Leibniz. Yet, in the eyes of contemporary observers and present-day historians, all of them can be described as polymaths.

Although consensus on the precise meaning of concepts such as 'polymath' or 'polyhistor' thus remains elusive, there can be no doubt that the ideal and practice of polymathy in Europe reached its zenith in the late sixteenth and early seventeenth centuries, the very period when Simon Stevin was living and working. Like other polymaths, Stevin was active in a wide range of fields, but he *did* have distinct focal points of interest. He was more concerned with mathematics, music and engineering, for example, than with medicine or botany. Religion and metaphysics were largely absent from his writings. Arguably, Stevin was both an outstanding figure in the history of knowledge in his own right and an 'ordinary' polymath. Moreover, he had ideas of his own about the nature and structure of knowledge, introducing the term 'wisconst' for 'mathesis', the art of certainty (see below). Comparing Stevin with other polymaths and interpreting his work in the context of polymathy may help us, to some extent, to gain more insight into his thinking, aspirations and actions. On the other hand, studying the particular case of Stevin may also contribute to our understanding of polymaths and polymathy in general, even if there is no widespread agreement on the precise meaning of these terms. We can learn much by taking such a dual perspective, and that is exactly what this volume of essays sets out to do.

3 Polymaths and Polymathy in Comparative Perspective

The first part of this book contains three studies on 'polymaths' in different countries in early modern Europe, allowing us to view Simon Stevin's activities in comparative perspective. Pietro Daniel Omodeo discusses the courtly identity of the Italian mathematical philosopher Giovanni Battista Benedetti, Jan Lazardzig studies the oeuvre of the mid-seventeenth-century German architect, engineer and merchant Joseph Furttenbach the Elder, and David Dunér analyzes the genesis of a Swedish polymath around 1700, Christopher Polhem. Like Stevin, all of these figures engaged to a greater or lesser extent in mathematics and practical mathematics, but they also concerned themselves with a wide range of other fields.

In the light of the continuing differences of opinion about the meaning of 'polymaths' and 'polymathy' outlined above, one of the questions that is often debated is the extent to which, and in what ways, the label 'polymath' is really distinctive or appropriate. Stephen Johnston has suggested that 'polymathic authorship' in Stevin's time may have been more 'routine', and thus less extraordinary, than we as historians living in an age of specialization perhaps tend to assume. Thomas Blundeville, for example, did not differ greatly from Simon Stevin in the range of his interests and pursuits, yet he did not fashion himself as an exceptionally learned human being.[17] Lazardzig argues that according to the characteristics identified by Hole Rössler, Joseph Furttenbach, despite his training as a 'Vitruvian polymath', did not entirely qualify as a 'polymath' himself. After all, Furttenbach lacked proficiency in Latin, mathematics and philosophy, and he did not transcend the disciplinary borders of knowledge.

Which conditions facilitated the rise of the polymath in early modern times? Johnston has drawn attention to the 'normal' features of the cultural background to polymathy.[18] In his opinion, polymathic ambitions were advanced by ideas that were quite broadly shared in contemporary societies, such as the view that mathematics was both a practical and an intellectual pursuit (and that science, technology and mathematics were not separate categories), the belief that methodology should show a balance between reason and experimentation, and the conviction that an 'active life in service of the commonwealth' was an eminently laudable goal. By contrast, in his contribution to this volume, Pietro Daniel Omodeo stresses the differences in the 'intellectual positioning' of polymaths, despite their shared areas of interest, which in his view arose from a fundamental divergence in social backgrounds and political contexts. Even though Benedetti and Stevin worked on 'similar problems of mathematics, physics and technology', he argues, their self-perception and ambitions were very different. The former liked to see himself as a court philosopher, the latter took pride in his status as an engineer. These contrasting identities, Omodeo suggests, were related to differences in the dominant economic and social forces in the societies in which Benedetti and Stevin lived and worked – the landed aristocracy in Piedmont and the mercantile bourgeoisie in the Dutch Republic, respectively – and this disparity, in turn, resulted in differences in their epistemic values and styles of thinking. Lazardzig and Dunér likewise underline how different social and political environments shaped the ways in which polymaths could rise and thrive. A polymath in an urban, commercial community such as Ulm evolved very differently from a

17 In a paper entitled 'Fellow travellers? Polymathy and community in the English mathematical arts', presented at the conference 'Stevin Inside Out' in Amsterdam, December 2016, which is not included in this volume.

18 Ibidem.

polymath in a society such as Sweden, where the centralized state had a profound impact on the development of the economy.

In an essay on the Scottish polymath Duncan Liddel (1561–1613), John Henry suggested that 'polymathy may have seemed the best course to pursue in order to try to make a decent living', especially for those who 'wanted to pursue a career in the mathematical sciences'. Mathematics alone would not have been lucrative enough.[19] Whether polymathy at this time was really a strategy born of necessity is somewhat questionable, however, considering the fact that most mathematicians did not fashion themselves as 'polymaths' and were not viewed as such by contemporaries, either. The Dutch Republic in the late sixteenth and early seventeenth centuries was home to countless mathematicians who, unlike Simon Stevin, did not evolve into polymaths. And why would they have needed to do so, in any case? The 'mercantile' and 'bourgeois' society of the Northern Netherlands, as Omodeo reminds us, offered sufficient opportunities for mathematicians or engineers to make 'a decent living'. There was a rapidly-growing demand for knowledge and skills in surveying, navigation, fortification, gauging and bookkeeping.[20] As Lazardzig points out, in the case of Joseph Furttenbach in Ulm, having a self-assumed identity as a 'universal' expert in all matters relating to architecture was not a means of generating income, either; nor would this have been necessary, as a well-established and well-connected merchant. As Dunér makes clear, Christopher Polhem should not primarily be seen as a mathematician branching out into other fields of knowledge, but as a man of 'indecisive social identity, between craftsman and nobleman, between illiterate and savant', who used the opportunities offered by the mercantilistic policies of the Swedish state to fashion himself into a polymath. As all four case studies demonstrate, what a polymathic identity could help to improve, at least, were the chances of a talented individual obtaining patronage or climbing the social ladder.

How did one become a polymath? The few scraps of information relating to the education of Simon Stevin – his training as a clerk in Bruges, his travels, his enrollment as a student at Leiden University – give us an inkling as to the sheer variety of ways in which a polymath could be formed. Lazardzig documents how Joseph Furttenbach's knowledge of mechanics, perspective and the art of civil and military architecture grew out of a mixture of very different sources. Although this Swabian merchant-author-architect lacked a grounding in Latin, he received an education at a German school, travelled

19 Henry, 'A Pragmatic Aspect of Polymathy', 93.

20 Davids, *Rise and Decline*, 426–431.

extensively in Italy, served an apprenticeship with a merchant and a master in artillery, and attended a private academy in Florence, where he learned about the theory and practice of mechanics and architectural design and mingled with artists and noblemen's sons. The English polymaths Thomas Harriot and Edward Wright did not only follow the 'conventional route to learning' by going directly from school to university, but also, unusually, combined this academic training with practical experience at sea. The full range of possible ways to make a polymath is shown in Dunér's essay on Christopher Polhem. Elementary schooling, practical experience in artisans' workshops, self-study by reading books, private tutoring in Latin, attendance at university, extensive travel abroad, reverse engineering of mechanical objects – all of these steps helped to hone Polhem's cognitive abilities as a versatile and prolific inventor and natural philosopher.

The Renaissance did not only create a favorable environment for polymaths, but it also saw a significant change in the formation and expression of identity, which Stephen Greenblatt has called 'self-fashioning'. Self-fashioning can be defined as 'an increased self-consciousness about the fashioning of human identity as a manipulable, artful process'.[21] Polymathy and self-fashioning were not unrelated. Girolamo Cardano's autobiography *De propria vita,* for example, illustrates this precise phenomenon: his self-fashioning by boasting about his productivity as a polymath and explaining 'his state of mind, his motivations, and his most mature reflections on life'.[22] In a sense, the case studies by Omodeo, Lazardig and Dunér in this volume confirm this connection between polymathy and self-fashioning, too. Benedetti portrayed himself as a 'court philosopher', Furttenbach presented himself in his publications and *Kunstkammer* 'as a person who [was] universally knowledgeable in architecture and engineering', and Polhem fashioned himself in his autobiography 'as the poor orphan, who with ingenuity, talent and diligence alone, worked hard on his own in a grudging and hostile environment'. Compared to these individuals, Simon Stevin was perhaps an unusual polymath, in the sense that he did not expressly construct an image of himself for the outside world. He did not write an autobiography. Stevin's relative lack of 'self-fashioning' is one of the very reasons why we know so few details about his life. And yet, he was perfectly aware of what he was doing and what he was worth. This Dutch polymath displayed a high level of reflection on his thoughts and actions, and it is to this aspect of Simon Stevin that we now turn.

21 Greenblatt, *Renaissance Self-fashioning*, 1–2.

22 Maclean, 'Girolamo Cardano', esp. 595.

4 Stevin Thinking: Searching for Keys in the *Consten*

The essays in the second part of the volume take a closer look at Simon Stevin's ideals, motives and methodologies from the perspective of the *consten*, a term that literally means 'the arts' (but see further below). Approaching Stevin from the perspective of 'present-day mathematics', which was the main focus of the *Principal Works* and affiliated publications, emphasized his originality and contribution to the sixteenth-century revival of Archimedes. This raised interesting questions about engineering, theory-practice relationships and so forth. These questions were often informed by modern concerns, however, and were less sensitive to the specific historical context in which Stevin himself was thinking, writing and acting.[23]

Approaching Stevin's oeuvre from the perspective of the *consten*, by contrast, emphasizes sixteenth-century debates on scholarship and contested conceptions of learning. *Const* (art) can and should be understood in a broad and integrated sense: broad, because it spans modern disciplinary distinctions, as exemplified by Stevin's simultaneous writing and publishing in *wis-* and *rede-const* (from mathematics to dialectics); and integrated, by starting from the category of 'const' and inquiring how Stevin understood this, and how this positioned him in sixteenth-century debates on and the development of the arts, *artes* and knowledge. As we shall see, *consten* are not identical to the arts or *artes*. Simon Stevin occupied a delicate position, because he was searching for a language to understand and interpret *consten* as well as practicing some of them himself. The importance of this concept for Stevin is clear from the various titles of his works that include the term *consten*. Indeed, his oeuvre can be seen as a program for the *consten*. In this way, the polymathic character of Stevin's oeuvre can be turned inside out, starting from the 'vernacular scholar' aspect rather than mathematics. Mathematics was only one of Stevin's many interests as a polymath. The *consten* perspective thus sheds new light on the questions and themes that played a role in the older historiography on Stevin, such as the rise of engineering science and the relationship between theory and practice.

Not only does the *consten* perspective offer a key to understanding Stevin, it also provides a link with recent historiography on early modern knowledge cultures. Stevin has always been hailed for combining theory and practice,

---

23 Verwey, 'Hoe zag Coornhert zichzelf?', makes a similar point for Dirck Coornhert, who moved in similar circles to Stevin and engaged in similar projects. The historiography on Coornhert has tended to focus on his particular case, with the result that his work has not been connected to the wider context of sixteenth-century political and ethical thinking.

anticipating the modern integration of science and technology, and illustrating the artisanal roots of modern science.[24] Recent historiography has made the use of modern categories such as 'science' and 'technology' problematic when writing about the early modern period, and has tried to transcend dichotomies such as 'scholar and craftsman' as well as modern disciplinary categories.[25] These accounts investigate the place and meaning of material and embodied practices with respect to cognitive and formal ones, approaching the history of early modern knowledge as a theatre of culturally and socially-embedded knowledge practices that interact and compete with one another.[26] The dichotomy of theory and practice is thus largely superseded by questions regarding vernacular epistemologies and knowledge claims from the arts.[27] By emphasizing the epistemic and cultural ambitions of practitioners and non-academic writers, the mathematics of a figure such as Stevin can be placed alongside natural history, anatomy and natural magic.[28] Early modern articulations of the nature and role of knowledge from the arts provide important clues for this new historiography of early modern knowledge.[29]

Stevin's reflections on the *consten* are interesting, but he hardly features in recent literature. This volume makes a case for Stevin as a significant and enlightening example. We employ new perspectives on early modern knowledge history to re-examine Stevin's legacy, asking how his 'wisconsten' can be understood as *consten* and how his ambitions can be understood in the context of Renaissance learning in the vernacular. Both Stevin's oeuvre and the theme of early modern *consten* are complex historiographical issues. Rather than claiming to provide definitive answers, this volume offers a collection of original essays intended to stimulate historical thinking. There are some overlaps in content between the chapters and some differences of opinion between the authors, which we believe enriches the academic debate.

The second part of the volume starts by taking an integrative look at the material by and about Stevin himself. This reconsideration uncovers the blind spots of the *Principal Works*: texts by Stevin that have been left out of or overlooked in the discussion until now. Some of the essays address the neglected

---

24 Struik, *Land of Stevin and Huygens*.
25 Roberts, *Mindful Hand*.
26 Long, *Artisan/Practitioners*; Klein, *Nützlich*.
27 Smith, *Body of the Artisan*; Valleriani, *Structures*.
28 Johnston, 'Mathematical Practitioners' and *Making Mathematical Practice*; Cormack, *Mathematical Practitioners*; Vermij, *Kleine geschiedenis*.
29 Glatigny and Vérin, *Reduire en Art*; Davids, *Rise and Decline*.

works, such as the *Eertclootschrift* [Geography] (Vermij) and the *Bewysconst* [Dialectics] (Buning). Starting from the *consten,* other essays take a fresh look at more well-known works, such as the *Weeghconst* [Statics] (Van Dyck) or the *Singconst* [Art of Singing] (Cohen and Kursell), and place them in the context of Stevin's oeuvre – the *Wisconstighe ghedachtenissen* (Van den Heuvel) – and the ideas of his time – the *Wijsentijt* [period of pristine knowledge or 'age of the sages'] (Dijksterhuis and others) and the *Burgherlick Leven* [Vita Politica] (Secretan). The chapter on Stevin's 'Singconst', which concludes this volume, is literally a retake in the form of a debate, in which Julia Kursell responds to Floris Cohen's earlier account.

An integrative approach is taken to explore new connections between Stevin's various writings and activities. The meditation on the 'Age of the Sages' in the 'Wijsentijt', which has usually been considered an oddity, is presented as a methodological reflection that provides important clues to Stevin's ideas. The puzzle of the 'Wijsentijt' has yet to be solved completely, but the chapters offer some new clues towards its meaning and purport. Van den Heuvel focuses on the place of the 'Wijsentijt' in Stevin's oeuvre as a whole, and Dijksterhuis on the intellectual context of the sixteenth century. Vermij draws attention to the fact that Stevin's discourse on the 'Wijsentijt' was included in his geography and inquires into its function in that context. Likewise, the logic of *Bewysconst* – which is studied here for the first time – turns out to be a key to the *Weeghconst*. Stevin's *wis-* and *redeconst* can be understood as parts of a broader project that had a strong programmatic character. It is tempting to read the *Weeghconst* as an 'essay' on Stevin's program for a vernacular epistemology.

Taking a comprehensive look at Stevin's works quickly yields results, in particular regarding the coherence of the sciences and humanities, or the 'quadrivium' and 'trivium', to use contemporary terms. Buning explains that in 1585, Plantin used a printing privilege from the Council of State for the *Bewysconst*, *L'Arithmetique*, and *De Thiende* [The Tenth]. In Stevin's publication activities, the *redeconst* and the *wisconst* belonged together. Van Dyck clarifies that the contents were also connected: in the *Weeghconst,* Stevin used the concepts and methods of inference that he had elaborated in the *Bewysconst*. Thus, as early as 1585, Stevin seems to have had a coherent vision of the *consten*. Such coherence was also maintained over the years: for example, in the role of the concept of number from the early *L'Arithmetique* to the later 'Singconst' (Cohen and Kursell). Van den Heuvel brings the diachronic element into his discussion of the *Wisconstighe ghedachtenissen*. Building upon his earlier inquiry into the *Huysbou,* he sketches a reconstruction of this complex publication and planned treatises connected to it, as well as its place in Stevin's oeuvre as a whole.[30] Van den Heuvel argues that the *Wisconstighe ghedachtenissen* was the product of the pedagogical turn that Stevin made in the 1590s, moving from the methodological emphasis of his early work to concerns about didactics and the transfer of knowledge. This may have been stimulated by his new role as an advisor to Stadtholder Maurits.

All of the *const* books that Stevin published during his lifetime raise the question of how he understood the *consten*. The chapters offer some new clues for moving towards a reinterpretation. First of all, Stevin's interests were quite focused: singing rather than *musica*, perspective rather than painting. In their discussion of the 'Singconst', Kursell and Cohen emphasize that Stevin was not a practitioner in the modern sense. In addition, he digressed in a curious way from the tradition of *musica* by proposing a correction of the conception of number in the classical theory of consonance. As in the visual arts – the topic of the *Deursichtinge* [Perspective] – he was less interested in the art itself than in reflecting on it. The 'Singconst' did not aim at codification; the goal was to explain the art in a way that artists themselves were unable to. In this regard, Stevin acted as an intermediary between the artist and the patron – Maurits, in this case. This is consistent with the view that Van den Heuvel developed earlier, regarding his architectural works.[31] A similar reading of Stevin's better known works in the mathematical sciences may also help to put his alleged practitioner's background into context. The 'Wijsentijt' and *Eertclootschrift* suggest that we should broaden the category of 'mathematician' to cover a more inclusive group of Archimedean 'ingenuous persons' (Vermij, Dijksterhuis).

If a *const*, for Stevin, was an explanation rather than a codification, the meaning of his well-known adage of combining theory and praxis can be reconsidered as well. Historical and conceptual prudence are needed when interpreting his words – 'spiegheling en daet' (reflection and act). Rather than simply reading them in modern terms of theory and practice (although justified in some cases), we should ask what exactly is being juxtaposed and how the *const* functioned in this respect. Discussing the mechanical works, Dijksterhuis points out parallels with the early modern epistemic distinction between pure and mixed mathematics, whilst Van Dyck notes a comparable disciplinary distinction between Archimedean mathematics and Aristotelian physics. The main point is that *const* should not be placed on either side of the dichotomy, but understood in terms of the process that was known at that

30 Van den Heuvel, *Huysbou*.

31 Van den Heuvel, 'Wisconstighe Ghedachtenissen'.

time as 'reduction to art'.[32] As in the case of the 'Singconst', this was intended to subject practical operations to reflection and bring them into an art.

Van Dyck offers further insights into Stevin's understanding of explanation by highlighting the role of proportional reasoning. Referring to the double meaning in Dutch of 'reden' as both proportion and cause, he argues that the key to causal inference lies in the concept of proportionality. Stevin introduced and systematically used the word 'evenredenheyd' in his analyses. Cohen explains that for Stevin, this Dutch word provided the link between 'ratio' and 'proportio' that Latin could not. Language was central to Stevin's conception of the *consten*. His reflection upon the dignity of Dutch is well-known and formed part of the widespread Renaissance promotion of the vernacular. Marijke van der Wal has placed his ideas in the broader context of sixteenth-century linguistic developments and the renewal of Dutch in particular. Stevin was building upon ideas about monosyllabism and composability to develop concepts such as 'evenredenheyd'.[33]

Even though Stevin could be critical of the methods of humanist scholars, his exposition on the Dutch language makes reference to what was an established humanist discourse at that time, especially that of the scholar Goropius Becanus, who worked mostly in Antwerp a generation before Stevin discussed the properties of an ideal language. Drawing on intricate etymological and linguistic arguments, Goropius argued that this ideal language was actually Dutch. His works were widely known among linguistic scholars in the Netherlands, and although Stevin does not refer to them, it seems obvious that he would have been influenced by them.[34] Like Goropius, Stevin saw language as the key to reality and therefore searched for the language that could function best. Again, this happened to be Dutch. The central role of language in Stevin's epistemology is also clear in his logic and its applications. Thus, Stevin's introduction of new Dutch words to logic, mathematics and so forth primarily served a cognitive goal. Definition was the key factor in the *consten*, being a way to stake out distinct concepts (Van Dyck, Buning). In his epistemological reflection on the 'Wijsentijt', Stevin again emphasized the importance of 'elementating', and promoted Euclid as an exemplar of learning (Dijksterhuis, Van den Heuvel). In his account of the *Weeghconst*, Van Dyck makes it clear that as early as 1585, Stevin's conceptions of inference and causality were systematic, profound and original.

---

32 Van Dyck refers to the work of Verin.

33 Van der Wal, 'Simon Stevin'; Van der Wal, 'Logic, Linguistics, and Simon Stevin'.

34 Frederickx and Van Hal, *Johannes Goropius Becanus*, 187–188 for Becanus' influence on Stevin.

Lying at the heart of Stevin's epistemic considerations is the question of what can be learned from practice: what knowledge can be conveyed from artefacts, operations and skill, and how can this be done? *Const* constitutes the transformation of tacit knowledge into understanding that can provide an artful basis for the *scientia* (Vermij). Conversely, the textual practices of the *const* offered conceptual tools to advance embodied practices. The *Bewysconst* provided understanding of the natural faculties of reasoning in order to control and improve them (Buning). The educational turn that Van den Heuvel identifies in Stevin's later *Wisconstighe ghedachtenissen* pointed towards this issue and its institutional aspects. In the plan for the 'Duytsche Mathematique', he sketched out how the *consten* could be learned and what practitioners could learn from them.[35] In this regard, the *Burgherlick Leven* offers an interesting counterpoint to traditional, learned treatises, in that it presents an 'art' of social living; a mirror of proper conduct, as Secretan points out, albeit one aimed at the citizen rather than the prince.

5 Conclusion

With his program for the *consten,* Stevin formed part of the Renaissance trend towards civil science and vernacular learning. This volume provides a framework for putting his work in historical context, and also suggests some starting points for further research. Secretan builds upon her earlier account of Stevin's politics and its connections to debates on civil society in the budding Republic. Likewise, the dialogue between Stevin and Maurits can be understood in the context of the Republic as a whole, including Friesland, where the stadholder Willem Lodewijk and mathematics professor Adriaan Metius gave shape to the integration of 'academics and idiots' in an idiosyncratic way.[36] Dijksterhuis connects the 'Wijsentijt' to the teachings of Ramus and their reception. The fact that Stevin did not explicitly engage in debates about such doctrines may have contributed to his invisibility in the historiography. Such leads can be explored further, for example in the context of projects on universal language as well as the uses of Dutch in learning. A critical study of his prolific publishing from the perspective of the history of the book may provide insights into the way he mobilized publishers and employed books to achieve his ambitions.

---

35 See also the recent accounts by Wiesenfeldt of the 'Duytsche Mathematique' and Leiden University.

36 Dijkstra, *Between Academics and Idiots.*

The integrative approach to Stevin's oeuvre that we take in this volume also raises questions about the way he can be positioned in his time. Stevin had broad and penetrating interests with specific foci, but this was not academic and merely erudite. If we wish to portray him as a polymath, the emphasis should be on the 'math', and on the particular interpretation Stevin gave to understanding. Rather than knowing a lot, he was interested in the quality of knowledge, in particular the added value that could be gained from the *consten* with respect to practical pursuits. Stevin translated 'mathesis' into 'wisconst' – the art of certainty. As such, he was an artful polymath. His program for the *consten* suggests some further lines of inquiry. Although theological and metaphysical considerations are clearly absent from his works (a fact that is quite interesting in itself), it may be worthwhile investigating the ethical dimensions of his vernacular epistemics (Vermij). Wisdom was generally seen as a key to virtue, and it was also starting to gain a vernacular setting. In the period when Stevin published his *Weegh-* and *Bewysconst,* the Dutch humanist Dirck Volkertsz. Coornhert published a *Zedekunst* in which he elaborated *eudaimonia* as an art that could be acquired by exercise and education.[37]

By rethinking Stevin, we can ask once again whether, and how, he might be considered to exemplify Dutch science. Struik and Van Berkel used him as a linchpin for the history of science in the Netherlands. Building on the *Principal Works,* Stevin was portrayed as a scientist who stood out for his orientation towards practice and for his pioneering work in 'engineering science'.[38] In line with the traditional history of science, content and method played a central role in defining Stevin as a model Dutch scientist. Aside from the image of Stevin as ingenious mathematician, no new synthetic interpretation had been developed to date. This introduction has tried to outline a new approach to Stevin from the perspective of the cultural history of early modern knowledge. The themes of the polymath and the *consten* serve as pointers for rethinking Stevin, in comparative fashion and based on his body of work as a whole, as a cultural actor.

Bibliography

Andersen, Kirsti, "Stevin's Theory of Perspective: The Origin of a Dutch Academic Approach to Perspective", *Tractrix,* 2 (1990) 25–62.

37 Gruppelaar in Coornhert, *Zedekunst.*
38 Struik, *Land*; Van Berkel, *Voetspoor*; Van Berkel, *History of Science in the Netherlands.*

Athanasius Kircher. The Last Man Who Knew Everything, ed. Paula Findlen, New York and London: Routledge, 2004.

Berkel, Klaas van, *In het voetspoor van Stevin. Geschiedenis van de natuurwetenschap in Nederland 1580–1940*, Meppel: Boom, 1985.

Berkel, Klaas van, A. van Helden and L. Palm eds., *A History of Science in the Netherlands*, Leiden and Boston: Brill, 1999.

Blair, Ann M., *Too Much to Know. Managing Scholarly Information before the Modern Age*, New Haven and London: Yale University Press, 2010.

Boer, Pim den, ed., *Simon Stevin. Het burgherlick leven. Vita Politica*, Utrecht: Erven Bijleveld, 2001.

Bos, Henk J.M., "Simon Stevin, wiskundige", in: Vanden Berghe et al., *Simon Stevin 1548–1620: De geboorte van de nieuwe wetenschap*, Turnhout: Brepols, 2004, 49–61.

Bryce, Judith, *Cosimo Bartoli (1503–1572). The Career of a Florentine Polymath*, Geneva: Librairie Droz, 1983.

Burke, Peter, *A Social History of Knowledge*, vol. II, *From the* Encyclopedie *to* Wikipedia, Cambridge: Polity Press, 2012.

Burke, Peter, *A Social History of Knowledge. From Gutenberg to Diderot*, Cambridge: Polity Press 2000, 84–85.

Cohen, H. Floris, *Quantifying Music. The Science of Music at the First Stage of the Scientific Revolution, 1580–1650*, Dordrecht: Reidel, 1984.

Coornhert, Dirck Volckertsz., *Zedekunst, dat is wellevenskunste* (1586). Translated, introduced and annotated by J. Gruppelaar, Hilversum: Verloren, 2015.

Cormack, Lesley B. et al. (eds.), *Mathematical Practitioners and the Transformation of Natural Knowledge in Early Modern Europe*, Cham: Springer, 2017.

Davids, Karel, *The Rise and Decline of Dutch Technological Leadership. Technology, Economy and Culture in the Netherlands, 1350–1800*, Leiden and Boston: Brill, 2008.

Devreese, Jozef T. and Guido Vanden Berghe, *"Magic is No Magic": The Wonderful World of Simon Stevin*, Southampton: WIT, 2008.

Diderot, Dennis and Jean le Rond d'Alembert, *Encyclopédie ou dictionnaire raisonné des sciences, des arts et des métiers*, Paris: Le Breton, Durand, Briasson, David, 28 vols., 1751–1772.

Dietz, Luc, "Ioannes Wower of Hamburg, Philologist and Polymath. A Preliminary Sketch of his Life and Works", *Journal of the Warburg and Courtauld Institutes*, 58 (1995) 132–151.

Dijksterhuis, Eduard Jan, *Simon Stevin*, The Hague: Martinus Nijhoff, 1943.

Dijksterhuis, Eduard Jan, *Simon Stevin. Science in the Netherlands around 1600*, The Hague: Martinus Nijhoff, 1970.

Dijksterhuis, Fokko Jan, "Duytsche Mathematique and the Building of a New Society: Pursuits of Mathematics in the Seventeenth-Century Dutch Republic," in: eds. Lesley B. Cormack, John A. Schuster and Steven A. Walton, *Mathematical*

Practitioners and the Transformation of Natural Knowledge in Early Modern Europe. Studies in History and Philosophy of Science; Vol. 45. Berlin: Springer, 2017, 167–181.

Dijkstra, Arjen, *Between Academics and Idiots: A Cultural History of Mathematics in the Dutch Province of Friesland (1600–1700)*, PhD dissertation, Enschede, University of Twente, 2012.

Elkhadem, Hossam and Wouter Bracke eds., *Simon Stevin 1548–1620: L'émergence de la nouvelle science*, Turnhout: Brepols, 2004.

Frederickx, Eddy and Toon van Hal, *Johannes Goropius Becanus (1519–1573). Brabantse arts en taalfanaat*, Hilversum: Verloren 2015.

Grafton, Anthony, "The World of the Polyhistors: Humanism and Encyclopedism," *Central European History*, XVIII (1985) 31–47.

Grafton, Anthony, *Leon Battista Alberti. Master Builder of the Italian Renaissance*, New York: Hill & Wang, 2000.

Greenblatt, Stephen, *Renaissance Self-fashioning. From More to Shakespeare*, Chicago and London: The University of Chicago Press, 1980.

Henry, John, "A Pragmatic Aspect of Polymathy: The Alliance of Mathematics and Medicine in Liddel's time," in: ed. Pietro Daniel Omodeo in collaboration with Karin Friedrich, *Duncan Liddel (1561–1613). Networks of Polymathy and the Northern European Renaissance*, Leiden and Boston: Brill, 2016, 91–112.

Heuvel, Charles van den, *'De Huysbou': A Reconstruction of an Unfinished Treatise on Architecture, Town Planning and Civil Engineering by Simon Stevin*, Amsterdam: KNAW Edita, 2005.

Heuvel, Charles van den. " 'Wisconstighe Ghedachtenissen'. Maurits over de kunsten en de wetenschappen in het werk van Stevin," in: ed. Kees Zandvliet, *Maurits Prins van Oranje*, 245–257. Amsterdam/Zwolle: Rijksmuseum/ Waanders, 2000, 106–121.

Hockney, Mike, *The Last Man Who Knew Everything*, Hyperreality Books, s.l. 2013.

Isaacson, Walter, *Leonardo da Vinci. The Biography*, New York: Simon & Schuster, 2017.

Jaumann, Herbert, "Was ist ein Polyhistor? Gehversuche auf einem verlassenen Terrain," *Studia Leibnitiana*, XXII (1990) 76–89.

Johnston, Stephen, "Mathematical Practitioners and Instruments in Elizabethan England," *Annals of Science* 48.4 (1991) 319–344.

Johnston, Stephen, *Making Mathematical Practice: Gentlemen, Practitioners and Artisans in Elizabethan England*, PhD dissertation, Cambridge University, 1994.

Klein, Ursula, *Nützliches Wissen. Die Erfindung der Technikwissenschaften*, Göttingen: Wallstein, 2016.

Long, Pamela O., *Artisan/Practitioners and the Rise of the New Sciences, 1400–1600*, Corvallis: Oregon State University Press, 2011.

Maclean, Ian, "Girolamo Cardano: The Last Years of a Polymath," *Renaissance Studies*, 21 (2007) 587–607.

Mare, Heidi de, *Huiselijke taferelen. De veranderende rol van het beeld in de Gouden Eeuw*, Nijmegen: Vantilt, 2012.

Murphy, Kathryn, "Robert Burton and the problems of polymathy," *Renaissance Studies*, 28 (2014) 281–297.

Oomes, R.M.Th.E., "Een onbekend werk van Simon Stevin," *Vereniging an van Hout. Nieuwsbrief*, 7 (1996), 2–4.

Roberts, Lissa, Simon Schaffer, and Peter Dear eds., *The Mindful Hand: Inquiry and Invention from the Late Renaissance to Early Industrialisation*, Amsterdam: Edita KNAW, 2007.

Robertson, Andrew, *The Last Man Who Knew Everything. Thomas Young, the Anonymous Polymath Who Proved Newton Wrong, Explained How We See, Cured The Sick and Deciphered the Rosetta Stone*, New York: PI Press, 2005.

Romein-Verschoor, Annie, and G.S. Overdiep eds., *Het burgherlick leven. Herdruk naar de eerste uitgave van 1590*, Amsterdam: Wereldbibliotheek, 1939.

Rössler, Hole, "Polyhistorie und Polymathie," in *Neue Diskurse der Gelehrtenkultur in der Frühen Neuzeit. Ein Handbuch*, eds. Herbert Jaumann and Gideon Stiening. Berlin: De Gruyter, 2016, 635–676.

Schilt, Kees, *Simon Stevin en het hermetisme. Een cultuurhistorische studie naar zijn methodologie* (unpublished PhD thesis, Vrije Universiteit Amsterdam, 2013).

Schouteet, A., "Een brief over het werk van Simon Stevin door zijn zoon Hendrik, 1636," *Handelingen van het Genootschap Société d'Émulation te Brugge*, 101 (1964) 236–239.

Schwartz, David N., *The Last Man Who Knew Everything. The Life and Times of Enrico Fermi, Father of the Nuclear Age*, New York: Basic Books, 2017.

Secretan, Catherine, and Pim den Boer eds., *Simon Stevin. De la vie civile 1590*, Lyon: ENS éditions, 2005.

Smet, Rudolf de, "Simon Stevin et le paradoxe de l'Humanisme fragmenté," in: Hossam Elkhadem and Wouter Bracke eds., *Simon Stevin 1548–1620: L'émergence de la nouvelle science*, Turnhout: Brepols, 2004, 27–34.

Smith, Pamela, *The Body of the Artisan: Art and Experience in the Scientific Revolution*, Chicago: Chicago University Press, 2004.

Stevin, Hendrick, *Wisconstich filosofisch bedrijf*, Leiden: De Croy, 1667.

Stevin, Hendrick, *Materiæ politicæ. Burgherlicke stoffen*, Leiden: Iustus Livius, 1649.

Stevin, Simon, *The Principal Works of Simon Stevin*. 5 vols., Amsterdam: C.V. Swets & Zeitlinger, 1955–1966.

Strotmann, Vivian, *Majd al-Dīn al-Fīrūzābādī (1329–1415). A Polymath on the Eve of the Early Modern Period*, Leiden and Boston: Brill, 2016.

Struik, Dirk J., *The Land of Stevin and Huygens. A Sketch of Science and Technology in the Dutch Republic during the Golden Century*, Dordrecht: Reidel, 1981.

Taverne, Ed R.M., *In 't land van belofte: in de nieue stadt. Ideaal en werkelijkheid van de stadsuitleg in de Republiek* 1580–1680, Maarssen: Gary Schwartz, 1978.

Valleriani, Matteo (ed.), *Structures of Practical Knowledge*, Cham: Springer, 2017.

Vanden Berghe, Guido et al., *Simon Stevin (1548–1620). De geboorte van de nieuwe wetenschap*, Turnhout: Brepols, 2004.

Varro Varius, the Polymath of the Roman World, ed. D.J. Butterfield, Cambridge: The Cambridge Philological Society, 2015.

Vérin, Hélène, "Rédiger et réduire en art: un projet de rationalisation des pratiques," in: Glatigny, Pascal Dubourg, Hélène Vérin, *Réduire en Art. La Technologie de la Renaissance aux Lumières*, Paris: Editions de la Maison des sciences de l'homme, 2014, 17–59.

Vermij, Rienk, *The Calvinist Copernicans. The Reception of the New Astronomy in the Dutch Republic, 1575–1750*, Amsterdam: Edita KNAW, 2002.

Vermij, Rienk, *Kleine geschiedenis van de wetenschap*, Amsterdam: Nieuwezijds, 2005.

Verwey, Gerlof, "Hoe zag Coornhert zichzelf? De idee van *paideia Christi* als matrix van zijn zelfinterpretatie," in: ed. Jaap Gruppelaar and Jürgen Pieters, *'Un certain Hollandois' Coornhert en de vragen van zijn tijd*, Hilversum: Verloren, 2014, 13–33.

Wal, Marijke J. van der, "Logic, Linguistics, and Simon Stevin in the Context of the Sixteenth and Seventeenth Centuries," in: ed. K.R. Jankowski, *History of Linguistics*, Amsterdam: John Benjamins, 1993, 147–156.

Wal, Marijke J. van der, "Simon Stevin: Taalbeschouwer en taalgebruiker." In *Simon Stevin 1548– 1620. De geboorte van de nieuwe wetenschap*, ed. Wouter Bracke, Leuven: Brepols, 2004, 170-77.

Warren, Leonard, *Joseph Leidy. The Last Man Who Knew Everything*, New Haven: Yale University Press, 1998.

Wiesenfeldt, Gerhard, "The 'Duytsche Mathematique' and Leiden Family Networks, 1600–1620," in: eds. Fokko Jan Dijksterhuis, Andreas Weber, Huib Zuidervaart, *Locations of Knowledge in Dutch Contexts*, Leiden and Boston: Brill, forthcoming 2020.

Wiesenfeldt, Gerhard, *Leerer Raum in Minervas Haus: Experimentelle Naturlehre an der Universität Leiden, 1675–1715*, Amsterdam: Edita KNAW, 2002.

CHAPTER 1

The Engineer and the Philosopher

Reflections on the Culture and Economy of Mechanics in Court Society

Pietro Daniel Omodeo

Throughout his life, Simon Stevin was proud of his status as an 'ingenieur'.[1] By contrast, his Italian contemporary, Giovanni Battista Benedetti, would have considered the title 'ingegnere' to be a denigration of his status as a 'patrician' and 'court philosopher'.[2] Yet, the work, interests and methods of these two men were so close that such a divergence of identities presents the social historian of science with a curious puzzle. What accounts for the remarkable difference in the intellectual positioning of these two protagonists of Renaissance science? An inquiry into the social and epistemological intersections and interrelationships between the activities of the early modern engineer and the philosopher, who worked on similar mathematical, physical and technological problems, promises to be a fruitful means of pinpointing the individual and contextual differences between this Dutch proto-scientist and his Italian counterpart. In addition, it may also help to highlight the processes underlying the hierarchization of knowledge and division of intellectual labor that characterized the emergence of modern science.

Above all else, the problem in question relates to a practical-theoretical form of polymathy – the technical versatility (or *polytechny*) typical of so-called 'scientist-engineers' – and its relation to higher forms of legitimate scholarship, especially the literary and artistic culture of the Renaissance courts. In the case of practical mathematics, a new class of practitioners, proud of their professional skills, emerged out of the context of competing republics and small states troubled by constant political and economic turmoil, rivalry and conflict. Against this socio-political background, architecture, the mechanical arts and the exact sciences became increasingly relevant for military and civil purposes, transportation and commerce, and, to a lesser extent, production. In their seminal works on the socio-economic history of mechanics,

1 Dijksterhuis, *Simon Stevin: Science in the Netherlands*, 130. In fact, the most prominent indication attached to his name in publications is the toponymic 'of Bruge'.

2 Omodeo, 'Social Position and Intellectual Identity of the Renaissance Mathematician-Physicist Giovanni Battista Benedetti'.

 | DOI:10.1163/9789004432918_003

Boris Hessen and Edgar Zilsel mapped out the *social function* of 'higher artisans' and 'scientist-engineers'.[3] The new society showed a new appreciation of once-discredited arts and crafts; professions such as 'low mechanics' won repute, whilst practical knowledge acquired unprecedented epistemological value.[4] Correspondingly, the reputation of scientist-engineers was enhanced.[5] As a 'mathematically trained engineer',[6] Stevin belonged to the new group of practical mathematicians who addressed theoretical-technological problems linked to navigation, surveying, shipbuilding, fortification, gunnery, and other fields of evident economic or military relevance.[7] The spheres of economy and war were irrevocably intertwined at this time of constant external and internal European conflict, on the two fronts of colonial expansion and religious civil wars. For practical mathematicians and experts in technology, warfare was a lucrative field and presented an opportunity for social advancement.

Before he became famous for his astounding telescopic discoveries, Galileo Galilei gained a reputation for the invention of a military compass with significant applications, for instance to ballistics. Attention has been recently drawn to such aspects of his scientific activity, as well as his ties to practical mathematicians and artisans. This has led to a re-evaluation of Galilei and earned him the appellation 'engineer' in a recent book.[8] During the Renaissance, an accomplished engineer could hope to exchange his position on the battlefield for a more comfortable position at court. Such a transformation was actually quicker than it might appear at first glance, given that in the Iron Century both warfare and the court were the domain of the aristocracy.[9] The social and epistemological implications of such a promotion could be far-reaching, however. Thus, the interpretations of 'Galilei the courtier' and 'Galilei the engineer' do not necessarily conflict; instead, they are complementary, and the tension between the two identities is worthy of special consideration. A certain proximity between the two types could also be observed in the Netherlands: in the 1590s, Stevin's social ascent was crowned when he entered the service of the stadholder, Prince Maurice of Nassau.

3 Hessen, 'Social and Economic Roots', and Zilsel, 'Sociological Roots of Science'.

4 Smith, *Body of the Artisan* and Long, *Artisan/Practitioners and the Rise of the New Sciences*.

5 Cf. Lefèvre, 'Galileo Engineer: Art And Modern Science'.

6 Alan F. Chalmers, 'Beyond Archimedes: Stevin's *Elements of Hydrostatics*', Chap. 3, in *One Hundred Years of Pressure: Hydrostatics*, 27.

7 Cf. Matthias Schemmel's definition in *English Galileo*, 15.

8 Matteo Valleriani has stressed the importance of the military context for Renaissance engineering and science. See *Galileo Engineer*, 197.

9 Kamen, *Iron Century*.

Giovanni Battista Benedetti (Venice 1530 – Turin 1590), court mathematician to the dukes of Savoy, was in many respects Stevin's Italian counterpart and constitutes the main focus of this essay. Hailing from a noble Venetian family of Spanish origin, he learned mathematics with passion as an autodidact – although he acknowledged that he had received some lessons on Euclid from the famous mathematician Niccolò Targaglia. The well-known astrologer Luca Gaurico was one of the first to recognize Benedetti's mathematical talent, publishing the horoscope that Benedetti had cast for himself in his early twenties in his *Tractatus astrologicus* (1552).[10] Benedetti affirmed his status as a mathematician and mathematically-inspired natural philosopher with the publication of his *Resolutio omnium Euclidis problematum* (Venice 1553), which offered a 'solution to all' geometrical problems in the Euclidean tradition by using a compass with a fixed opening, and of his *Demonstratio proportionum motuum localium contra Aristotilem et omnes philosophos* [Demonstration Concerning the Proportions of Local Motions against Aristotle and All Philosophers] (Venice 1554). In the latter work, he presented a new theory of falling bodies, according to which, and contrary to the famous Aristotelian dictum, bodies of the same material fall at the same speed independently of their weight, if the medium is the same. From 1558 to 1567 he was a courtier to Ottavio Farnese in Parma, and between 1567 and 1590 he served the dukes of Savoy in Turin. In this courtly context, he acted as a mathematical advisor and published works on various topics, such as gnomonics and calendar reform. Most importantly, he published a large folio of miscellaneous 'mathematical and physical speculations', *Diversarum speculationum mathematicarum et physicarum liber* (Turin 1585), which is particularly well-known among historians of science for its book on mechanics (the third book). It also comprised a book of arithmetic theorems, a book on perspective, a critique of Aristotle's physics (which also tackled problems in post-Copernican astronomy), a brief discussion of Euclid's theory of proportions in the fifth book of the *Elements*, and a large collection of epistles on a diverse range of scientific and natural matters, which Benedetti approached and solved using mathematics.

Both Benedetti and Stevin wrote on geometry and arithmetic, and contributed to the advancement of mechanics, hydrostatics, perspective and mathematical astronomy, including the study of Copernican theory; and both dealt with problems linked to navigation, military technology and

10 Luca Gaurico, *Tractatus astrologicus in quo agitur de praeteritis multorum hominum accidentibus per proprias eorum genituras ad unguem examinatis* (Venice: Curtius Troianus Navò 1552), f. 76*r*.

perspective.[11] Nevertheless, Benedetti would have considered it offensive to have been called an 'ingegnere'. On the title pages of his works, he proudly referred to himself as a 'patritius' and 'philosophus', and used corresponding Italian expressions such as 'filosofo del sereniss[imo] duca di Savoia' [Philosopher to the Most Serene Duke of Savoy].[12] As a Venetian gentleman who had been granted the privileges of the imperial nobility by his sovereign, Duke Emanuele Filiberto,[13] he certainly did not wish to lower his status to that of an artist, engineer or practitioner, although he did not disrespect those professions or the intellectual and practical value of their knowledge and expertise. Apart from the differences between the two individuals' biographies and social origins, however, other questions arise when attempting to interpret the science of the time. Did Benedetti's sense of superiority towards engineers and practitioners shape his science: its content, form or epistemology? And more generally, did the introduction of technological knowledge of the practical arts to the courts of proto-absolutist states transform the latter in any significant manner? To what extent was the relationship between the engineer and the philosopher in court society relevant to the birth of modern scientific culture and the specific class of intellectuals who represented it? In what follows, I offer some considerations on the figure of the *court intellectual* as compared to that of the *engineer*.

1 The Codification of an Aristocratic Discipline

The scientific culture surrounding Benedetti and Stevin had slowly emerged from the practical knowledge and theoretical efforts of the preceding generations, at a time of profound social and political change. It is expedient to start by giving an overview of this historical setting. Beginning in the Italian towns of the Quattrocento, with their political autonomy, strong guild system and affirmation of new entrepreneurial interest groups, there arose a proud class of technicians and engineers who shared the utilitarian values of

11 Dijksterhuis' monograph on Stevin is still a valid reference point for an overview of Stevin's achievements; for a recent overview of Benedetti's work, cf. Omodeo and Renn, *Science in Court Society*.

12 Benedetti appears as 'patritius Venetus philosophus' in the titlepages of his major work, *Diversarum speculationum mathematicarum et physicarum liber* (1585) and that of his work on gnomonics, *De gnomonum umbrarumque solarium usu liber* (1574). In vernacular publications, such as the *Consideratione. … d'intorno al discorso della grandezza terra et dell'acqua* (1579), he correspondingly appears as 'filosofo del sereniss. duca di Savoia'.

13 Bordiga, *Giovanni Battista Benedetti, filosofo e matematico veneziano*, 752.

the mercantile bourgeoisie, but did not disregard ecclesiastical or aristocratic patronage.[14] Leonardo Olschki depicted these artist-artisans in detail in his seminal studies on early scientific literature written in the vulgar tongue. According to his interpretation, the artists and technicians demonstrated their social and intellectual pride by choosing to write in the vernacular.[15] These men practiced a specific form of polymathy, which corresponded to the social need for and potential application of their artistic, technical and architectonic knowledge. Olschki identified Filippo Brunelleschi as one conspicuous embodiment of the Renaissance artist-technician *typus,* who united mathematical expertise, technical knowledge and artistic creativity.[16] As his case shows, it was not unusual if 'the same master constructor [*Baumeister*] who was entrusted with churches, hospitals and public palaces, also had to build bridges, fortresses, war machines, city walls, canals and dams, as a civil and military engineer'.[17]

Among the examples of the celebration of the noble art of mechanics during the Renaissance, the letter to the reader of the Italian edition of Guidobaldo Del Monte's *Le mechaniche* [Mechanics] (1581) by the soldier, traveler and diplomat Filippo Pigafetta stands out for its incisiveness. Aiming to dispel widespread prejudices, it addresses problems of knowledge and social hierarchy:

> The present book [...] is entitled *The Mechanics*. However, many will probably misunderstand the correct meaning of this word, i.e. 'mechanics'. We can easily meet those who regard this term as an insult. In fact, in many parts of Italy it is usual to call somebody 'mechanic' in order to mock or offend him. Some others, who call themselves 'engineers', would show indignation. In consideration of this, it is very appropriate to emphasize that, according to Plutarch, 'mechanic' is a very honored term as it refers to the military profession. It suits the illustrious man who is capable of realizing, with his hands and ingenuity, marvelous works, which are extremely useful and agreeable in life.[18]

---

14 Lefèvre, *Naturtheorie und Produktionsweise*, 96.

15 Olschki, *Geschichte der neusprachlichen wissenschaftlichen Literatur*, Vol. 3, 70: 'Dieser Stand der Künstler und Techniker bekundete zuweilen mit Stolz und Affekt diese seine Selbständigkeit mit dem Gebrauch der Volkssprache'.

16 Cf. Renn, 'Florenz: Matrix der Wissenschaft'.

17 Olschki, *Geschichte der neusprachlichen wissenschaftlichen Literatur*, Vol. 1, 36 (all translations are my own, unless stated otherwise).

18 Del Monte, *Le mechaniche*, f. b3*r*. On the age-old 'prejudice' concerning 'vile mechanics', cf. Rossi, *Birth of Modern Science*, 15–17.

In order to assert the prestige of the 'art' of machines, Pigafetta had to reverse the age-old disdain for the practical arts, a lack of recognition that resulted from the stigma attached to manual work in general. The *practical* arts had long been contrasted with the *liberal* arts, those marked by the disinterestedness of men free of 'vulgar interests'.

Pigafetta indirectly informs us about the established social distinction between the lower 'mechanico' and the superior 'ingegnere', whilst simultaneously trying to establish an even higher position for the scholar of mechanics. In order to make his argument, he does not explicitly rehabilitate work done by the hand, but rather appeals to humanistic literacy, cultural values and classicist bias. Thus he strengthens his positive evaluation of mechanics with a reference to an acknowledged classical authority such as Plutarch. Moreover, he emphasizes the nobility of the art by disconnecting it from civil utility and the social status of its masters. He prefers to accentuate the connection of this art to the most aristocratic of all activities: the *art* of war. Indeed, it appears from Pigafetta's biography that he himself combined a military career with a literary one, including the study of architecture and engineering. The general of artillery of the Venetian republic, Giulio Savorgnan, commissioned him to translate the *Mechanics* into Italian.[19]

Del Monte could be seen as an example of aristocratic engagement with the discipline at a level that surpassed the activities of a common mechanic or engineer. He in fact belonged to one of the most prominent families in the belligerent Duchy of Urbino, and was very close to the Della Rovere rulers. The author of the *Mechanics* had himself received an aristocratic military education, including training in fencing and horse-riding. He had taken part in a campaign in Hungary in the retinue of the Grand Duke of Tuscany in 1566. Later, he would have participated in the expedition against the Turks leading to the Battle of Lepanto if a serious back injury had not hindered him from continuing the *vita activa* of a soldier. Nevertheless, his commitment to the study of the art of machines shows that aristocratic and military interests could be sublimated into intellectual engagement with practical mathematics. Pigafetta alluded to this continuity between the active and intellectual life in a letter addressed to Del Monte on 5 November 1580:

> Everybody celebrates Your very noble nature, Your capacity to use weapons and ride, as well as the profound knowledge of every science. They

19 For an introduction to Pigafetta's life and work, cf. Pozzi, *Filippo Pigafetta consigliere del Principe*, vol. 1.

> are such that You excel in comparison to any other Signore and You are not lesser than any man of letters.[20]

The ability to master weapons and riding formed part of the standard image of a noble education – together with hunting, the chivalric quest for glory, and a refined education in humane letters. Baldassar Castiglione had codified this model in *Il libro del Cortegiano* [The Book of the Courtier] (1528) by means of an idealization of the very court of Urbino that Del Monte would serve some decades later.[21] In that book, the arts – especially music, sculpture and painting – were elevated from being the mere application of technical skills to becoming an intellectual ornament of the courtier and his milieu:

> I would like to address another thing which I deem of great relevance and therefore believe our courtier ought in no wise neglect. And that is the cunning in drawing and the knowledge of the art of painting. Do not wonder if I wish [him to have] this skill, which today is perhaps considered to be mechanical [*mecanica*] and of little convenience to a gentleman, because I remember that I read that the ancients, especially all over Greece, wanted the noble children to learn to paint in the schools as a matter both honest and necessary; it was even admitted in the first degree among the liberal arts [...].[22]

One should bear in mind that the modern conception of the artist, gradually distinguished from the artisan, is the product of the courtly environment and patronage system of the Renaissance, even though the term itself was established later; Giorgio Vasari called the authors of his biographies *artefici del disegno*, or 'drawing art-makers'.[23] Titian, to give one example of aristocratic art, had noble origins and was at his ease at court, whilst Raphael was praised for his art as well as for his costumes and grace.[24] But what about the art of machines, which Castiglione seems to exclude from the *liberal* arts as something improper to a gentleman? Could mechanics be elevated to the higher level of the other courtly arts, thereby undergoing a similar promotion to painting, to become an area of activity worthy of a gentleman? Castiglione's contempt for

20 Quoted from Martin Frank, *Guidobaldo del Monte's Mechanics in Context*, 28, n. 7.

21 Barberis, 'Baldassar Castiglione', lxiv.

22 Castiglione, *Il libro del Cortegiano* I 49, 103–104.

23 Chastel, 'L'artista', 239.

24 Burke, 'Il cortigiano', 156.

mecanico contrasts to the later efforts to rehabilitate mechanics by aristocrats such as Del Monte and Benedetti.

Del Monte's legitimation strategy was linked to the codification of mechanics as a *science*, namely a deductive science following a rather Archimedean and Euclidean ideal. Countering the disparagement of the practical arts by many, Del Monte's work can be read as an effort to reverse the denigration of mechanics – as an art for the subaltern classes – and turn it into an aristocratic discipline. In fact, his *Mechanics* was not a *practical* treatise in the popular sense of the term. Rather, it dignified mechanics through an *epistemological* transformation of the art of machines from a practical to a theoretical science. Del Monte simplified and generalized mechanics: he reduced the five simple machines (lever, pulley, winch, wedge and screw) to the lever; began his treatment with a definition (that of the center of gravity), then moved to the most general notions (or *communes notiones*) and theses (or *suppositions*) from which the rest descended; and addressed simple machines, one by one, in the form of a series of *propositiones*. In summary, he conferred a deductive structure upon the discipline of mechanics, and established its mathematical and systematic order.[25] Pigafetta's exaltation of Del Monte's achievement emphasized that a 'distilled' version had been produced of what was originally a *practical* art. Its promotion to a military and aristocratic science, by means of reforming its methods and concepts, was intended to obliterate its roots.

In late fifteenth and early sixteenth-century Italy, when important centers such as Florence were undergoing a shift from republican to proto-absolutist forms of centralized power, the practical arts were transformed upon their introduction to the courts. The peer relationship between the high engineer and the theoretician – as exemplified by the interaction in Florence between Brunelleschi and the mathematician Paolo dal Pozzo Toscanelli – became vertical. The court intellectual appropriated the knowledge of the subaltern classes in order to advise, supervise and direct court society from its center of power. Eugenio Garin has convincingly argued that the political shift that occurred with the affirmation of De Medici rule in Florence caused a move in philosophy from the discussion of ethics and politics to the theologizing metaphysics of the Platonic Academy.[26] Similar to this philosophical shift in the passage from the *Quattrocento umanistico* to the *Signoria*, a process of courtly appropriation could be witnessed in the natural sciences as well. This process can be regarded as a theoretical and philosophical codification and grounding

25 See the open access edition by Renn and Damerow, *Guidobaldo del Monte's Mechanicorum Liber*.

26 Garin, *Cultura filosofica del Rinascimento italiano*, 71.

of knowledge rooted in, and alienated from, the practical realms of handicraft, technology and engineering.

2 Benedetti's Physico-Mathematics: a Courtly Path to Science

Although practical mathematics enjoyed a higher status during the Renaissance than previously, its introduction to the court was by no means an obvious development. Castiglione neglected to include mathematics while admitting other arts, which were praised for their capacity to elevate the spirit and taste of the accomplished *cortigiano*. By contrast, Del Monte and Benedetti were committed to introducing practical mathematics and mechanics into the exclusive circles of Renaissance courts such as Urbino and Turin, whose rulers were deeply committed to military enterprises for different reasons. Whilst Urbino had secured a key role in the turbulent political landscape of Renaissance Italy, thanks to its army and strategic alliances in the continuous conflicts of the time, the restored Duchy of Savoy had to invest in fortifications and the army in order to survive the proximity of France and Spanish Lombardy.[27] Architecture and mechanical knowledge constituted essential technological assets in warfare.

Emanuele Filiberto, famous for his military skills and the endurance he demonstrated on the battlefield as the chief commander of Charles V's imperial army in the Netherlands, was himself keenly interested in mathematical disciplines. As the Venetian ambassador Giovanni Correr reported in 1566,

> That Duke is no man of letters but he loves the virtuosi. Hence, he has many of them by him; he likes to listen to their reasoning and he asks them questions. However, there is no subject that delights him more than mathematics, as this science is not only apt but also necessary to the profession of military commander.[28]

Benedetti bears witness to the duke's bias. In the dedicatory letter of his major work, the *Diversarum speculationum mathematicarum et physicarum liber* [Book on Various Mathematical and Physical Speculations] (Turin, 1585), he

27 The geopolitical situation of the restored duchy is described well in Barberis, *Armi del Principe*, Chap. 1. For a comprehensive historical reconstruction of Turin in the time of Emanuele Filiberto and his successor, see Ricuperati (ed.) *Storia di Torino: Dalla dominazione francese alla ricomposizione dello Stato*.

28 Firpo, *Relazioni di ambasciatori veneti al Senato*, 123.

reminded his protector and the dedicatee of the book, Emanuele, of his father Emanuele Filiberto's passion for mathematics.[29]

Benedetti's scientific epistolary reveals how his work was embedded in the court. His *Physica et Mathematica responsa per Epistolas* [Epistolary Answers on Physics and Mathematics] constitutes the final, very extensive part of the *Diversae speculationes mathematicae et physicae*. The letters, on the most varied scientific topics, contain the expert advice he gave to the dukes, courtiers and foreign correspondents, most of whom were in Venice. The most evident function of their publication was to show off the 'social capital' of the court *philosopher*. The selection of epistles and their disposition is the *mise-en-scène* of the asymmetric network emanating from the court.[30] As a courtier, Benedetti was expected to honor the dukes and the high court dignitaries. Emanuele Filiberto, Carlo Emanuele of Savoy and the most powerful military commander Andrea Provana of Leynì – one of the 'heroes' of Lepanto – are the addressees of the first letters in the collection.[31] The relationship with other court intellectuals is symmetrical. For instance, Benedetti discussed meteorological and natural issues in a friendly and respectful tone with the court physician, the Aristotelian Francesco Arma, as well as with the other court peripatetics, even though he disagreed with them on almost everything, beginning with respect for Aristotle's authority. In addition, he considered himself superior to architects, artists and other technicians. If any subaltern showed a lack of respect, his response was fierce. When a lesser-known censor, Benedetto Altavilla, attacked the reliability of the astrological prognostics cast for the dukes in 1580, Benedetti promptly responded with a sharp booklet discrediting his adversary. The court intellectual emphasized his provocateur's lack of mathematical preparation while cautiously avoiding addressing the ethical implications of the polemic, that is, the controversial belief in astrological influence.[32]

As for epistemology, the social differences between those who made things and those who thought had to be represented as the subordination of practice to theory. Benedetti and Del Monte can be seen as representatives of the shift from practice-oriented knowledge, the *polytechny* of the Renaissance engineer, to a kind of aristocratic *polymathy* more suited to the court.[33]

29 Benedetti, *Diversarum speculationum mathematicarum et physicarum liber*, f. A2r.

30 See Cecchini and Roero, 'Corrispondenti di Giovanni Battista Benedetti'.

31 Benedetti, *Diversarum speculationum mathematicarum et physicarum liber*, 205–210 (to Duke Emanuele Filiberto, on calendar reform), 211–213 (to Carlo Emanuele on geometrical problems), 214–224 (to Andrea Provana de Leyní on navigation instruments).

32 See Omodeo, 'Efemeridi e critica all'astrologia tra filosofia naturale ed etica'. On Renaissance court astrology, cf. Azzolini, *Duke and the Stars*.

33 On Del Monte's work and environment, see especially Becchi et al., *Guidobaldo del Monte*.

Although their work on mechanics, practical mathematics and physics would be inconceivable without the technical and practical context of their time, their scientific activity was marked by a degree of speculative generalization that went beyond the technical and the practical. Del Monte envisaged a *theoretical* science of mechanics. It was informed by the humanistic ideals of the *restaurator mathematicarum*, Federico Commandino, who fostered a mathematical purism inspired by Hellenistic models to be appropriated with philological acumen.[34] Within such a cultural project, mechanics could be raised above the 'vulgar' and made suitable for the refined culture of the court. Yet, it should be noted that mathematical ideals radiated beyond humanistic circles and the courts that hosted them. Even Stevin's 'Archimedean and Euclidean ideal' shared similar methodological and formal characteristics, to an extent, although he did not intend to widen the gulf between theory and practice.[35]

Benedetti chose a different path towards establishing mechanics and practical mathematics at court. Rather than being simply mathematical, his project can be labeled 'physico-mathematical', or, to use his own language, 'speculatively mathematical and philosophical'. In the works of his youth, he had already attacked Aristotelian natural philosophy for the purposes of an improved theory of motion, but it was only in his mature work that he sought to draft a new conception of natural philosophy.[36] Book four of the *Diversae speculationes* addressed motion and Copernican astronomy, among other things, in order to dispute the validity of Aristotle's doctrines. The section is entitled *Disputationes de quibusdam placitis Aristotelis* [Disputations on Some Opinions Held by Aristotle]. It is here that Benedetti discussed the necessity of the void to ensure the very possibility of motion, which implicitly rehabilitated atomistic conceptions of matter. He recast central concepts of natural philosophy, such as space and time, outside of an Aristotelian framework. Rather than 'place' and the 'measure of motion', he conceived of them as an absolute frame for physics. Considered as a whole, Benedetti's *Various Mathematical and Physical Speculations* presented a novel conception of nature. He strove for a Pythagorean ideal, a philosophy offering the *theory* behind the mathematical arts and mechanics. 'The book of nature is written in mathematical language' could be seen as Benedetti's dictum before Galilei famously formulated it in *Il Saggiatore* [The Assayer] (1623). His goal, in fact, was to raise mechanics from being

34 Cf. Rose, *Italian Renaissance of Mathematics*, especially Chaps. 9–11.

35 Cf. Van Dyck, 'Motion and Proportion in Simon Stevin's Mechanics'.

36 Maccagni, *Speculazioni giovanili 'de motu'*.

the art of machines to a general vision; that is, to a mathematical philosophy of nature.[37]

This project is perfectly consistent with Benedetti's self-fashioning as a 'court philosopher' instead of a 'court mathematician' (not to speak of 'engineer'), before Galileo made a similar move at the De Medici court.[38] The high status of the philosopher was mirrored by the construction of the theory of a physico-mathematical science. Del Monte and other scholars in Urbino disdained the lack of elegance of Benedetti's conflation of physics and mathematics, and opposed his intellectual project in the name of an Hellenophile mathematical purism.[39] Yet, in spite of the differences, these courtiers shared his vision of a scientific approach – either philosophical or purely mathematical – that distanced itself from the popular handicraft in which this knowledge had emerged.

3 The Socio-Economic and Political Terrain of Court Science: the Case of Benedetti

As I have argued so far, the Renaissance court served as a catalyst that transformed the art of mechanics and practical mathematics into theoretical disciplines, mathematical and physical. As important as this step was for the historical development of physics and the establishment of the intellectual forebears of the modern scientist, our consideration of epistemological shifts and rhetorical strategies will be insufficient unless it is integrated with a specific discussion of the socio-political coordinates of Renaissance scientific culture. This implies that we have to reassess the 'Hessen question' about the socio-economic roots of modern mechanics.

As far as the economic impulses underlying the establishment of practical mathematics and technological knowledge are concerned, the thesis of this founder of socio-economic historiography is well known: 'economics [...] *present*[*s*] demands, which *pose* technical problems, which *generate* scientific problems'.[40] According to Hessen, the economic roots of early modern science,

37 Thus, Benedetti's program has to be seen as part of a more general trend. Cf. Renn and Damerow, 'Transformation of Ancient Mechanics into a Mechanistic World View'.

38 Biagioli, 'Social Status of Italian Mathematicians', 49–50. Also, see idem, *Galileo, Courtier*.

39 Bernardino Baldi's 'Urbinate scorn' regarding Benedetti is telling of the distance between the two approaches. In his *Cronica de matematici* he depicted the adversary in quite negative terms. Cf. Baldi, *Cronica de matematici*, 140.

40 This formulation is to be found in Freudenthal and McLaughlin, *Social and Economic Roots*, 4.

particularly the development of mechanics in the age of Newton, lay in commerce and transportation, industry and warfare. The seventeenth-century English economy was based on growing merchant capital and manufacturing, linked to early global colonialism. The needs of communication and transport, which were mainly maritime, posed technical problems such as the need to improve vessels and navigation techniques (especially on the high seas), and the construction of canals and locks. Moreover, crucial sectors of production, such as mining and the war industry, posed technical challenges for chemistry and mechanics in a variety of forms, ranging from the construction of machines to hydrostatics and aerostatics and the science of materials. The introduction of firearms required the study of ballistics and appropriate fortification techniques, which in turn stimulated dynamics and architecture. As a result, certain areas of scientific inquiry became crucial: simple machines, inclined planes and statics; free fall and projectile trajectories; hydrostatics and aerostatics; celestial mechanics and tidal theory; and optics and magnetism.[41]

Benedetti addressed all of these topics before Galileo and Newton, with the exception perhaps of magnetism. Because his context was profoundly different from the proto-capitalistic one on which the socio-economical historiography has tended to focus, however, there is an urgent need to address the following questions: what specific socio-economic characteristics of his context, the Duchy of Savoy, account for Benedetti's engagement with his subjects of inquiry? How did their socio-economic relevance relate to the forms of court culture and the appropriation of the practical arts by that milieu?

Early modern Piedmont was in fact far from a throbbing center of mercantile culture, financial speculation and large-scale manufacturing, and its character was thus very different from the mercantile societies of England and the Netherlands. Its main source of wealth was agriculture, especially the abundant production of cereals in the Po Valley and wine on the hillsides. Large landowners, the aristocratic *latifondisti*, owned the land, which was cultivated by sharecroppers and later by tenants. Emanuele Filiberto largely supported mulberry cultivation to increase silk production, the enterprise that was most generously funded by the state. He and his successor, Carlo Emanuele, also invested in the mining industry, which mainly extracted iron and lead; attempts to obtain silver in large quantities failed. Moreover, despite Emanuele

41 This is not the place to discuss the useful insights in and limitations of Hessen's socio-economic history of science. For the present discussion, it is sufficient to outline the economic drivers of early modern science and refer, for a discussion of explanatory models of the history of science, to the extensive literature on Hessen and Cold War HPS paradigms, as summarized in Omodeo, 'Socio-Political Coordinates of Early-Modern Mechanics'.

Filiberto's efforts to develop infrastructure, commerce remained local, a situation that remained unchanged for centuries.

The fundamental social and economic force in the country was that of the landed aristocracy, which also constituted the main hindrance to transformation in the direction of more industry and commerce. 'In the state [the Duchy of Savoy] the feudal nobility, which certainly constituted a very strong economic force, was almost exclusively directed toward land and money investments, not to commerce and industry'.[42] Thus, despite a sizeable silk and wool industry, no urban economy and mercantile class emerged, unlike in other parts of northern and central Italy. In his *Relationi universali* [Universal Reports], Giovanni Botero treated Piedmont as an 'underdeveloped' periphery, an appendix of Italy comparable to Friuli. His assessment of its human resources and economic activities is pitiless:

> Indeed, the Piedmontese have an open and simple nature. They engage little with weapons and even less with commerce. They like amusements and idleness. They are especially devoted to agriculture. Altogether the population of the country surpasses 700,000 souls.[43]

Botero might have been right concerning the inertia of an economy that had remained agrarian and had been largely eroded by early sixteenth-century French-Spanish conflicts. Still, he failed to acknowledge the potential of the military reforms that Emanuele Filiberto introduced after the reconquest of his territories from the French in 1559 and the reconstitution of the state and its new capital, Turin, in 1562. The duke introduced obligatory conscription for all men aged between 18 and 50 years, following the Swiss model celebrated by Niccolò Machiavelli.[44] In the long run, this would constitute the basis of Piedmontese strength, in accordance with a military project shared by the rulers and the landed aristocracy. As representatives of the state's hegemonic social and economic class, the nobles lent money to the ruler for war. Throughout the early modern period, Savoy military expenditure was exceptionally high. In the eighteenth century, its relative weight in the state economy was second only in comparison to Prussia; it reached a peak of 54% of state expenditure at the time of the War of Spanish Succession, and even 68% at the time of the War of Polish Succession.[45]

42 Woolf, 'Sviluppo economico e struttura sociale in Piemonte', 25.

43 Botero, *Relationi universali*, f. 27r.

44 The difficulties of the project have been stressed by Barberis, *Armi del Principe*, Chaps. 1 and 2.

45 Woolf, 'Sviluppo economico e struttura sociale in Piemonte', 49.

Andrea Palladio, who visited Turin in 1568, gave Emanuele Filiberto expert advice on the organization of the popular militia based on the model of the Roman army.[46] In addition, he dedicated to him the third of his most celebrated series of books on architecture, *Quattro libri dell'architettura* (Venice, 1570). It is the volume that deals with public constructions, streets, bridges, squares, basilicas and gymnasia:

> As your Highness is familiar with the most noble arts and sciences related to these issues [concerning architecture], you will have much pleasure and relief by considering the subtle and beautiful inventions of humankind as well as the true science of this art, which you understand very well and which has been brought to the most rare and almost absolute perfection. This can be seen in the illustrious and royal buildings that have been constructed in many parts of your large and happiest state.[47]

Palladio's architecture was an expression of a classicist project that desired the elevation of architecture to a noble science. Those who strove to raise mechanics to similar heights could look to him as a paradigm.

Palladio pointed out to Emanuele Filiberto that a ruler could secure popular consent through public buildings, which displayed his 'greatness and generosity'. Upon closer examination, these words might sound ironic, given that the duke's largest projects concerned the fortification of the capital and the territories. When Palladio visited Turin, the new citadel had just been built. For the construction, Emanuele Filiberto had recruited the engineer Francesco Paciotto of Flanders. As Martha Pollak has observed, the purpose of the citadel was twofold: 'Paciotto proposed a five-sided fortress, with three bastions oriented towards the countryside, defending the approach to the city from the west, and two bastions facing the city, ready to bring it under control in case of riotous uprisings against the duke'.[48] The implementation of a new *Signoria* over Turin was intended to discourage the subjects from giving into the temptation of too much autonomy. Such was the climate in which the high culture of the court developed; it was an absolutist endeavor in which distances needed to be emphasized in order to strengthen and legitimize new power hierarchies. These were not only aesthetic or rhetorical distances, but also had a social and political impact in terms of enforcing and securing structures of power and dominion.

---

46 Tessari, 'Sul soggiorno di Andrea Palladio a Torino'.

47 Palladio, *Quattro libri dell'architettura*, III, 3.

48 Pollak, *Turin 1564–1680*, 16.

Land and war were the two tenets of the economy and socio-political system of Savoy in the time of Benedetti. Both were in the hands of one and the same class. The interests of the ruling family, those of the landed aristocracy and those of court scientists converged in the duchy's agrarian-military economy. It was an economic system far from the proto-liberal bourgeois model addressed by scholars investigating the link between the inception of capitalism and the Scientific Revolution, who only considered the English (or the English-Dutch) case. In order to understand the rise of modern science, alternative social arrangements and economies should also be taken into consideration. Court society and later absolutist monarchies intersected with the proto-capitalistic world in many ways through conflicts and exchanges in the complex geopolitics and economy of early modern Europe. Did they not *present* specific demands, which *posed* technical problems, which *generated* scientific problems? In fact, many of the scientific problems Benedetti addressed as a courtier to Emanuele Filiberto in Turin coincided with those outlined by Hessen, among others. His treatment of simple machines and statics is well known and celebrated in the history of physics.[49] Moreover, he has been acknowledged for his studies on free-falling bodies, and he dedicated part of his major work to projectile trajectories. His considerations on the possible motion of the Earth, which he equated with that of a turning wheel, went in the direction of a mechanical treatment of celestial motion.[50] The fact that all of these contributions by an Italian *court philosopher* are close to the work of the Dutch *engineer* Stevin suggests that mercantile and agrarian-military societies were interlinked, despite irrevocable political and cultural differences.

It should be emphasized, however, that there is a series of themes in Benedetti's work that are specific and directly connected with the political interests associated with knowledge in a court society. To give an example, consider the calendar reform on which he advised the dukes; this was a sort of technical duty of his. His reflections on this issue, to which the Catholic rulers had been asked to contribute by the Roman Pontiff, were printed separately and then as a letter to Emanuele Filiberto in the *Diversae speculationes*. Additionally,

49 See, among others, Drake and Drabkin, *Mechanics in Sixteenth-Century Italy* and Renn and Damerow, *Equilibrium Controversy*.

50 Although Jochen Büttner has cautioned against rashly interpretating Benedetti's discussion of the turning wheel as connected to the discussion of terrestrial motion, Benedetti's dual treatment of physics and Copernican astronomy certainly suggested that his readers should consider the two topics as related, and did in fact pave the way to Galileo's physical inquiry into the consequences of Copernican theory. Büttner, 'Big Wheel Keep on Turning'.

his service as a court astrologer, including much of his work on astronomical observations and calculations, should be read as an integral part of his function as a court expert.[51] Third, his epistles on navigation techniques were connected to Provana's maritime war expeditions. He also wrote on perspective in the service of painting, and discussed problems of literary exegesis for which astronomical knowledge was essential.[52] His scientific production that was linked directly to the court tended to be of an occasional and fragmentary character (the variety stressed in the title of the *Diversae speculationes*), as his writings depended on circumstances and commissions rather than on an 'autonomous' scholarly agenda.

Moreover, we can discern a tension between seriousness and lightness, and between necessity and superfluity, in Benedetti's studies. Looking more closely, we see that this polarity lay at the core of courtly life. The court was at once both a literary and refined center, in which luxury and culture were displayed, and the main political quarter in which the most important decisions had to be taken. 'Reality and imagination – as has been remarked – prescribed that at court the weight of the duty of government be mirrored by the lightness of amusement'.[53] The glamour of the court, the 'natural habitat' of the *cortigiano*, has been described at length by microhistorians, but it is to a sociologist, Norbert Elias, that we owe our understanding of the *seriousness* of flattery in that milieu. It was not a space of pure divertissement, but rather the command center of a specific societal formation, which Elias called the *court society*.[54] Similar to other expressions of court culture, court science exhibited the superfluity of its speculative nature whilst remaining connected to the most serious of all business. This dual abstraction and interest holds particularly true for science in the service of a state such as Savoy, whose economy and politics were determined by military necessity and interests. At the headquarters of proto-absolutist states, science had to be at once useful and pleasant, practice-oriented but not purely empirical, witty and sensitive to court sociability. It displayed an abstract disinterestedness whilst securing the vital interests of the ruling elites. In accordance with his cultural values, Benedetti elevated mathematics from being a practical discipline of scientist-engineers to being a refined cultural activity.

The courtly nature of Benedetti's science was at once its strength and its weakness, however: it was relatively free from the bookish tradition, but not

51 Omodeo, 'Polemiche astronomico-astrologiche'.

52 Idem, 'Renaissance Science and Literature'.

53 Barberis, 'Baldassar Castiglione', XVII.

54 Norbert, *Höfische Gesellschaft*.

from the contingencies of courtly life. As it rested on such a volatile institution as the court, it ultimately depended on the will of the sovereign. These conditions could not guarantee any continuity in scientific activity, nor did they allow the founding of a school. The limitations of early modern court society would be overcome in the course of the seventeenth century, when the creation of scientific societies secured continuity for patronized science. The founding of the *Académie royale des sciences* in France is the most prominent case. It was a means to patronize *all* of the sciences at once, and for Louis XIV it meant the conquest of a new kingdom, *la république des lettres tout entière*.[55] On the other hand, the institutionalization of the *savant* within a bureaucratic and meritocratic system integrated into the absolutist state gave birth to the profession of the modern scientist. Benedetti's culture was distinct (though not disconnected) from that of the engineers and the mechanics; he was not the organic intellectual of modern bourgeois society, but rather that of the emerging absolute state. His profile anticipated that of the *savant d'ancien régime*, just as Renaissance courts can be seen as prefiguring later absolutist courts and state apparatuses.[56] Benedetti's environment shares many characteristics with the environment that culminated in absolutist France, which, under the banner of a centralized national monarchy, facilitated the exceptional flourishing of eighteenth-century science.

4 Concluding Remarks

Stevin may have been acquainted with Benedetti's work or some of his achievements in various ways. He could have accessed Benedetti's considerations on mechanics through the plagiarism of the Flemish scholar Jean Taisner.[57] His hydrostatics also share points of contact with some of Benedetti's technical solutions.[58] His work on mechanics shows a mathematical order that is not only reminiscent of Commandino's school, but also makes use of a distinction between theory and practice that is akin to that of contemporary Italian scholars.[59] The word Stevin used to refer to theory, *spiegheling* as opposed to *daet* (effect, action or practice), is the literal Dutch translation of Benedetti's

55 Biagioli, 'Prince et les savants', 1438.

56 For a portrait of such figures, read Ferrone, 'Età dei lumi', 67.

57 Taisner, *Opusculum*. See the discussion in Maccagni, 'Contributi alla bibliografia', 344–455, n. 13.

58 Cf. Chalmers's remarks in *One Hundred Years of Pressure*, 94.

59 As discussed by Van Dyck, 'Motion and Proportion in Simon Stevin's Mechanics'.

most cherished *speculatio*.[60] Unlike his Italian colleague, however, Stevin never obscured the connection between theory and practice. Rather, he keenly maintained that continuity, *'t menghen der spiegheling en daet*.[61] His desire to remain anchored in the terrain of technological practice typical of an engineer represents the main identity-related difference between the Dutch proto-scientist and his Italian counterpart.[62]

Practice and engineering are much more visible in Stevin's work and activities than in those of Benedetti.[63] Unlike the Savoy courtier, Stevin published works that were aimed at commerce, such as the *Tables of interest,* and contributed to pedagogical projects such as the reform of mathematical teaching for engineering schools. The differences between the court philosopher and the engineer undoubtedly reflected their different social backgrounds. Stevin started his career by engaging in activities that were very far from those of a patrician and an aristocrat. He was first a cashier and bookkeeper in Antwerp, established a reputation as a windmill engineer and public hydraulic engineer in Delft, and then became a teacher of mathematics at Leiden, before being promoted to quartermaster general of the Dutch armies and councilor to Prince Maurits of Nassau. When making sense of the different identities of the philosopher and the engineer, however, I regard the differences in political context as more important than biographical details. Stevin authored a booklet of political reflections, *Vita politica: het burgherlick leven* (1590), which bears witness to his commitment to a model of republican citizenship opposed to absolutism. Stevin emphasized the civic virtues of Dutch *burgherlickheid*, seen as an anti-monarchic model of political freedom.[64] In this light, one can consider his role as an engineer to be a dignified one involving public utility. In fact, Stevin directed his scientific-technological efforts towards societal 'progress', particularly to the benefit of the urban classes. This is particularly evident in his work on functional architecture and what we would now call urban studies,

60 The English word 'speculation' fails to capture the mirror metaphor in the original term, which brings together the ideas of reflection and abstraction.

61 See the chapter by Dijksterhuis in this volume.

62 I write 'proto-scientist' bearing in mind the controversial problem of the emergence of the figure of the 'modern scientist'. The seminal work on this issue is Zilsel, 'The Sociological Roots of Science', which identifies the various cultural elements (derived from scholasticism, humanism and craftmanship) that the generation of Gilbert, Galilei and Bacon brought together.

63 Even more than in the standard monograph on Stevin by Dijksterhuis, the preeminence of engineering and practical mathematics in Stevin's work emerges from Sarton's old but still valuable overview, 'Simon Stevin of Bruges'.

64 See the chapter by Secretan in this volume.

which is attentive to the specific life of the *burgher* class to which he himself belonged.[65]

The intellectual identity of an engineer in the bourgeois context of the United Provinces did not undergo the same aristocratic transformation that occurred in many Italian provinces, as an epistemological consequence of the passage from republicanism to court society in line with European tendencies toward absolutism. This produced a change in the self-perception and social ambitions of modern proto-scientists, particularly a dynamic interplay between the *social position* of the engineer and the *intellectual identity* of the court intellectual, which resulted in a change of epistemic values and *Denkstil*.[66] A new image of the 'mathematician' and the 'scientist' was slowly established; in fact, it has been argued that these two professions (or intellectual identities) assumed their modern forms at the beginning and at the end of the so-called Scientific Revolution, respectively.[67]

Benedetti's case illustrates this socio-political and epistemological shift from Renaissance *polytechny* towards court *polymathy*. Humanistic literacy and rhetoric were deployed to elevate the status and perception of mechanics, and make it acceptable to court intellectuals. In this context, the problem of the codification of practical and technical knowledge took the form of a social and intellectual appropriation by the new literate elites, at the expense of lower-class artisans and practitioners. This had consequences for the form and content of science, as Del Monte's mathematical mechanics and Benedetti's mathematical philosophy document in different ways. In the codification of court physico-mathematics, experience and know-how, the legacy of the artisans, lost its centrality. In addition, the specific contribution made by higher-status artisans and engineers through their methods and techniques was raised to the speculative heights of theory – *speculation*, as Benedetti claimed for himself, more precisely *speculatio mathematica et physica*. This surplus of theory and philosophy was the courtier's contribution to the codification of science, secured through the epistemological triad of theory, method and practice. As for the content, the new science was expanding its boundaries; the age of the mathematical conquest of the world had just begun.

In conclusion, I should summarize the issue of the ethos and culture of the emerging court society, and how it paradigmatically affected the enterprise of science. Courtly values entered the epistemology of court science in terms of

---

65 De Mare, 'Rule Worth Following'.

66 See Omodeo, 'Social Position and Intellectual Identity'.

67 Cf. Høirup, 'Practitioners', and Rossi and Ferrone, *Lo scienziato nell'età moderna*.

theoretical abstraction and knowledge hierarchization. The shift in epistemic values from practice to speculation is best captured if one considers the contrast between the engineer and the philosopher (or the mathematical purist). In court society, the social pyramid of the *signoria* and the absolutist monarchy was reflected in its cultural expressions, including science. A comparison between an Italian *signoria* and the Dutch case, however, suggests that social and epistemological differentiations were not only a matter of diachronic change, but also of geopolitical differentiation. The Seven United Provinces and the Duchy of Savoy are representative of different and in many ways opposite political models, with corresponding intellectual *ethoi*. The 'engineer' and the 'courtier' stand out as the 'organic intellectuals' of two different societal projects.

Acknowledgment

This essay was written in the framework of the ERC consolidator project *Early Modern Cosmology*, which received funding from the European Union's Horizon 2020 Research and Innovation Programme (GA 725883).

Bibliography

Azzolini, Monica, *The Duke and the Stars: Astrology and Politics in Renaissance Milan*, Cambridge, MA: Harvard University Press, 2013.

Baldi, Bernardino, *Cronica de matematici overo epitome dell'istoria delle vite loro*, Urbino: Monticelli, 1707.

Barberis, Walter, "Baldassar Castiglione: Gli ultimi bagliori dell'Umanesimo," in Baldassar Castiglione, *Il libro del Cortegiano*, Turin: Einaudi, 2017, v–lxviii.

Barberis, Walter, *Le armi del Principe: La tradizione militare sabauda*, Turin: Einaudi, 1988.

Becchi, Antonio, Domenico Bertoloni Meli and Enrico Gamba (eds.), *Guidobaldo del Monte (1545–1607): Theory and Practice of the Mathematical Disciplines from Urbino to Europe*, Berlin: Edition Open Access, 2013.

Benedetti, Giovanni Battista, *Diversarum speculationum mathematicarum et physicarum liber*, Turin: Haeredis Nicolai Bevilaquae, 1585.

Biagioli, Mario, "Le prince et les savants: La civilté scientifique au 17e siécle," *Annales: Historie, Sciences Sociales* 6: 1417–1453.

Biagioli, Mario, "The Social Status of Italian Mathematicians, 1450–1600," *History of Science* 27/1 (1989) 41–95.

Biagioli, Mario, *Galileo, Courtier: The Practice of Science in the Culture of Absolutism*, Chicago-London: The University of Chicago Press, 1993.

Bordiga, Giovanni, *Giovanni Battista Benedetti, filosofo e matematico veneziano nel secolo XVI con un aggiornamento bibliografico ragionato di Pasquale Ventrice*, Venice: Istituto Veneto di Scienze, Lettere ed Arti, 1985.

Botero, Giovanni, *Relationi universali*, Vicenza: heredi di Parin librario, 1595.

Burke, Peter, "Il cortigiano," in *L'uomo del Rinascimento*, ed. Eugenio Garin, Bari–Roma: Laterza, 1988, 133–166.

Büttner, Jochen, "Big Wheel Keep on Turning," *Galilaeana* 5 (2008): 33–62.

Castiglione, Baldassar, *Il libro del Cortegiano*, Turin: Einaudi, 2017.

Cecchini, Michela and Clara Silvia Roero, "I corrispondenti di Giovanni Battista Benedetti," *Physis* 41/1 (2004) 31–66.

Chalmers, Alan F., *One Hundred Years of Pressure: Hydrostatics from Stevin to Newton*, Cham: Springer, 2017.

Chastel, André, "L'artista," in *L'uomo del Rinascimento*, ed. Eugenio Garin, Bari–Roma: Laterza, 1988, 237–270.

D. Pollak, Martha, *Turin 1564–1680: Urban Design, Military Culture, and the Creation of the Absolutist Capital*, Chicago-London: The University of Chicago Press, 1991.

De Mare, Heidi, "A Rule Worth Following in Architecture? The Significance of Gender Classification in Simon Stevin's Architectural Treatise (1548–1620)," in *Women of the Golden Age*, ed. Els Kloek et al., Hilversum: Verloren, 1994: 103–120.

Del Monte, Guidobaldo, *Le mechaniche ... tradotte in volgare dal Sig. Filippo Pigafetta*, Venice: Francesco di Franceschi Sanese, 1581.

Dijksterhuis, Eduard J., *Simon Stevin: Science in the Netherlands around 1600*, The Hague: Nijhoff, 1970.

Drake, Stillman and Israel Edward Drabkin, *Mechanics in Sixteenth-Century Italy: Selection from Tartaglia, Benedetti, Guido Ubaldo and Galileo*, Madison: University of Wisconsin Press, 1969.

Elias, Norbert, *Die höfische Gesellschaft*, Frankfurt a.M.: Suhrkamp, 2002.

Ferrone, Vincenzo, "L'età dei lumi," in Vincenzo Ferrone and Paolo Rossi, *Lo scienziato nell'età moderna*.

Firpo, Luigi, *Relazioni di ambasciatori veneti al Senato: tratte dalle migliori edizioni disponibili e ordinate*, vol. XI, Turin: Bottega di Erasmo, 1983.

Frank, Martin, *Guidobaldo del Monte's Mechanics in Context: A Research on the Connections between His Mechanical Work and His Biography and Environment*, PhD Thesis (Pisa 2011).

Freudenthal, Gideon and Peter McLaughlin (eds.), *The Social and Economic Roots of the Scientific Revolution: Texts by Boris Hessen and Henryk Grossmann*, Dordrecht: Springer, 2009.

Garin, Eugenio, *La cultura filosofica del Rinascimento italiano*, Milan: Bompiani, 1994.

Hessen, Boris, "The Social and Economic Roots of Newton's Principia," in *Science at the Cross Roads*, Londen: Kniga 1931, 147–212.

Høirup, Jens, "Practitioners – School Teachers – 'mathematicians': The Divisions of Pre-Modern mathematics and Its Actors" (2003): http://akira.ruc.dk/~jensh/Publications/2003%7BK%7D04_Divisions%20of%20pre-Modern%20mathematics.PDF (accessed 18 January 2018).

Kamen, Henry, *The Iron Century: Social Change in Europe 1550–1660*, Worcester-London: The Trinity Press, 1971.

Lefèvre, Wolfgang, "Galileo Engineer: Art and Modern Science," in *Galileo in context*, ed. Jürgen Renn, Cambridge: Cambridge University Press 2001, 11–27.

Lefèvre, Wolfgang, *Naturtheorie und Produktionsweise, Probleme einer materialistischen Wissenschaftsgeschichtsschreibung: Eine Studie zur Genese der neuzeitlichen Naturwissenschaft*, Darmstadt-Neuwied: Hermann Luchterhand, 1978.

Long, Pamela O., *Artisan/Practitioners and the rise of the new sciences, 1400–1600*, Corvallis: Oregon State University Press, 2011.

Maccagni, Carlo, "Contributi alla bibliografia di Giovanni Battista Benedetti," *Physis* 9/3 (1967): 338–364.

Maccagni, Carlo, *Le speculazioni giovanili "de motu" di Giovanni Battista Benedetti*, Pisa: Domus Galilaeana, 1967.

Olschki, Leonardo, *Geschichte der neusprachlichen wissenschaftlichen Literatur*, 3 vols., *Galilei und seine Zeit*, Halle Saale: Niemeyer, 1919–1927.

Omodeo, Pietro Daniel and Jürgen Renn, *Science in Court Society: Giovanni Battista Benedetti's* Diversarum speculationum mathematicarum et physicarum liber (*Turin, 1585*), Berlin: Edition Open Access, 2019.

Omodeo, Pietro Daniel, "Efemeridi e critica all'astrologia tra filosofia naturale ed etica: La contesa tra Benedetti e Altavilla nel tardo Rinascimento torinese," in *Preprints of the Max Planck Institute for the History of Science* 458 (2014).

Omodeo, Pietro Daniel, "Polemiche astronomico-astrologiche nella Torino di Benedetti," *Galilaeana: Journal of Galilean Studies*, 11 (2014): 73–103.

Omodeo, Pietro Daniel, "Renaissance Science and Literature: Benedetti, Ovid and the Transformations of Phaeton's Myth after Copernicus," *Science & Education* 21 (2012), 1–8.

Omodeo, Pietro Daniel, "Socio-Political Coordinates of Early-Modern Mechanics: A Preliminary Discussion," in *Emergence and Expansion of Pre-Classical Mechanics, Boston Studies in the Philosophy and History of Science*, ed. Rivka Feldhay, Jürgen Renn, Mattias Schemmel, and Matteo Valleriani, Cham: Springer, 2018, 55–78.

Omodeo, Pietro Daniel, "The Social Position and Intellectual Identity of the Renaissance Mathematician-Physicist Giovanni Battista Benedetti: A Case Study in the Socio-Political History of Mechanics," in *Emergence and Expansion of Pre-Classical Mechanics*, ed. Rivka Feldhay, Jürgen Renn, Mattias Schemmel and Matteo Valleriani (Cham: Springer, 2018), 181–213.

Palladio, Andrea, *I quattro libri dell'architettura*, Venice: Dominico de' Franceschi, 1570.

Pozzi, Mario, *Filippo Pigafetta consigliere del Principe*, Vicenza: Biblioteca Civica Bertoliana, 2004.

Renn, Jürgen and Peter Damerow, *Guidobaldo del Monte's* Mechanicorum Liber, Berlin: Edition Open Access, 2010.

Renn, Jürgen and Damerow, Peter, *The Equilibrium Controversy: Guidobaldo del Monte's Critical Notes on the Mechanics of Jordanus and Benedetti and their Historical and Conceptual Background*, Berlin: Edition Open Access, 2012.

Renn, Jürgen and Peter Damerow, "The Transformation of Ancient Mechanics into a Mechanistic World View," in *Transformationen antiker Wissenschaften*, ed. Georg Toepfer and Hartmut Böhme, Berlin-New York: de Gruyter, 2010, 243–267.

Renn, Jürgen, "Florenz: Matrix der Wissenschaft," in *Florenz! Die Stadt der Medici kommt an den Rhein*, Bonn: Hirmer, 2014, 101–112.

Ricuperati, Giuseppe (ed.), *Storia di Torino: Dalla dominazione francese alla ricomposizione dello Stato (1536–1630)*, vol. 3, Turin: Einaudi, 1998.

Rose, Paul Lawrence, *The Italian Renaissance of Mathematics: Studies on Humanists and Mathematicians from Petrarch to Galileo*, Geneva: Droz, 1975.

Rossi, Paolo, *The Birth of Modern Science*, Oxford: Blackwell, 2001.

Sarton, George, "Simon Stevin of Bruges," *Isis* 21/2 (1934): 241–303.

Schemmel, Matthias, *The English Galileo: Thomas Harriot's Work on Motion as an Example of Pre-Classical Mechanics*. Dordrecht: Springer, 2008.

Smith, Pamela, *The body of the artisan: art and experience in the Scientific Revolution*, Chicago: The University of Chicago Press 2004.

Taisner, Jean, *Opusculum perpetua memoria dignissimum, de natura magnetis, et eius effectibus*, Cologne: Johann Birckmann, 1562.

Tessari, Antonio Secondo, "Sul soggiorno di Andrea Palladio a Torino per le questioni militari di Emanuele Filiberto," *Studi Piemontesi* 22/1 (1993) 9–20.

Valleriani, Matteo, *Galileo Engineer*, Dordrecht: Springer, 2010, 197.

Van Dyck, Maarten, "Motion and Proportion in Simon Stevin's Mechanics," in *Eppur si muove: Doing History and Philosophy of Science with Peter Machamer: A Collection of Essays in Honor of Peter Machamer*, ed. Marcus P. Adams *et al.* (Cham: Springer, 2017), 21–37.

Woolf, Stuart J., "Sviluppo economico e struttura sociale in Piemonte da Emanuele Filiberto a Carlo Emanuele III," *Nuova Rivista Storica* 46/1 (1962) 1–57.

Zilsel, Edgar, "The Sociological Roots of Science," *American Journal of Sociology* 47/4 (1942) 544–562.

CHAPTER 2

Vitruvian Universalism

On the Order of Mechanical Knowledge in the Works of Joseph Furttenbach the Elder (1591–1667)

Jan Lazardzig

Can there be a template or model for being a polymath? Contesting long-held ideas about the new 'art of knowing' in the early modern period, one might ask how certain ideas and concepts of polymathy materialized, played out and were transformed in accordance with existing models or concepts of knowledge. Did polymaths such as Simon Stevin find an epistemic template for their engagement in diverse fields and matters, including mathematics, mechanics, hydrostatics, astronomy, geography, navigation, technology, military science, book-keeping, architecture, music and logic? And how did these templates relate to social patterns and the local conditions of polymathy? In this chapter, I want to examine a specific variant of mechanical universalism in the works of a contemporary of Simon Stevin (1548–1620), the Swabian architect, engineer and merchant Joseph Furttenbach the Elder (1591–1667). Furttenbach's universalism emerged and operated in the context of patronage, provided by networks of merchants and craftsmen. Unlike Stevin, who enrolled as a student at Leiden University, Furttenbach derived his knowledge from self-study, artisan workshops, travel and private tutoring. In addition to his almost two dozen publications on matters such as artillery and fireworks, and civil and military architecture (fortifications), Furttenbach published a travel guide to Italy and texts on palace and garden architecture, grottoes, theaters and machinery, ships and, last but not least, mechanical instruments. Taking the devastation and desolation caused by the Thirty Years' War as a point of departure, Furttenbach's mechanical universe was cast in a postlapsarian light, imbued with biblical references. His celebrated *Rüst- und Kunstkammer*, a sought-after cabinet of mechanical curiosities with a vast number of architectural models that he erected in his private house in Ulm, was a material manifestation of what I would call *Vitruvian universalism*, or *polymathy*. As Paolo Sanvito and others have shown, Vitruvius' *De architectura libri decem* [Ten Books on Architecture] offered an epistemic template for mathematicians, engineers and architects between

 | DOI:10.1163/9789004432918_004

1500 and 1800, linking mathematics and mechanics with a diverse range of knowledge.[1]

Vitruvius' *De architectura* was first printed around 1487 in Italy and subsequently translated into German (Basel, 1514), Italian (Milan, 1521), Spanish (Toledo, 1542) and French (Paris, 1539/1547).[2] Furttenbach, who lacked the linguistic proficiency to read the *Libri decem* in Latin, might have had access to the Italian translation by Daniele Barbaro (Venice, 1556) during his stay in Genoa. Later, the German translation by Walter Ryff/Rivius (Nuremberg, 1548, 1575, 1614) may have entered Furttenbach's library, which contained (by his own account) some 140 books on architecture and mechanics. Furttenbach rarely gave indications of his theoretical reading on architecture and mechanics. Time and again, Furttenbach instead stressed his practical experience, his apprenticeship with (Italian) masters, and his training at academies of architecture, engineering and artillery. Though largely absent as a theoretical voice, Vitruvius, who is mentioned on several occasions with regard to his polymathy, seems to be present in the very concept of the universal architect that Furttenbach strove to embody as an architect-engineer and city master-builder of Ulm. Hence, Vitruvianism is primarily present in Furttenbach's work not as an aesthetic ideal but, as I shall argue, as an epistemic imperative for being universally knowledgeable in architecture, engineering and all related matters.

At the beginning of book I, Vitruvius famously offers an account of the universal knowledge that an architect should acquire. The opening of the *Libri decem* imbues everything that follows with a sense of exemplarity or ideality, comparable to a template or even a textbook or guideline for future architects. To begin with, the knowledge of the architect and master-builder is characterized by the interrelation of theory and practice, *fabrica* (craftsmanship) and *racionatio* (mental activity):

> THE architect should be equipped with knowledge of many branches of study and varied kinds of learning, for it is by his judgement that all work done by the other arts is put to test. This knowledge is the child of practice and theory. Practice is the continuous and regular exercise of employment where manual work is done with any necessary material according to the design of a drawing. Theory, on the other hand, is the

1 Sanvito, 'Introduction'.

2 See Fensterbusch, 'Einleitung', 13–14.

ability to demonstrate and explain the productions of dexterity on the principles of proportion.[3]

According to Vitruvius, the master-builder has to acquire proficiency in writing as well as in design and painting. Furthermore, he has to gain knowledge in various fields such as geometry, perspective and arithmetic, history and mythology, philosophy, music, medicine, law and astrology.[4] Vitruvius emphasizes the length and duration of these studies and compares them to a kind of step-ladder which the apprentice has to climb. Although they may appear heterogeneous to the layman, the integrality of these diverse fields of knowledge becomes evident in the task of the architect itself.

> Consequently, since this study is so vast in extent, embellished and enriched as it is with many different kinds of learning, I think that men have no right to profess themselves architects hastily, without having climbed from boyhood the steps of these studies and thus, nursed by the knowledge of many arts and sciences, having reached the heights of the holy ground of architecture. But perhaps to the inexperienced it will seem a marvel that human nature can comprehend such a great number of studies and keep them in the memory. Still, the observation that *all studies have a common bond of union and intercourse with one another*, will lead to the belief that this can easily be realized. For a liberal education forms, as it were, a single body made up of these members. Those, therefore, who from tender years receive instruction in the various forms of learning, recognize the same stamp on all the arts, and an

---

3 Vitruvius, *The Ten Books on Architecture*, 1.1.1. (my italics) In Rivius' translation (1st ed., 1548, 2v): 'Die Architectur .../ ist ....mit vilfeltigen andren künsten und wissen geziert/ dadurch der erfarne Bawmeister auch die werck aus dem grundt andrer kunst volbracht urtheilen und approbirn kan. Und hat dise kunst iren grundt und fundament in zweyerley sachen .... Ein ding .... in das werck künstlichen zu bringen/ erfordert ernstliche und fleissige nachtrachtung mit gantz fleissiger auffmerckung teglicher übung/ .../ Aber ursach dar zuthun solcher werck/ ist das man aus verstandt und wissen anzeige/ ire proportion'.

4 Idem, 4. 'Darumb ist von nöten / das ein füurtrefflicher berhümbter vollkommener Bawmeister fur das aller erst gelehrt sey / und der Schrifft wol erfahren/ Er soll auch des Reissens und Malens einen guten bericht haben/ auch der Geometrei/ das ist deß Zirckels und Richtscheidts gerechtigkeit / deßgleichen der Perspectiua und Rechenkunst / der Historien oder alten geschicht sol er vorauzimlich Wissen haben / auch der Philosophei etwas bericht seyn / die Musica können / und der Reglen der Ertzney etwas erfarnen sein/ wie auch der Satzungen der Juristen/ in der Astrologeiist im auch nit wenig von nöten / daser ein wissen hab des Himels lauff'.

intercourse between all studies, and so they more readily comprehend them all.[5]

The idea of interconnectedness and mutual benefit between the different fields or branches of knowledge, epitomized in the image of a chain – *encyclios omnium doctrinarum disciplina* (Vitruvius) – is not restricted to architecture or technology, but can be found in rhetoric as well (Cicero, Quintillian).[6] The metaphor of a *chain of knowledge* promotes the idea of eclectic learning, but also the idea of order, necessity and specialization. It thus became popular in the sixteenth and seventeenth centuries, promoting education in the liberal arts as well as the ideal of the *uomo universale*; and it was central to early modern notions of polymathy.[7]

According to Hole Rößler, there is significant overlap between *polymathia* as a concept and *polyhistoria* and *topica universalis*.[8] *Polymathia* literally signifies a comprehensive, transdisciplinary inventory of knowledge particles, the cohesion of which cannot necessarily be deduced from its individual elements. It describes an attitude towards knowledge whereby disciplinary borders were often merged or transcended, questioned or displaced. Among the 'polyhistors' (polymaths) identified by Rößler are well-known persons such as Giovanni Battista Benedetti (see the chapter by Pietro Daniel Omodeo), Hermann Conring, Conrad Gesner, Gottfried Wilhelm Leibniz, Daniel Georg Morhof, Caspar Peucer, Vincentius Placcius, Conrad Samuel Schurtzfleisch, Gerhard Vossius and Johann Christoph Wagenseil. Here it becomes clear that access

5 Vitruvius, *The Ten Books on Architecture*. I.1.11 and I.1.12 (my italics). In Rivius' German translation (4v/5r): 'Dieweil aber nun die Architectur oder kunst des Bawwercks / mit solchen / und auch so vielen furnemen und trefflichen künsten geziert ist / mag nicht wol müglich sein / das einer dieser kunst in eyl / und also stümpflichen gnugsamen bericht empfahe / es sey dann sau/ *das er in solchen künsten und wissen von jugendt auff / von tag zu tag auffsteige / so lang daser zu oberst komen / und den höchsten spitz und besten kern erlange*. Es möcht aber die unverstendigen wunder nemen / wie es der Natur müglich / das ein mensch so vielerley Künst möcht erlernen / und in gedechtnusfassen oder behalten. Solche aber wo sie vernemmen oder mercken / wie alle diese obgemelte gute freye oder Mathematische kunst / dermassen einander verbunden unnd angehörig seind / werden sie uns solchs leichtlichen zugeben müssen / das es wol müglich / dann sie alle wie ein ketten aneinander gehefft / unnd dermassen verschlossen / das sie schwerlich getheilt oder abgesondert werden mögen'.

6 Cicero, *Pro Archia Poeta*, II, 2; Quintilian, *Institutio Oratoria*, 12, 3–4.

7 On the early modern concept of polymathia, see Rößler, 'Polyhistorie und Polymathie', 660.

8 Rößler states that what appears as *polyhistoria* in German territories qualifies as *polymathia* elsewhere. See ibid., 659.

to Latin was a condition *sine qua non* for the polyhistor or *multiscius*, who, according to Nicolas du Mortier's *Etymologiae Sacrae Graeco-Latinae* (1703), knew and remembered much ('qui multa novit et memorat').[9] Thus, unlike Simon Stevin, who could be listed among the learned men mentioned above, Furttenbach would not necessarily have qualified as a polymath. He did not go beyond the disciplinary borders of knowledge, but interpreted an existing topos of universal knowledge, the Vitruvian ideal of the universal architect. But the differences are less clear than one might think, in particular with regard to Stevin's concern with architecture.

Among the Renaissance architects north of the Alps who published written works, including Hans Blum (1520/27-after 1562), Philibert De l'Orme (1512/15–1570), Jacques Androuet Du Cerceau the Elder (ca. 1520–1585/86), Hans Vredeman de Vries (1526–1609), Daniel Specklin (1536–1589), Simon Stevin, Wendel Dietterlin (1550/51–1599), Heinrich Schickhardt (1558–1634) and Elias Hall (1573–1646), Furttenbach's theoretical work stands out for its preoccupation with the ordering and displaying of mechanical knowledge. Unlike other writers on architecture, Furttenbach develops his idea of universal architecture not by interpreting Vitruvius' order of pillars, but by re-thinking the encyclopedic order or connectedness of architectural and technical knowledge in the broadest possible sense.

In what follows, I shall analyze Furttenbach's self-fashioning as a person with a universal knowledge of architecture and engineering, first by examining his publications, in particular the frontispieces of three of his publications, *Architectura Universalis* (1635), *Mechanische ReißLaden* (1644) and *Mannhaffter Kunst-Spiegel* (1663). Furttenbach systematically used the frontispieces of his books to reflect upon the knowledge presented to the reader. I shall then focus on Furttenbach's Vitruvian universalism as it was displayed in his *Kunstkammer*, which not only presented models, sketches, drawings, instruments and objects of all kinds, but also set up a specific order of mechanical knowledge for his visitors. The *Kunstkammer* and Furttenbach's theoretical publications were imbued with a tightly-knit system of mutual references. Visitors entered an exhibition space for the epistemic objects (models, instruments, sketches, paraphernalia from his travels, etc.) that Furttenbach referred to in his books (and vice versa). Readers of Furttenbach's books were likewise bombarded with references to other texts and objects from the architect's universe of mechanical knowledge.

---

9 Ibid., 643.

1 Peregrinatio Mechanica and Militaris

Joseph Furttenbach the Elder (1591–1667) was born in 1591 in the imperial city of Leutkirch in Württemberg, the twentieth child of the patrician Hieronymus II Furtenbach (1539–1596) and his second wife, Clara Föls of Constance.[10] His father, a forestry superintendent, was the head of the city council, the city treasurer and master mason of Leutkirch. At the age of ten, Joseph's mother sent him to the German school in Isny in the Allgäu, although he moved back to the city chancery in Leutkirch only a few years later. Around the age of sixteen, Joseph traveled to Italy in order to complete an apprenticeship as a merchant there. The members of his family maintained exceptionally close trade relations with Upper Italy. After a two-year stay in Milan, where he learned Italian (amongst other things), Joseph moved to Genoa, where he would remain for seven years (albeit interrupted by lengthy travels that would take him to Rome and to Florence, where he stayed for at least a year).

Italy inspired Furttenbach's interest in architecture and construction, and his stay there – originally conceived as a mercantile peregrination – was transformed into a *peregrinatio mechanica et militaria* (Johann Heinrich Alsted).[11]

10 On Joseph Furttenbach's biography, see: Berthold, *Joseph Furttenbach (1591–1667)*. A condensed version of this study appeared in essay form: Berthold, 'Joseph Furttenbach von Leutkirch'. See also the new edition of the second (and only extant) part of Furttenbach's diary-like 'Lebenslauf' ('curriculum vitae'), based on recent scholarship: Furttenbach. *Lebenslauff 1652–1664*.

11 Following Johann Heinrich Alsted (1588–1638), Furttenbach's travels can be described as a *peregrinatio mechanica et militaris*. In his *Encyclopaedia* (2nd ed. 1649), Alsted combines the apprentice's journey to Italy to learn civil technology with the journey to learn military technology, since in both one studies technology by means of a concrete object, while also considering the context of its application. He explains that 'the mechanic carefully weighs which art is to be learned, and which examples are to be regarded, in which place. For the mechanical arts are not the same in each of the cities. Thus, e.g., architects, sculptors, painters, weavers, which are generally respected, live in Italy, potters and carpenters of all kinds in Germany'. ('Mechanicus diligenter expendat, quae ars discenda, quaeque exempla spectanda, quo loco. Non enim mechanicae artes in singulis urbibus sunt eaedem. Sic v.g. architecti, statuarii, pictores, textores, quae observent, habebant in Italiam: figuli & fabri omne genus in Germaniam'.) Alsted, *Scientiarum Omnium Encyclopaediae*, 250. The not uncommon figure of the dual war- and theater-engineer shows that the course of Furttenbach's own education was in no way unusual for architect-engineers. On this topic, see also Adami, 'Tra guerra e teatro'. On the early modern education of architects, see Binding's survey work *Meister der Baukunst*; Johannes (ed.), *Entwerfen.* (unfortunately, this does not include a discussion of architects' journeys). For Florence and Upper Italy specifically, see: Kiene, 'Das Berufsbild des Architekten im 16. Jahrhundert'; Lorenzo, 'Being an engineer and being an architect'. On early modern educational journeys in

Furttenbach, who had attended neither Latin school nor university, sought access to 'the most renowned engineers and their academies' ('zu den vornembsten Ingegnieri und in dero *Achademien*'), in order, as he formulated it in the foreword to his 1627 travel handbook *Newes Itinerarium Italiae*, 'not only to hear the learned discourses there, but also to see if these would enable me to carry out and complete the work myself'.[12] In Genoa, he was introduced to the (Italian) arts of fortification architecture, the design of *castra doloris* (decorated catafalques), and perspectival construction by Paolo Rizio – 'an excellent engineer whom God has gifted with understanding, whom I can boast is my honored patron and instructor'.[13] Moreover, in Genoa he also acquired a certificate of apprenticeship in the *Arte de la Bombardieri*, this time from the artillery and pyrotechnics master Hans Veldhausen of Augsburg.

Most decisive for Furttenbach's training in mechanics, perspective and the art of civil and military architecture was probably his stay in 1617 at the private academy of Giulio Parigi (1571–1635) in Florence, which lasted approximately one year. Giulio, son of Alfonso Parigi the Elder (†1590) and student of Bernardo Buontalenti (1531–1608), came from a family of architects. He worked as an architect, painter and festival organizer for the Medici.[14] In 1594 he became a member of the *Accademia del Disegno*, which was supported by the Medici and dedicated to the education of artists and architects. On several occasions, he held the academy's offices of consul and *festaiolo*, the director of courtly festivities.[15] The little we know about Parigi's *Accademia* comes from scattered remarks in the artists' biographies by Filippo Baldinucci (1625–1696). Regarding the syllabus, we read: 'Giulio Parigi had opened a school – or perhaps we should say an academy – in which one reads Euclid and is instructed in

---

Germany: Berns, 'Peregrinatio academica und Kavalierstour'. (This publication mainly discusses the literary results of travel.)

12 '[a]llda nit allein ihre vernünfftige discursi zuhören/ sonder auch ein Muht zu fassen/ ob ich durch derselben befürderung/ die Werck selber führen unnd zu end bringen möchte'. Joseph Furttenbach. '*Newes Itinerarium Italiae: In welchem der Reisende nicht allein gründtlichen Bericht, durch die herrlichste namhaffteste örter Italiae sein Reiß wol zubestellen, sonder es wirdt jhme auch* [...] *beschrieben, was allda* [...] *an fürstlichen Hoffhaltungen* [...] *denckwürdig zu sehen* [...]', Furttenbach, *Newes Itinerarium Italiae*, foreword, n.p.

13 '[e]in von Gott mit Verstand hochbegabter trefflicher Ingegnier, den ich für mein hochgeehrten Patron und Lehrmeister berühme'. This is probably the engineer and cartographer Pier Paolo Rizzio, who is little-known today. See Quaini, 'Cartographic Activities', 868 (with references). Reference is made to a drawing of Rizzio's in: Graziani (ed.), *Vistighe Corse*, 240.

14 See Negro Spina, *Giulio Parigi e gli incisori della sua cerchia*; Smalzi, 'Giulio Parigi architetto di corte'.

15 See Schlimme, 'Die frühe Accademia et Compagnia dell'Arte del Disegno'.

mechanics, perspective, and the art of civil and military construction'.[16] Alongside artists, Parigi's private academy admitted the sons of the Italian nobility. Their education was oriented less towards an artistic-artisanal praxis – that is, sketching and modelling, engraving and etching – than toward a theoretical and aesthetic education that would later allow them to assess architectural and technical designs adequately. The academy thus possessed some traits of a knight's academy.[17] Coming from a family of merchants, Furttenbach's affinity with practical mathematics might have helped him to meet the requirements. As is evident from his later accounts, Furttenbach clearly took pride in having studied engineering. Alongside the Italian and Florentine architects, artists and scenographers who attended Parigi's academy were artists such as Jacques Callot and Remigio Cantagallina, who contributed to the school's international renown. Furttenbach collected a large number of engravings and models at Parigi's academy that would later form the basis for his own design practice (and eventually find their way into his private *Kunstkammer*).[18] Whereas from a courtly perspective engineering and architecture in particular might then still have been seen as 'lower' forms of (practical) knowledge, from the perspective of a young merchant from Ulm, architecture and engineering offered a pathway to recognition and social status. As Omodeo has shown in his contribution to this volume, Giovanni Battista Benedetti contrasted himself, as a courtier and polymath, to the more lowly engineer.

In January of 1620, the now 29-year-old Furttenbach returned to Upper Swabia. In the following year, he settled in Ulm, where he became a manager of sorts for a local merchant.[19] In 1627 Furttenbach began his publishing career, which would make him famous in the German-speaking world as an architectural theorist who was equally well-versed in ship-building and theater construction, garden design and the construction of fortifications and artillery. By the time of his death in 1667, Furttenbach had authored more than twenty richly illustrated treatises, which were printed first in Ulm and later in Augsburg and disseminated widely throughout the German-speaking world.[20]

---

16 'Aveva Giulio Parigi eretta in casa sua una scuola, o vogliamo dire accademia, nella quale leggeva Euclide, insegnava le meccaniche, prospettiva, architettura civile e militare'. Quoted following ibid., 339f.

17 See Conrads, *Ritterakademien der Frühen Neuzeit*.

18 Furttenbach, *Architectura Privata*, 37, 46. For his training in scenography and theater machinery at Parigi's Florentine academy, see Lazardzig/Rößler, 'Joseph Furttenbach and the Transfer of Mechanical Knowledge'.

19 Greyerz, 'Joseph Furttenbach', 18.

20 Furttenbach's description of his Italian journey (*Newes Itinerarium Italiae*, 1627), a book on artillery (*Halinitro Pyrobolia*, 1627), and treatises on civil architecture (*Architectura Civilis*,

When compared with contemporary writings on architectural theory, Furttenbach's books are captivating due to their decidedly non-academic approach. They neither perform Vitruvius' exegesis nor do they speculate on the ordering of columns. Instead, Furttenbach favors experience and the knowledge of craftsmen. Against the background of the Thirty Years' War, his discussions are always directed toward the practical, contextually-determined practicability and immediate applicability of mechanical and architectural knowledge. As a Lutheran, Furttenbach equipped his architectural program with the religiously motivated promise of wide-ranging *recreation* (in the dual sense of reconstruction and recovery) for a society damaged by war. Here, there are strong parallels between his mechanical universalism and the biblical story of the creation, or rather recreation. In 1631, Furttenbach was named master mason of the city of Ulm, and in 1636 he was elected to the city council. As the city's master builder (and not as a 'professional' architect), Furttenbach was not only responsible for much repair work, but also for the construction of a new hospital *extra muros*, a military hospital and a German school. He also designed his own town house and transformed the former Dominican cloister into a permanent theater. Furttenbach achieved renown beyond the German-speaking world for his cabinet of curiosities or *Kunstkammer*, which was frequented by countless locals and foreigners. It was located in the attic of his town house in Ulm, which he had built in 1640.

A half-length portrait of Furttenbach (figure 2.1), drawn by Johann Jakob Campanus and engraved by Matthäus Rembold in 1635, presents Furttenbach at the height of his career, fashioned as a man of noble descent and bearing the insignia of universal architecture.[21] In the foreground on the right, Mars

---

1628), shipbuilding (*Architectura Navalis*, 1629) and fortress construction (*Architectura Martialis*, 1630) appeared in quick succession. After becoming the master mason of Ulm in 1631, he published a universal architectural text on civil and military construction (*Architectura Universalis*, 1635), a further treatise on civil architecture (which also contained his well-known designs for a *periaktoi* stage – *Architectura Recreationis*, 1640), a text on the house he built in Ulm in 1640 (*Architectura Privata*, 1641), a further book on artillery (*Büchsenmeisterey-Schul*, 1643), a book on instruments (*Mechanische ReißLaden*, 1644), and finally his most comprehensive work (*Mannhaffter Kunst-Spiegel*, 1663), in which he recapitulated all of his learning. Links to high-quality digital versions of all of Furttenbach's writings (and to the ten architectural treatises that he published under the name of his son, Joseph Furttenbach the Younger), as well as to biographical sources have been collected by Hole Rößler, are available at http://www.holeroessler.de/furttenbach.html. Until now, only a few passages on scenography and theater construction have been translated into English and made available to non-German-speaking theater scholars. See Hewitt (ed.), *The Renaissance Stage*.

21 Sebastian Fitzner notes that Furttenbach's posture and gestures reflect his rank and status. The branch of the Furttenbach family from Leutkirch had been knighted ('Reichsritter')

FIGURE 2.1 Matthäus Rembold (engraver), portrait of Joseph Furttenbach the Elder, copperplate, 36.8 x 21.4 cm, 1635. Joseph Furttenbach. *Architectura Universalis: das ist: von Kriegs-, Statt- u. Wasser-Gebäwen.* [*Architectura Universalis: that is, of war, city and water buildings.*] Ulm: Meder, 1635, frontispiece

SOURCE: COPYRIGHT UNIVERSITY LIBRARY HEIDELBERG (HTTPS://DIGI.UB.UNI-HEIDELBERG.DE/DIGLIT/FURTTENBACH1635/0007)

embodies the *architectura militaris*, bearing armor, sword and lances. Behind him is a canon, symbolizing artillery. On the left, we see an allegorical personification of the *architectura civilis* with circle and staff, sitting on a scale model of a palazzo-like building. Above the female figure are two tables depicting a floor plan and an order of columns. In the center, we see a ship which illustrates the *architectura navalis*. The two putti holding the cartouche with the coat of arms of the Furttenbach family are equipped with circle, staff and plumb-line, referring to geometry as a prerequisite of all mechanical arts, symbolized by various paraphernalia behind the putti. Furttenbach's motto *Con la Patienza/S'aquista Scienza* ('Knowledge is acquainted with patience') evokes Vitruvius' account of the time and patience needed to become an architect. According to the caption, Furttenbach is the 'most experienced friend' of the technical, i.e., mechanical arts ('viro spectatissimo, technophilo pertissimo'). A later and more somber portrait of Furttenbach from 1652 omits most of the symbols of universal architecture. Instead, Furttenbach is presented as the author of the *Mannhaffter Kunst-Spiegel*, his last book, which was not published until 1663.

2 Self-Fashioning: the Locality and Universality of Knowledge

Furttenbach's publications not only address fellow engineers and architects, craftsmen and master builders, but are also directed towards broader technophile audiences (wealthy citizens, noblemen) who share his fascination for technical creation as well as for architectural fantasies. In this regard, his books have much in common with other image-savvy technical books on early modern engineering, such as the encyclopedic *Theatrum machinarum* treatises. Here too, technological inventions such as machines and instruments are presented as both useful and playful, adhering to the juxtaposition of reason and entertainment. But in contrast to the technical books by Italian Renaissance engineers, who promoted their ingenuity to attract the attention of prospective commissioners and customers, Furttenbach's writings seem to promote the idea and ingenuity of architectural creation, not so much for prospective building owners or commissioners, but for the recreation-seeking readers of his books. Hence, his books are not a means to an end, but an end in themselves.

---

by Emperor Ferdinand II in 1623. See Furttenbach, *Mechanische ReißLaden*, ed. by Fitzner, 23.

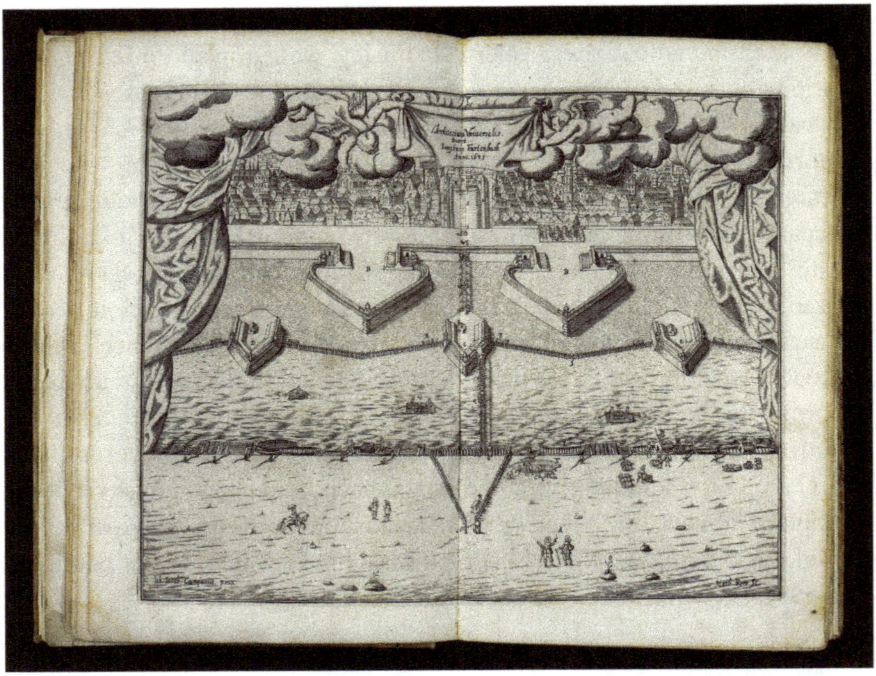

FIGURE 2.2 Joseph Furttenbach. *Architectura Universalis: das ist: von Kriegs-, Statt- u. Wasser-Gebäwen.* [*Architectura Universalis: that is, of war, city and water buildings.*] Ulm: Meder, 1635, frontispiece
SOURCE: COPYRIGHT UNIVERSITY LIBRARY HEIDELBERG (HTTPS://DIGI.UB.UNI-HEIDELBERG.DE/DIGLIT/FURTTENBACH1635/0025)

In Furttenbach's skillfully-illustrated publications, the frontispieces play a highly significant role. They indicate not only the content of the book but also an order of knowledge, presenting the architect-theorist as universally knowledgeable. In accordance with a common practice of his time, most of these frontispieces are accompanied by an interpretive poem.

In his *Architectura universalis* (1635), the frontispiece (figure 2.2) – designed by Furttenbach, drawn by Johann Jakob Campanus and engraved by Matthias Rembold – offers a bird's-eye (albeit strictly compartmentalized) view on a well-fortified city. The picture is dominated by an extensive fortification system (fences, water pits, glacis, bastions, ravelins, walls). A draped curtain on both sides creates the impression of a proscenium stage. Thick cumulus clouds define the upper frame. The putti on the upper frame present a cloth bearing the title and author of the book. In the lower section of the print, workers, merchants and carpenters go about their business. Letters and symbols are allocated to the different elements of the fortification system, which bears striking similarities to the fortifications in Ulm depicted in Merian's atlas of Swabia and Württemberg.[22]

The explanatory poem on the following four pages ('Reimenw: Erclärung') is largely written as a discourse between two *cavaliers* (who are depicted in the foreground of the image, letter A). In the course of the poem, these two men (one is meant to be the architect or city master-builder, presumably Furttenbach himself) enter the city and discuss the different types of architecture (civil and military architecture as well as artillery), offering an introduction to the book's content. In line with Vitruvius' *Libri decem*, the assessment of the city (as the epitome of universal architecture) is carried out from the outside to the inside, affirming the city's fortifications as a prerequisite for the existence of its civil architecture. The societal function of universal (military and civil) architecture is compared to the function of medicine or medical knowledge as being *praeservativ* (preservative – also in the sense of being preventive) and *curativ* (curative – healing, also in the sense of correcting or recreating):

> Architecture can aptly be compared to *medicine*. For the latter partly pre-empts vicious plagues through *preservatives*, and partly banishes them by certain *curatives*. Similarly, architecture partly prevents a conquest, so that it protects against a siege; partly often provides help and good advice: one can indeed comprehend its usefulness. And as the physician often *cures* an old vicious illness with great mastery, through burning, cutting and bathing, so the artful architect corrects the flaws and *defects* of old buildings. As *medicine* upholds the dignity of the body, so this art provides an ornament to countries, protection to cities, and stability to castles.[23]

Whilst the frontispiece of the *Architectura Universalis* promotes the integrity of architectural knowledge and its connectedness to the urban space, the next example, the frontispiece of the *Mechanischer ReißLaden* (1644), appears to be slightly more abstract in its visual argumentation.

The book introduces an invention by Furttenbach, a universal instrument, bringing together in a box (*Laden*) 'all' the geometrical instruments needed to project and draw architectural inventions (as described in Furttenbach's

22 See Zeiller, *Topographia Sueviae*.
23 Furttenbach, *Archictura Universalis*, Reimenw. Erclärung.

publications). In his *ReißLaden*, Furttenbach discusses a device popular among architects and engineers.[24] Apart from Furttenbach's *ReißLaden* (which was reportedly built at least once for the duke of Wolfenbüttel), three comparable devices are still extant: Christoph Schissler's (the Elder) *Instrumentenkasten* (late sixteenth century), Benjamin Bramer's *Instrument-Kästlein* (1628), and Johann Heinrich Burchart's *Mathematischer Besteckkasten* (1686).[25] The frontispiece (designed by Furttenbach) depicts a pyramidal structure, featuring allegorical personifications of the theoretical and practical branches of the mechanical arts (figure 2.3).[26] The image, which was also on display in Furttenbach's *Kunstkammer*, evokes Vitruvius' image of a stepladder, emphasizing the lengthy acquisition of both theoretical and practical knowledge. The frontispiece emphasizes the *artes mechanicae*, i.e., the synthesis of empirical-practical and theoretical knowledge (and not the study of theoretical architecture treatises).[27] In the image, the synthesis of practical knowledge on the one hand and theoretical knowledge on the other is portrayed as constitutive of mechanics through the masculine (practical) and feminine (theoretical) figurations of the arts. The practical arts encompass shipbuilding (*Architectura Navalis*), civil architecture (*Architectura Civilis*), military architecture (*Architectura Militaris*), artillery (*Büchsenmeisterey*), fireworks (*Feuerwerck*), water pipes (*Wasser-Laitung*) and grottoes (*Grottenwerck*). The theoretical personifications are perspective (*Prospectiva*), navigation (*Navigatio*), astronomy (*Astronomia*), geography (*Geographia*), planimetry (*Planimetria*), geometry (*Geometria*) and arithmetic (*Arithmetica*). The central personification of *Mechanica*, probably a portrait of Furttenbach seated upon a hoist, iconographically reaches up into the divine sphere: one's technical abilities draw one into the realm of divine creation. As a brief dedicatory poem informs us on the next page, underneath *Mechanica* we see the latter's antagonist, envy (*Neid*). *Fama* (on the upper left) and *Gloria* (on the upper right) herald fame and glory for mechanics.

---

24 See for example Albrecht, *Eygendliche Beschreibung*; Faulhaber, *Newe Geometrische vnd Perspectivische Inventiones*; Pomodoro, *Geometria prattica*.

25 See Furttenbach, *Mechanische ReißLaden*, ed. by Fitzner, 236–238.

26 Furttenbach's inspiration for both the instrument and the frontispiece might have come from the Ulm mathematician Johannes Faulhaber (1580–1635), who invented a *Perspectivisches Instrument* (1625). In his *Ingenieurs Schul* (1631), the frontispiece depicts a gathering of twelve allegorical personifications of the mathematical and mechanical arts: *Architectonica, Astronomia, Logarithmographia, Mechanica, Musica, Gnomonica, Trigonomitria, Optica, Geometria, Topographia, Geographia,* and *Arithmetica,* all under the auspices of wisdom (*Sapientia*). Furttenbach mentions six of these twelve fields of knowledge. See Furttenbach, *Mechanische ReißLaden*, ed. by Fitzner.

27 On the significance of the *artes mechanicae* in the sixteenth and seventeenth centuries, see Bacher, 'Artes Mechanicae', 35–49.

FIGURE 2.3 Joseph Furttenbach. *Mechanische ReißLaden, Das ist, Ein gar geschmeidige, bey sich verborgen tragende Laden.* [Mechanical box, that is, a very flexible box to be carried hidden on one's person.] Augsburg: Schultes, 1644
SOURCE: UNIVERSITY LIBRARY HEIDELBERG (URN:NBN:DE:BSZ:16-ARTDOK-53832)

The title page of the book presents the above fields of knowledge as *Recreationen*, drawing on the dual meaning of the word (to recreate and to recover/recuperate). Whilst the *Architectura Universalis* – written *during* the war – highlights the preventive quality of the fortification system and the curative function of civil architecture, the *Mechanische ReißLaden* takes a quasi-postdiluvian perspective, emphasizing the rebuilding of society following the destruction of war. The poem is written from the perspective of *Mechanica*, addressing her 'sons' and 'daughters': 'I have raised you all with great fear and care: I have carefully considered what you should learn: that above all God's glory and the prosperity of the polity shall grow and blossom in the entire land'.[28] As such, the interconnectedness of all fields of knowledge is expressed by the genealogical concept of family ties.

The last example originates from Furttenbach's best-known book, *Mannhaffter Kunstspiegel* (1663). A manuscript version of Furttenbach's encyclopedic work had formed part of the *Kunstkammer* for at least a decade before its publication. The title page distinguishes sixteen 'recreationen' or branches of knowledge, adding *Architectura insulata* to the sub-disciplines known already in 1644. One chapter is dedicated to each discipline. The frontispiece

28 Furttenbach, *Mechanische ReißLaden*, ed. by Fitzner, n.p. 'Mit grosser Angst und Sorg hab ich euch all erzogen; Was ihr lernen sollt/ hab ich gar wohl erwogen: Daß hierdurch Gottes Ehr zuforderst / und Wolstand Deß gmeinen Wesens wachs/ und grün in allem Land'.

FIGURE 2.4 Joseph Furttenbach. *La Riviera di Genova*. Panorama of the Riviera between Antibes and the Gulf of La Spezia. Six pages glued together (tables 31, 3, 1 and 10 from the *Mannhafften Kunst-Spiegel*, plus two handwritten titles above tables 31 and 10). Insert by Joseph Furttenbach. *Mannhaffter Kunst-Spiegel/ Oder Continuatio, vnd fortsetzung allerhand Mathematisch- vnd Mechanisch-hochnutzlich-Sowol auch sehr erfrölichen delectationen*. [The manly mirror of art or *continuation* and sequel of all kinds of mathematical and mechanical, highly useful as well as very pleasant delights.] Augsburg: Schultes, 1663
SOURCE: BY COURTESY OF HÖLE RÖSSLER. COPYRIGHT HERZOG AUGUST LIBRARY WOLFENBÜTTEL, SIGNATURE: M: UF 4° 51

(figure 2.4) offers a bird's eye view of the city of Genoa, dramatically beleaguered by a fleet of warships.

Having been Furttenbach's home for almost seven years during his *peregrinatio mechanica et militaria*, Genoa is presented as a place of apprenticeship in the accompanying poem. Letters and symbols mark buildings and sites in the (urban) topography, representing the sub-disciplines of mechanical and architectural knowledge. Furttenbach, who designed the frontispiece, blends the order of knowledge present in his book with the historical and biographical context of its origin (almost half a century beforehand). He situates his encyclopedic knowledge in the topography of Genoa, taking the reader from the harbor to the recreational gardens, from military buildings to civil architecture. The universal architectural realm of the city guarantees the connectedness of the diverse disciplines presented in the book.

As we shall see, Furttenbach's *Kunstkammer* serves a similar purpose: reminding the visitor of Furttenbach's peregrination, it combines the ordering of mechanical and architectural knowledge with the individual trajectory of its acquisition.

3 Armory and the Cabinet of Curiosity

In his first publication, the travel book *Newes Itinerarium Italiae* (1627), Furttenbach already demonstrates in painstaking fashion his knowledge of the

Northern Italian cultural sphere, acquiring symbolic capital for his career as an architectural theorist.[29] Throughout his travel guide, Furttenbach enumerates numerous, mostly princely, cabinets of curiosity and natural history collections. His fascination for the cabinets he visited on many occasions during his Italian journey resonates with the central apodemic motif of *curiositas*.[30] Back in Swabian territory, Furttenbach used a unique combination of cabinet of curiosities and theater in his own design for his first architecture-theoretical publication, the *Architectura Civilis* (1628), which appeared a year after the *Newes Itinerarium Italiae*. One of Furttenbach's characteristic strategies is already visible in this book: the adaptation of the aristocratic architecture that he had encountered in Italy to the bourgeois milieu of his homeland. The case at hand is a design for a small indoor theater which combined a cabinet of curiosities and a theater in a single object. The front sides of this circus-shaped theater, conceived for a bourgeois town house, feature a grotto and a simple perspectival theater ('Sciena di Comedia'). Furttenbach was obviously seeking to echo in miniature the Gallery of the Uffizi, ringed with columns, which he had previously described with enthusiasm in his travel account.[31] This chamber theater was conceived by Furttenbach as a space of knowledge for architecture and engineering, intended to mobilize the imagination of both the layman and the experienced architect or designer.

Located along the long sides of the room are a total of nine wooden niches. Inside, hidden behind curtains, are models of maritime and civil constructions, astronomical instruments including a globe, an armillary sphere and a sundial, geometric and arithmetic instruments and tables, cannon and artillery instruments as well as a number of pyrotechnical devices. The full spectrum of architectural-mechanical creations is presented in model form. Precisely because of its theatrical organization, this chamber theater 'transforms the thoughts' of the observer, who 'speculates there', mentally conceives of new architecture, and 'has the models right before his eyes'.

> And if the building is constructed like this, the admirer of the art of engineering will have no small joy in *actually having it before his eyes and*

---

29 See Zaugg, 'bey den Jtalienern recht sinnreiche Gedancken [...] gespürt'. Joseph Furttenbach als kultureller Vermittler', 25–43.

30 See Furttenbach, *Newes Itinerarium Italiae*, 85ff., 89, 116, 191, 239–243. On Furttenbach's armory and cabinet of curiosities, see Lazardzig, *Theatermaschine und Festungsbau*, 123–132; as well as Siebenhüner, 'Entwerfen, Modelle bauen, ausstellen'.

31 On the Uffizi circuit, see Furttenbach, *Newes Itinerarium Italiae*, 85ff. The parallels between the *Kunstkammer* theater and the Uffizi have already been noted by Zielske, 'Die Anfänge einer Theaterbautheorie in Deutschland', 31.

> *also while contemplating it to modify his thoughts*, so that by speculating in front of it, time will pass quickly for him. The cases should be covered with curtains with graceful and beautiful geographical maps painted on them. Apart from that, a library might be arranged in this theater with benches, in such a way that in front of each case there would stand a bench with books related to the art that they are placed next to. This will afford the scholar great *recreation* [grosse *recreation*], when he stands there speculating and has the drafts or the model right in front of him. This library can also be removed at any time by taking the benches away. Then, one can hold a banquet or a dance in this theater, or put on a comedy.[32]

Furttenbach describes the recreational effect of speculation resulting from the theatrical arrangement and operationality of the objects. The models, plans, books and instruments were intended to stimulate the architectural and technical imagination to create new inventions. A readership is thus addressed that is composed both of admirers of the 'Arte Ingegnio', the engineer's art, and (future) architects and engineers ('studios[i]'). A correspondence on the spatial level is that the chamber theater, which initially takes the form of a workspace, can be quickly transformed into a ballroom, theater or banqueting hall. The connection between enjoyment and utility, an integral part of early modern engineering, is evident in the interplay of serious and recreational fireworks, machine arts and automatons. With Furttenbach's notion of *recreation* and its immediate reference to the destruction wrought by war, the link between pleasure and utility is presented in quite an original way.[33] The objective of the universal *Kunstkammer* theater in the *Architectura Civilis* was to serve as 'imagination technology'. This objective ultimately found its way into Furttenbach's

32 'Unnd wann diß Gebäw [...] also angestellt/ so wirdt Der Liebhaber der Arte Ingegnio dieses würcklich vor Augen zu sehen nicht geringe ergötzlichkeit empfinden/ auch in betrachtung derselben die gedancken also verendern/ daß ihme etlich stund darvor zu speculieren die zeit kurtz werden wird/ Anbelangt die Kästen/ da sollen dieselbige mit vorhengen bedeckt/ darauff aber von Geographischen mappen gar zierlich unnd lustig gemahlt [...]. Sonsten aber mag in Disem Theatro ein Bibliotheca, oder Liberey doch allain mit hinein gestelten bencken gericht/ Also daß vor einem jeden kasten ein banck mit bücher verordnet/ welche samentlich von derselben Arte, warbey sie am negsten stehn tractieren. Das macht dem studioso wann er alda Speculiert und die visierungen oder Modelli gleich im gesicht hat/ grosse *recreation*. Dise Liberey kan aber zu jederzeit behend abgehöbt/ die benck weg genommen/ Alßdann nach belieben in disem Theatro ein Comedien Agiert/ oder ein Pangett/ oder Dantz darinnen gehalten werden'. Furttenbach: *Architectura Civilis*, 54.

33 On this topic, see Rößler, 'For lack of a site, and also to save on expenses'.

own armory and cabinet of curiosity, which he began to assemble shortly after his return from Italy and completed in 1620. In multiple rooms in the attic of his town house in Ulm, he assembled the instruments, models, designs, maps, engravings, manuscripts, books, material samples and curiosities that he had brought back from Italy. In the years that followed, Furttenbach continued to expand his collection. His armory and cabinet of curiosities soon acquired widespread and later even international renown.[34]

In his *Architectura Privata* of 1641, Furttenbach depicts the layout of the collection rooms and the manner of their use (figure 2.5). The space is shaped like a circuit (he writes – as he had previously done, with reference to the path through the Uffizi – of a 'walk' ('Spaziergang')), by means of which the visitor (and reader) can relive the most important stations of Furttenbach's *peregrinatio* and gain some understanding of the various sub-disciplines of the *artes mechanicae*.[35] At the center of the cabinet room stood a glass-doored cabinet known as the 'model case' ('Modell Kasten'), which contained Furttenbach's architectural models. These were arranged according to the subfields of architecture – from *architectura civilis, navalis* and *martialis* to pyrotechnics.

The latent references to salvation history in Furttenbach's cabinet were symbolized by a model of Noah's ark (Latin *arca* = chest/case, German *Kasten*) designed by Furttenbach according to biblical measurements and built by the master carpenter Johannes Nibling (also Nübling).[36] The ark was added to the collection in August 1657 (figure 2.6).[37] Furttenbach describes this model of the most resilient of all biblical constructions as an object of reflection, exchange and sense-making,

34 Kim Siebenhüner contextualizes Furttenbach's cabinet of curiosities within seventeenth-century Europe and Baroque collecting practices, while also offering a detailed analysis of local and foreign visitors. Siebenhüner: 'Entwerfen, Modelle bauen, ausstellen', especially 49. Siebenhüner emphasizes the originality of Furttenbach's cabinet, as it follows neither the model of the humanistic-artistic cabinet nor that of a natural history collection. Here further study is required, for Leonhard Christoph Sturm considers the model collection a self-explanatory element of an ideal cabinet of curiosities and natural history collection, which would seem to question the originality of Furttenbach's cabinet.

35 See Furttenbach, *Architectura Privata*, 20–52. On performative practice in early modern collections, see Welzel, 'Verhüllen und Inszenieren'. Welzel demonstrates the extent to which the ordering of cabinets of curiosity is to be understood as a space of dynamic action, which only discloses itself through 'performative receptive actions' (111).

36 According to Kaspar von Greyerz, the collection offered a mixture of natural history and biblical history typical of the *curiositas* of the intellectual visitor in the second half of the seventeenth century. See Greyerz, 'Religion und Wissenschaft im 16. und 17. Jahrhundert'.

37 Furttenbach, *Lebenslauff 1652–1664*, 250. The model was approximately 180 cm long, 28 cm wide and 28 cm high.

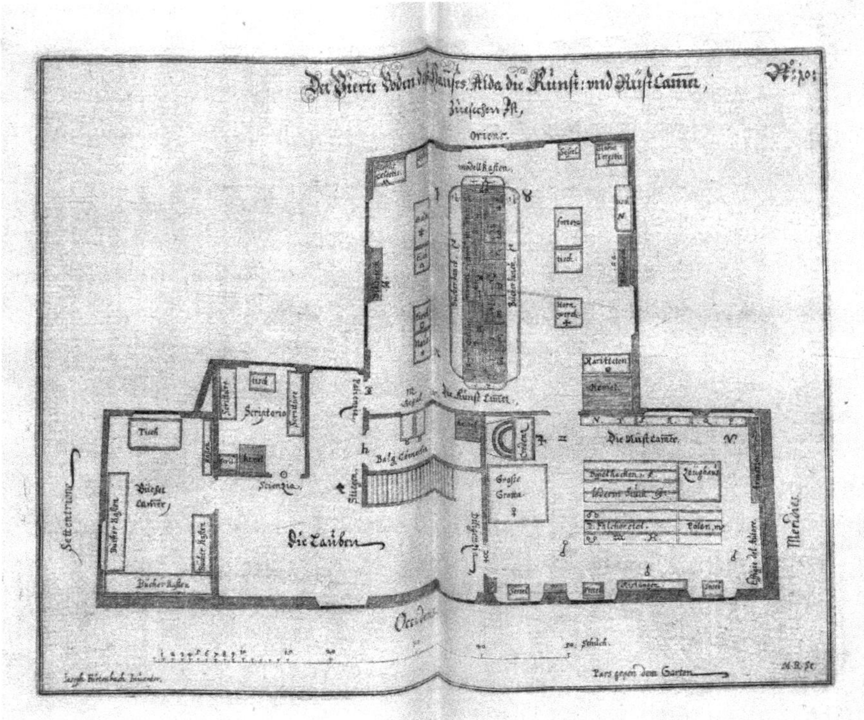

FIGURE 2.5 Joseph Furttenbach. *Architectura Privata*. Augsburg: Schultes, 1641, no. 10
SOURCE: SLUB DRESDEN (OAI:DE:SLUB-DRESDEN:DB:ID-265900999)

[…] displayed in Furttenbach's *Kunstkammer* for noteworthy *memoria*. Written on its roof are the words 'spiritual contemplation of Noah's Ark, how it was probably built according to God's command, following holy measurements' […]. A number of people, both of high and low standing, of the spiritual and the worldly orders, came to see this model, and to hold lofty debates about it, pro and contra. I tell them that it is only through the grace of God, to whom (and whom alone) the honor is due, that they are able to voice these opinions, just as our experienced carpenters shall not dare to reproduce it in full size.[38]

38 '[…] zu immer denckhwürdiger memoria in die furttenbachische kunstcammer aufgestellt. Auf seinem dach stunde geschriben, geistliches nachsinnen der archa Noe, wie dieselbige ungefehr nach Gottes beuehl, dem heiligen mass oder elen bogen nach möcht gebawt gewesen sein […]. Es erscheinen albereith vil geistlich- vnnd welltliche hoch- vnd niderstandts personen vnd herren, dises Modell zu sechen, vnd thun darüber hochvernünfftig discurß pro et contra halten, denen vnd aber allein durch die gnade Gottes, deme dann einig vnd allein die ehre gebürt, ich also begegnen, dz sie dise meinung

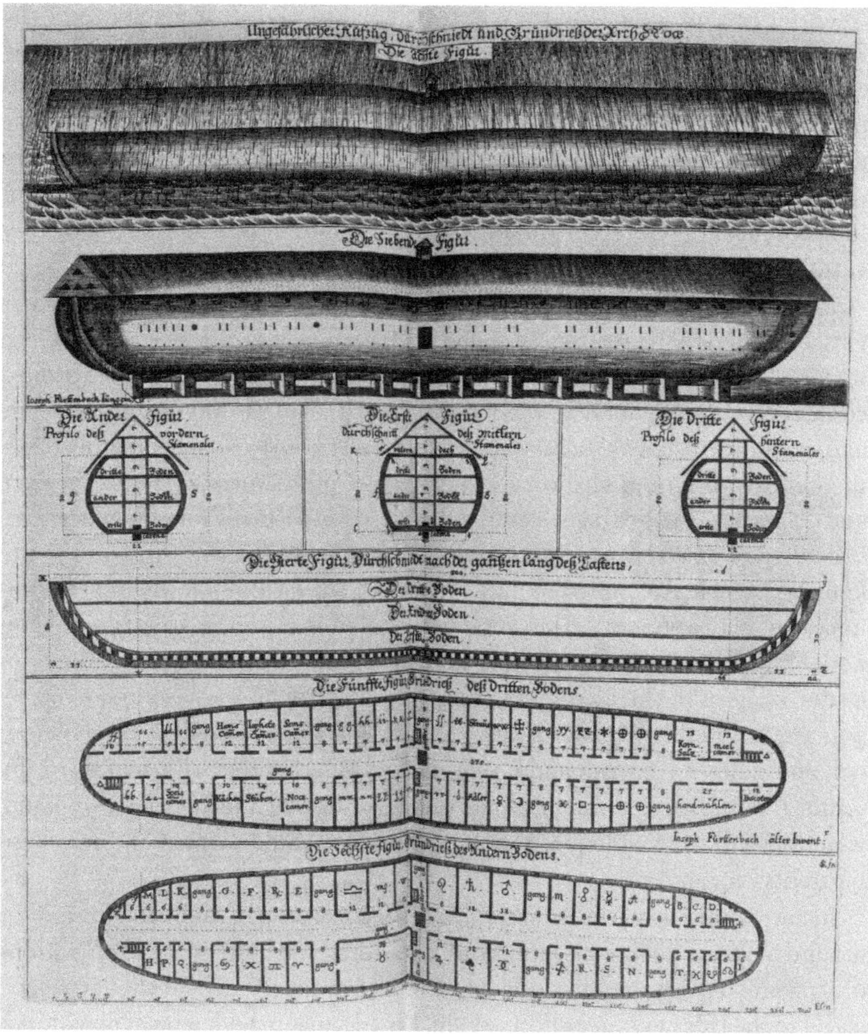

FIGURE 2.6 Joseph Furttenbach. 'Beschreibung der Archen Noae'. *Biblia, Das ist/ Die gantze H. Schrifft, Altes und Newes Testaments Teutsch/ D. Martin Luthers*. ['Description of Noah's arch'. *Biblia*, that is, the entire H[oly] Scripture, Old and New Testament in German by Dr. Martin Luther.] Nuremberg: Endter, 1662, n.p. [after p. 6]
SOURCE: COPYRIGHT SLUB / DEUTSCHE FOTOTHEK / RICHTER, REGINE (HTTP://WWW.DEUTSCHEFOTOTHEK.DE/DOCUMENTS/OBJ/80287611)

passieren lassen, wie dann auch unsere wolerfahrne zimmermeister sich vnderstehn wolten, also im grossen werckh selbste nachzumachen'. Ibid. On the function of biblical architecture as a model in the seventeenth century, see Bennett/Mandelbrote, *The Garden, the Ark, the Tower, the Temple*.

Noah's 'chest' or 'case', the only human construction to have survived the Flood, symbolically marks the starting point for reconstruction after the end of the war. Reconstruction and recreation are thus given a kind of postlapsarian significance. At the same time, the ark embodies an ordering principle for the world's knowledge. As such, this biblical reference can, in times of reconstruction, metonymically ennoble the knowledge system of the cabinet (of architecture and mechanics), thereby epistemologically universalizing it.

Alongside its function as a room of order and knowledge, the cabinet was closely related to the referential space of Furttenbach's work as an author. The material and medial transfer of technological knowledge was viewed within a context of authentication and validation. Nearly all of the models, instruments and machines displayed in the cabinet featured in Furttenbach's architectural-theoretical work (and vice versa). As a way of authenticating his experience and expertise, Furttenbach could refer to countless curious objects in his collection (models, instruments, drawings, engravings, etc.), testifying to the knowledge he had acquired in Italy. At the same time, this allowed him to capitalize on his 'experientz'. This network of references in the armory and cabinet of curiosities was underscored by a significant detail: in front of the door to the 'scriptorio', the architect's study and workroom, the visitor saw 'an oil painting entitled *La Dama Scienzia* [...] who has a white sheet of paper before her, and holds a compass and a ruler in her right hand'.[39] This figure led the visitor into the next room, in which 'books written by the author's own hand' were to be found;[40] here, visitors to the cabinet could inspect and purchase Furttenbach's books.

Just as the reader of the *Architectura Privata* is led by Furttenbach the author around a circuit filled with numerous anecdotes and stories, the explanations of Furttenbach the architect, drawing on his own experiences, must have made a visit to the cabinet into quite an event. Thanks to Furttenbach's explanations, many apparently nondescript objects become *mirabilia*, amazing objects and bearers of important knowledge.[41] Furttenbach recorded more than 700 visitors up until 1656. More than one third of them – as Furttenbach noted in his *Lebenslauff* for the period between June 1652 and January 1656 – were notable citizens; that is, patricians, city aldermen and merchants, doctors, architects,

39 '[v]on Oelfarb gemahlte Tafel nemblich la Dama Scienzia genant [...] welche vor ihr ein rein weissen Bogen Papier ligend/ in dero rechten Hand aber/ ein Zirckel vnnd Lineal haltet'. Furttenbach, 50f.

40 '[v]on deß Autoris eigen Handen geschribene[n] Bücher'. Ibid.

41 On this, see also Heesen, 'Geschlossene und transparente Ordnungen', especially 28f.

city jurists and office-holders.[42] Another third was made up of the military nobility and service gentry, young men on grand tours and court clerks. The remaining visitors were relatives, students and scholars, as well as prominent intellectual figures. Conversation and the exchange of information played a central role during these visits.[43]

Despite what the underlying didactic structure of the *Kunstkammer* and the nature of its visitors might lead us to believe, the cabinet rarely functioned as a place of instruction – or did not do so at all. In all likelihood, Furttenbach, who initially provided a sort of private education exclusively to his son Joseph – and, after the latter's untimely death in 1655, to his 'cousin' Gabriel Furttenbach (1640–1716) – simply lacked the time to run a private academy in the manner of his instructor Parigi, alongside his numerous professional and honorary duties.[44] Instead, Furttenbach's cabinet (like his *oeuvre*) functioned as a site and medium for the preservation, provision and promotion of knowledge. Against the backdrop of the Thirty Years' War, Furttenbach promoted an idea of universal mechanical knowledge, applying it to the local conditions and limitations of his hometown of Ulm. Referring to his training as a Vitruvian polymath in Italy and evoking Vitruvius' concept of the universally-trained architect, Furttenbach established his *Kunstkammer* as a showroom that promoted the 'recreation' of a neatly-built 'new world'.

4 Conclusion

To summarize, different but closely-connected functions can be attributed to Furttenbach's universal conception of architecture. First and foremost, Furttenbach, as an author, *promotes himself* as having a universal knowledge of both practical and theoretical matters related to architecture. This medial self-fashioning is not primarily aimed at winning architectural and engineering commissions, but rather at promoting the 'knowledge industry' of the 'merchant-architect' and 'collector-writer'.

This leads, secondly, to the intended readers of his writings. Furttenbach's architectural universalism is directed at a diverse audience. He

---

42 Here I am following the analysis of the visitor composition in Siebenhüner, 'Entwerfen, Modelle bauen, ausstellen', 61–63.

43 See ibid., 63.

44 He wrote in his diary that he did not have the time to take on students. It was only towards the end of his life that he also began to instruct his 'cousin'. Furttenbach, *Lebenslauff 1652–1664*, 232–235.

demonstrates his competence, presents his inventions and offers guidance for practitioners and fellow architect-engineers. Yet perhaps more importantly, he sparks constructivist fantasies in broader technophile audiences. Here, as I attempted to make evident, Furttenbach interprets Vitruvius' conception of the universal architect in the light of seventeenth-century mechanics and a biblical (Lutheran) concept of (re-)creation. Remarkably, the indebtedness to Vitruvius' *Libri decem* remains unmentioned, in this respect. For the learned reader, it must nonetheless have been palpable.[45] For the non-specialist, it would have seemed 'a marvel that human nature can comprehend such a great number of studies and keep them in the memory'.[46]

As we have seen, Furttenbach does not necessarily qualify as a polymath. As a 'merchant-collector' and 'merchant-writer', he lacked some foundations for polymathy, such as knowledge of algebraic mathematics and philosophy. Nonetheless, his mercantile family connections in Northern Italy granted him access to workshops and academies, where he acquired a theoretical knowledge of engineering and architecture. His subsequent career as a theoretical author on architecture demonstrated a growing bourgeois self-confidence, allowing him to offer a rather non-academic interpretation of Vitruvian universalism. Unlike his contemporary Simon Stevin, who excelled as a polymath, the mercantile Furttenbach, in diligently amassing mechanical and mathematical knowledge as well as instruments and objects, and in fashioning himself as universally knowledgeable via his publications, contributed to the creation of the image of the polymath per se.

Bibliography

Adami, Giuseppe, "Tra guerra e teatro: scienza e tecnologia militare al servizio dello spettacolo nell'Europa di Antico Regime," *Biblioteca Teatrale. Rivista trimestrale di studi e ricerche sullo spettacolo* 89–90 (2009), 13–45.

Albrecht, Andreas, *Eygendliche Beschreibung und Abriß eines sonderbaren nutzlich und nohtwendigen Mechanischen Instruments*, Nuremberg: Halbmayer, 1625.

Alsted, Johann Heinrich, *Encyclopaedia Scientiarum Omnium*, Leiden: Huguetan & Ravaud, 1649.

45 I think that this can safely be said, given the importance and significance of Vitruvius in all matters related to architecture and technology at this time. Cf. Sanvito, 'Introduction', 3–4.

46 Vitruvius, *The Ten Books on Architecture*, 1.1.12.

Bacher, Jutta, "Artes Mechanicae", in: ed. Hans Holländer, *Erkenntnis, Erfindung, Konstruktion. Studien zur Bildgeschichte von Naturwissenschaften und Technik vom 16. bis zum 19. Jahrhundert*, Berlin: Gebrüder Mann, 2000, 35–49.

Bennett, James Arthur and Scott Mandelbrote, *The Garden, the Ark, the Tower, the Temple. Biblical Metaphors of Knowledge in Early Modern Europe*, Oxford: Museum of the History of Science, 1998.

Berns, Jörg Jochen, "Peregrinatio academica und Kavalierstour. Bildungsreisen junger Deutscher in der Frühen Neuzeit", in: ed. Conrad Wiedemann, *Rom – Paris – London. Erfahrung und Selbsterfahrung deutscher Schriftsteller und Künstler in den fremden Metropolen. Ein Symposion*, Stuttgart: Metzler, 1988, 155–181.

Berthold, Margot, *Joseph Furttenbach (1591–1667). Architekturtheoretiker und Stadtbaumeister in Ulm*, unpublished dissertation, Munich, 1951.

Berthold, Margot, "Joseph Furttenbach von Leutkirch, Architekt und Ratsherr in Ulm (1591–1667)", *Ulm und Oberschwaben. Zeitschrift für Geschichte und Kunst. Mitteilungen des Vereins für Kunst und Altertum in Ulm und Oberschwaben* 33 (1953), 119–179.

Binding, Günther, *Meister der Baukunst. Geschichte des Architekten- und Ingenieurberufes*, Darmstadt: Primus, 2004.

Conrads, Norbert, *Ritterakademien der Frühen Neuzeit. Bildung als Staatsprivileg im 16. und 17. Jahrhundert*, Göttingen: Vandenhoeck & Ruprecht, 1982.

Faulhaber, Johannes, *Ingenieurs Schul*, Franckfurt am Mayn: Kieser, 1630–1637.

Faulhaber, Johannes, *Newe Geometrische vnd Perspectivische Inventiones, Perspektivisches Instrument*, Franckfurt am Mayn: Humm, 1625.

Fensterbusch, Curt, "Einleitung", in: Vitruv, *Zehn Bücher über Architektur.* Translated and annotated by Curt Fensterbusch, 5th ed., Darmstadt: Primus, 1996, 1–16.

Filippo, Tiziana De and Philipp Tscholl ed., *Architekt und/versus Baumeister. Die Frage nach dem Metier*, Zurich: gta-Verlag, 2009.

Fitzner, Sebastian, *Architekturzeichnungen der deutschen Renaissance: Funktion und Bildlichkeit zeichnerischer Produktion 1500–1650*, Cologne: Modern Academic Publishing, 2016.

Furttenbach, Joseph, *Architectura Civilis*, Ulm: Saur, 1628.

Furttenbach, Joseph, *Architectura Martialis. Das ist: Außführliches Bedencken/ uber das/ zu dem Geschütz vnd Waffen gehörige Gebäw*, Ulm: Saur, 1630.

Furttenbach, Joseph, *Architectura Navalis. Das ist: Von dem Schiffgebäw/ auff dem Meer und Seekusten zugebrauchen*, Ulm: Saur, 1629.

Furttenbach, Joseph, *Architectura Privata*, Augsburg: Rembold & Schultes, 1641.

Furttenbach, Joseph, *Architectura Recreationis. Das ist: Von Allerhand Nutzlich: vnd Erfrewlichen Civilischen Gebäwen*, Augsburg: Schultes 1640.

Furttenbach, Joseph, *Architectura Universalis. Das ist: Von Kriegs: Statt- und Wasser-Gebäwen*, Ulm: Meder 1635.

Furttenbach, Joseph, *Büchsenmeisterey-Schul*, Augsburg: Schultes 1643.

Furttenbach, Joseph, *Halinitro Pyrobolia: Beschreibung einer newen Büchsenmeisterey*, Ulm: Saur 1627.

Furttenbach, Joseph, *Lebenslauff 1652–1664*. Eds. Kaspar von Greyerz, Kim Siebenhüner and Roberto Zaugg. Cologne, Weimar and Vienna, 2013.

Furttenbach, Joseph, *Mannhaffter Kunst-Spiegel/ Oder Continuatio, vnd Fortsetzung allerhand mathematisch- vnd mechanisch-hochnutzlich- sowol auch sehr erfrölichen Delectationen* [...], Augsburg: Schultes 1663.

Furttenbach, Joseph, *Mechanische Reißladen/ Das ist/ Ein gar geschmeidige/ bey sich verborgen tragende Laden*, Augsburg: Schultes 1644. Ed. Sebastian Fitzner with essays, transcription and commentaries by Paul Brakmann, Sebastian Fitzner, Vera Henze-Mengelkamp, Elina Potratz, Lisa Rolle, Isabella Schneider, Viktoria Teske and Lars Weitemeier, FONTES 83, 2017, DOI: 10.11588/artdok.5383.

Furttenbach, Joseph, *Newes Itinerarium Italiae*, Ulm: Saur 1627.

Graziani, Antoine-Marie ed., *Vistighe Corse. Guide des sources de l'histoire de la Corse dans les archives génoises*, Vol. 2. Ajaccio: Piazzola, 2004.

Greyerz, Kaspar von, "Joseph Furttenbach: Autobiograph, frommer Lutheraner, kultureller Mediator, Kunstkammer-Patron". In: Joseph Furttenbach, *Lebenslauff 1652–1664*. Eds. Kaspar von Greyerz, Kim Siebenhüner and Roberto Zaugg, Cologne, Weimar and Vienna: Böhlau 2013, 9–24.

Greyerz, Kaspar von, "Religion und Wissenschaft im 16. und 17. Jahrhundert: Eine Einführung," in: ed. Kaspar von Greyerz, *Religion und Naturwissenschaften im 16. und 17. Jahrhundert*, Gütersloh: Gütersloher Verlagshaus, 2010, 9–32.

Heesen, Anke te, "Geschlossene und transparente Ordnungen. Sammlungsmöbel und ihre Wahrnehmung in der Aufklärungszeit". In: eds. Dürbeck Gabriele et al., *Wahrnehmung der Natur, Natur der Wahrnehmung. Studien zur Geschichte visueller Kultur um 1800*, Dresden: Verlag der Kunst, 2001, 19–34.

Hewitt, Bernard ed., *The Renaissance Stage. Documents of Serlio, Sabbattini and Furttenbach*, Coral Gables: University of Miami Press, 1958.

Johannes, Ralph ed., *Entwerfen. Architektenausbildung in Europa von Vitruv bis Mitte des 20. Jahrhunderts. Geschichte – Theorie – Praxis*, Hamburg: Junius, 2009.

Kiene, Michael, "Das Berufsbild des Architekten im 16. Jahrhundert. Bartolomeo Ammannati als letzter Dombaumeister und als Hofarchitekt in Florenz," *Marburger Jahrbuch für Kunstwissenschaft* 26 (1999), 157–187.

Lazardzig, Jan, *Theatermaschine und Festungsbau. Paradoxien der Wissensproduktion im 17. Jahrhundert*. Berlin: Akademie, 2007.

Lazardzig, Jan and Hole Rößler, "Joseph Furttenbach and the Transfer of Mechanical Knowlege. New Perspectives on Early Modern Theatre Cultures", in: eds. Jan Lazardzig and Hole Rößler, *Technologies of Theatre: Joseph Furttenbach and the Transfer of Mechanical Knowledge in Early Modern Theatre Cultures*. Zeitsprünge. Forschungen zur Frühen Neuzeit, vol. 20. Frankfurt a.M., 2016, 271–309.

Lorenzo, Renata De, "Being an engineer and being an architect in eighteenth-century Italy: professional identity as a reflection of political fragmentation," *Engineering Studies* 3.3 (2011), 171–194.

Negro Spina, Annamaria, *Giulio Parigi e gli incisori della sua cerchia*, Naples: Società Ed. Napoletana, 1983.

Pomodoro, Giovanni, *Geometria prattica,* Rome: Paulini, 1599.

Rößler, Hole, "For lack of a site, and also to save on expenses. Knowledge on Stagecraft. Joseph Furttenbach and the Limits of Cultural Translation", in: eds. Jan Lazardzig and Hole Rößler, *Technologies of Theatre: Joseph Furttenbach and the Transfer of Mechanical Knowledge in Early Modern Theatre Cultures*. Zeitsprünge. Forschungen zur Frühen Neuzeit, vol. 20, 3/4, 2016, Frankfurt a.M., 2016, 367–386.

Rößler, Hole, "Polyhistorie und Polymathie", in: eds. Herbert Jaumann and Gideon Stiening. *Neue Diskurse der Gelehrtenkultur in der Frühen Neuzeit. Ein Handbuch*, Berlin: De Gruyter, 2016, 635–676.

Sanvito, Paolo ed., *Vitruvianism. Origins and Transformations*, Berlin, Boston: de Gruyter, 2016.

Schlimme, Herrmann, "Die frühe Accademia et Compagnia dell'Arte del Disegno in Florenz und die Architektenausbildung", in: ed. Ralph Johannes, *Entwerfen. Architektenausbildung in Europa von Vitruv bis Mitte des 20. Jahrhunderts. Geschichte – Theorie – Praxis*, Hamburg: Junius, 2009, 326–343.

Schütte, Ulrich and Hartwig Neumann eds., *Architekt und Ingenieur. Baumeister in Krieg und Frieden. Katalog der Ausstellung der Herzog-August Bibliothek in Wolfenbüttel*, Wolfenbüttel: Herzog August Bibliothek, 1984.

Siebenhüner, Kim:, "Entwerfen, Modelle bauen, ausstellen. Joseph Furttenbach und seine Rüst- und Kunstkammer", in: Joseph Furttenbach, *Lebenslauff 1652–1664*, eds. Kaspar von Greyerz, Kim Siebenhüner and Roberto Zaugg, Cologne/Weimar/Vienna: Böhlau, 2013, 55–66.

Smalzi, Daniela, "Giulio Parigi architetto di corte: la progettazione dell'ampliamento di palazzo e piazza Pitti," in: ed. Mario Bevilacqua, *Architetti e costruttori del barocco in Toscana. Opere, tecniche, materiali,* Rome: De Luca Editori d'Arte (2004), 69–88.

Quaini, Massimo, "Cartographic Activities in the Republic of Genoa, Corsica, and Sardinia in the Renaissance". *Cartography in the European Renaissance* (*The History of Cartography*) 1, Chicago: University Press, (2007), 854–873.

Vitruvius, *Zehen Bücher von der Architectur und künstlichem Bawen: ein Schlüssel und eynleitung aller Mathematischen unnd Mechanischen Künst, Scharpffsinniger fleissiger nachtrachtung oder Speculation künstlicher Werck* [...], translated by Walther Hermann Ryff, Basel: Sebastian Henricpetri 1548.

Vitruvius, *The Ten Books on Architecture*. Translated by Morris Hicky Morgan, Cambridge: Harvard University Press, London: Humphrey Milford, Oxford University Press, 1914.

Welzel, Barbara: “Verhüllen und Inszenieren”, in: eds. Robert Felfe and Angelia Lozar, *Frühneuzeitliche Sammlungspraxis und Literatur*, Berlin: Lukas Verlag, 2006, 109–129.

Zaugg, Roberto, “ ‘bey den Jtalienern recht sinnreiche Gedancken [...] gespürt’. Joseph Furttenbach als kultureller Vermittler”, in: Joseph Furttenbach. *Lebenslauff 1652–1664*. Eds. Kaspar von Greyerz, Kim Siebenhüner and Roberto Zaugg, Cologne, Weimar and Vienna: Böhlau, 2013, 25–44.

Zeiller, Martin. *Topographia Sueviae, das ist Beschreib: und Aigentliche Abcontrafeitung Würtenberg Marggraffschafft Baden und andern zu dem Hochlöbl: Schwabischen Craiße gehörigen Landschafften und Orten*, Franckfurt am Mayn, 1643.

Zielske, Harald, “Die Anfänge einer Theaterbautheorie in Deutschland im 17. und 18. Jahrhundert”, in: eds. Rolf Badenhausen and Harald Zielske, *Bühnenformen – Bühnenräume – Bühnendekorationen. Beiträge zur Entwicklung des Spielorts. Herbert A. Frenzel zum 65. Geburtstag von Freunden und wissenschaftlichen Mitstreitern*, Berlin: Schmidt, 1974, 28–63.

CHAPTER 3

The Swedish Archimedes

The Formation of the Polymath Christopher Polhem

David Dunér

1 Introduction: Polhem's Headache

When the Swedish mechanic and natural philosopher Christopher Polhem (1661–1751) thought too much, his brain became sore and ached.[1] His body became weak and he felt powerless to think. Nevertheless, he constantly searched for new knowledge about the world: the laws of mechanics, the inner structure of matter, the movement of water, geological processes, and waves of sound and light. He tried to understand the world, interpret it, think about it, invent new things and build models of reality. He made contributions to engineering, mechanics, physics, the theory of matter, geology, mathematics, linguistics, music theory, economic theory and much more. He was a polymath.

As a polymath, Polhem transcended the established and imagined disciplinary borders of knowledge. What made him a polymath, however, was primarily a certain kind of combinatory quest; that is, the transdisciplinary cohesion of theory and practice. This 'Archimedean polymathy' is characterized by the striving and need for both theoretical understanding and practical experience, both mathematics and craftsmanship, rationalism and empiricism, in order to solve technological problems. The experience and knowledge of craftsmen should be combined with the mathematical, analytical mind of the learned natural philosopher. For a machine to be applicable, one needs the practical experience of the craftsman *and* the analytical insights of the mathematician. Time and again, Polhem himself emphasized the need for a unification of theory and practice. His compatriots called him 'the Swedish Archimedes', referring to his rare transdisciplinary skills as an inventor, a craftsman-mathematician who could combine the delicate theoretical sagacity of the mathematician with the broad, longstanding and tangible practical experience of the craftsman. In that sense, Polhem's 'Archimedean polymathy' was not a striving to be universally knowledgeable, to obtain encyclopedic

1 Polhem, *Christopher Polhems brev*, 36.

 | DOI:10.1163/9789004432918_005

knowledge for its own sake, but to combine practical experience with theoretical analysis to solve specific technological problems; to improve and enhance human creativity.

In this combined striving for theory and practice, Polhem reminds us to a great extent of Simon Stevin's blending of theory and praxis in his use of the mathematical sciences in technology, for example in statics and hydrostatics. For Polhem, taking a theoretical and mathematical approach to statics and hydrostatics was the key to developing practical mining machines, water mills, locks and dams, and many other technological enterprises. Like Stevin, Polhem focused on engineering and mathematics, but was also greatly interested in linguistics and music, much less so in medicine and botany. Polhem and Stevin alike were interested in the nature and structure of knowledge. Religion, however, is almost completely absent from both Stevin's and Polhem's works. Like Stevin, Polhem lacked the linguistic foundation for polymathy – proficiency in Latin – and similar to Stevin, he had the theoretical-mathematical skills of a true Archimedean polymath in the early modern period, in Polhem's case: skills in algebraic mathematics, an understanding of calculus, and a predilection for natural philosophical speculation. Polhem's Archimedeanism could also be compared to the Vitruvianism of Joseph Furttenbach the Elder:[2] being knowledgeable in both the practical and theoretical aspects of architecture in order to become the universal architect of the Vitruvian ideal. Polhem clearly transcended disciplinary borders, however, and did not seek universal knowledge for its own sake. Rather, the focus of his polymathy lay in finding the balance between theory and practice. Stevin, as E. J. Dijksterhuis put it, 'continually oscillated between what he called *spiegeling* (speculation, *i.e.* theoretical investigation) and *daet* (practical activity)'.[3] This oscillation between theoretical investigation and practical activity was likewise typical of Polhem's Archimedean exertions.

This chapter examines the formation of such an Archimedean polymath, the craftsman-mathematician. Polhem's case not only illustrates how polymathy could be achieved through education and the use of abilities such as perception and communication, but it also reveals the social factors that formed polymathy. The following sections discuss Polhem's professional formation through his formal education, perception and communication. Before addressing these sub-themes, we first present a recapitulation of Polhem's major technological achievements.[4]

2 See Lazardzig's chapter in this volume.

3 Dijksterhuis, *Simon Stevin*, 130.

4 Lindroth, *Christopher Polhem*; Bring, 'Contribution'; Dunér, 'Daedalus of the North'; Lindgren, *Christopher Polhems testamente*.

2 The Swedish Archimedes

Many of Polhem's most famous and innovative machines were constructed for the Stora Kopparberg mine in Falun, central Sweden. As the major copper mine in Sweden, technological improvements were essential for the economy of the Swedish state. Polhem became responsible for the machines at the mine, and was engaged in improving them for more than twenty years. Most notably, he designed and constructed machines for hoisting the ore from the pit (figure 3.1). Without previous knowledge and experience of mining, he approached the mechanical problems in a new fashion, independently of the long-lasting traditions upheld by miners and craftsmen. The phenomenon whereby an outsider tries to solve a problem despite lacking the knowledge of the experienced expert, and produces – contrary to all expectations – a successful solution, might be described as a sort of 'genial ignorance'. Ignorance sometimes liberates the mind, allowing the problem-solver to approach the task fearlessly with different, non-traditional methods. Polhem, for example, replaced costly and less durable leather ropes with pairs of wooden rods, with which the barrels were transferred from one pair of hooks to another.

Polhem lived at a time of increasing trade and more frequent international contact, when the Swedish state, in line with mercantilist economic doctrine, endorsed and subsidized domestic industry. In 1700, together with the lawyer Gabriel Stierncrona, Polhem established a mechanical manufacturing works in Stjärnsund in southern Dalecarlia. The proto-industrial Stjärnsund works embarked on the large-scale production of various utensils, cups, bowls and plates made of tinned iron plate; it manufactured bells, locks, agricultural equipment and tools for carpenters, tinsmiths, blacksmiths and other craftsmen. The idea was to operate the workshop using automatic machines, as far as possible. Instead of animal or human muscle power, the machines would be driven by ever-flowing, relentless hydropower – as Polhem said, a workforce that needed neither food nor hay. Polhem was also engaged by the Swedish king, Charles XII, to work on other projects in the country: the construction of a dock at the naval shipyard in Karlskrona, the building of sluices in the Göta River near Trollhättan, and the invention of war machines for the Swedish army.

When Polhem was not busy building models or constructing and maintaining machines, he was pondering the nature of the world. As a natural philosopher and mechanic, he viewed the world as one giant machine. Everything was mechanical. The human body, physical and geological processes and the starry sky all followed mechanical principles; according to Polhem,

FIGURE 3.1 Polhem's hoisting machine, Stora Kopparberg mine, Falun. Engraving made in Holland by Jan van Vianen after a drawing by Samuel Buschenfelt
SOURCE: *DAEDALUS HYPERBOREUS* (1716). PHOTO: LUND UNIVERSITY LIBRARY

even the soul and the mind could be explained using simple mechanics. There was no difference between the artificial mechanics of the engineer and the natural mechanics of God's creation.[5] Natural machines were subject to the same mechanical laws as those invented by man. He would return to this idea time and again. Another recurring idea concerned the relationship between theory and practice, linked to the 'Archimedean tradition' of the unity of theory and practice and the mutual interrelations of science, technology and mathematics, which became important in the Scientific Revolution; in the use of mathematical proofs of mechanical theories, for example, not least for Simon Stevin's practical application of mathematics and physics.[6] The Archimedean ideal also infused military practice with rigorous mechanistic reasoning.[7] According to Polhem, theory without practical knowledge and

5 Dunér, 'Christopher Polhem's Experiments'.
6 Kemp, 'Simon Stevin and Pieter Saenredam'; Koetsier, 'Simon Stevin"; Dunér, 'Science'.
7 Steele and Dorland, *Heirs of Archimedes*, 3.

craftsmanship without theoretical insight could not lead to any significant benefit for mankind. Rather, mathematics and crafts should be combined. Among his countless drafts, fragments and notes is an undated dialogue between Lady Theoria and the master builder Practicus.[8] The young lady explains to the builder how mathematical knowledge can be applied to the construction of watermills, and Practicus later declares his love to Theoria. The moral of the story is that the two would have extraordinary, beautiful children if they were to marry. Besides the evident gendered denotation of theory and practice as feminine and masculine, respectively, the dialogue reflects the social dimensions of theory and practice; noble theoretical studies versus ignoble practical labor. Theory and mathematics had a higher status, whereas the practical life belonged to more lowly carpenters and peasants. The dialogue also reflected Polhem's indeterminate social position; he was both a carpenter and a book-learned mathematician, and neither per se. As such, the fusion of theory and practice – Archimedean polymathy – could be seen as a strategy to achieve higher social status; to climb the social ladder from being an ignorant peasant, through the class of carpenters, to reach the elite of society. At the same time, the craftsman-mathematician marketed the practical utility of a learned, theoretical education, and how it could generate wealth and progress.

In 1697, Polhem founded a 'Laboratorium mechanicum' with the aim of developing mechanics, both theoretically and practically. The most advanced experimental machine developed within the framework of the laboratory was a hydrodynamic machine from 1702 that tested waterwheels of different types (figure 3.2). A key element in his education of young technicians was his 'mechanical alphabet', which consisted of simple, pedagogical wooden models that demonstrated the basic mechanical laws.[9] Over the years, he employed a number of disciples and assistants in his mechanical laboratory and for his work assignments. The more notable include Göran Vallerius, Emanuel Swedenborg,[10] Augustin Ehrensvärd and Carl Johan Cronstedt. His last undertaking was the construction of a new lock in Stockholm. By the time of his death on 31 August 1751, he was a celebrated man: a member of the nobility, Knight of the Polar Star, a fellow of the Royal Swedish Academy of Sciences, and known by his compatriots as 'the Swedish Archimedes' or the 'Daedalus of the North'.

---

8 Polhem, *Christopher Polhems efterlämnade skrifter I*, 277–307.

9 Dunér, 'Language of Cosmos', 59–62.

10 Dunér, *Natural Philosophy of Emanuel Swedenborg*.

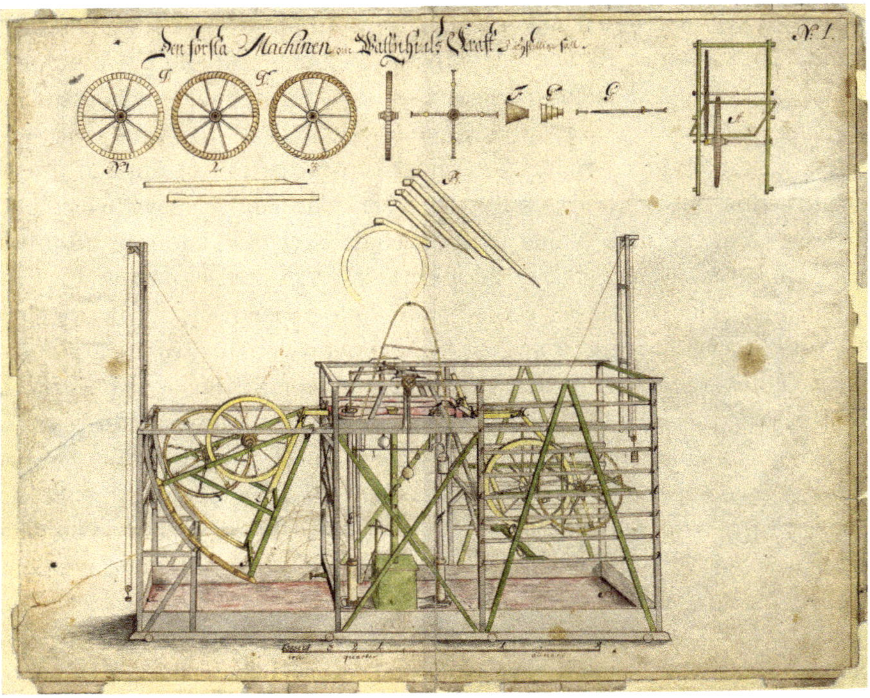

FIGURE 3.2 The hydrodynamic experimental machine from 1702
SOURCE: SKETCH BY GÖRAN VALLERIUS, 1705. PHOTO: ROYAL INSTITUTE OF TECHNOLOGY, STOCKHOLM

3 Mechanical Education

In the seventeenth century, many foreign professionals – German craftsmen, Dutch merchants and Walloon smiths – were recruited and attracted to economically buoyant but sparsely populated Sweden, to fill the needs of this expanding Northern European military power and its growing interest in cultural, scientific and technological development. With its expansionist policy, Sweden needed skilled workers and capital to strengthen the country's economy. New factories were established, water power was utilized, the iron industry was boosted, ports were re-equipped, natural resources as far north as northern Lapland were investigated and mapped, and new forms of capital accumulation and work organization came into use. All of this was connected to the war and foreign policy. Sweden had been at war for decades and vast territories had been conquered, but the country had a relatively underdeveloped economy as a mere producer of commodities such as iron, copper, tar and pitch.

Polhem's father, Wulf Christopher Polhammar, originally from Swedish Pomerania, was a merchant in Visby on the island of Gotland. After he died in 1669, Polhem's mother remarried a building contractor, who – as Polhem wrote later in his life – did not want to pay for his stepson's education.[11] Instead, he was cared for by his uncle in Stockholm. He moved to Stockholm in the autumn of 1671, and began studying at the German school. After just two years, however, his uncle also died. Now alone, Polhem had to take care of himself. He managed to get positions on various estates around Stockholm. He set up a carpentry workshop with a lathe and forge, where he could hammer, file, lathe, forge, fabricate and repair tools, knives and scissors. In his spare time, he made clocks and other more complicated mechanical devices.

His desire to study mathematics and mechanics was so great that he could hardly sleep at night, Polhem later revealed when thinking back on his poverty-stricken youth.[12] He harbored dreams of studying. His brief period at primary school in Visby and Stockholm had left him with major gaps in his knowledge. In order to develop his interest in mechanics and engineering, he understood that he needed theoretical knowledge to fertilize his practical mechanical skills. One day, surveyors came to the estate to make a two-dimensional, geometrical record of the meadows and arable lands on sheets of paper. Amazed, Polhem followed them around to learn their art. Keen to learn more, he took his savings and bought both a German and a Swedish textbook on surveying. At this point, the curious Polhem encountered a major obstacle to progressing in his theoretical studies: he was unable to read Latin. All of the amazing works on physics, mathematics and mechanics were written in a language he could not understand. He knew that priests had mastered this language, however. The parish minister lent him a Swedish-Latin dictionary and Polhem began to learn it by heart, but he soon realized that this method would not work. Another priest, Lars Olofsson Weldt, later came to his workshop to order a clock. Polhem spotted an opportunity: striking a deal with the priest, he paid for Latin lessons by making a clock showing the hours, days, dates, and the new and full moon. Once a month, when Weldt came to the Vansta Estate where Polhem was employed, he received instructions and new homework to do.

In order to develop his skills in mechanics even further, however, Polhem needed to study theory at university. His talent for mechanics was

11 Polhem, *Christopher Polhems efterlämnade skrifter IV*, 396.

12 Polhem, *Christopher Polhems efterlämnade skrifter IV*, 390.

recognized by the parish minister, the learned Erland Dryselius, and thanks to a letter of recommendation, Polhem enrolled at Uppsala University in the autumn of 1687. He was not yet 26 years old. On the way to Uppsala, he visited the surveyor's office in Stockholm, intending to become a surveyor's apprentice. While he was standing in the hall, waiting to be called up, he read through the requirements for enrolling as a surveying student. Realizing that he would fail, he soon left the building, and continued to Uppsala. He studied diligently at the university, especially mathematics and physics. He borrowed books on mathematics from the astronomy professor Anders Spole's library, and learned their contents by looking at the mathematical figures. He received a scholarship, and he also undertook a number of assignments in mechanics.

When Polhem was studying in Uppsala, one of the most divisive academic quarrels in the history of the university was taking place: a fierce debate between Aristotelians and Cartesians, which raged between 1686 and 1689. The defenders of the Cartesian or mechanistic worldview regarded the world as one gigantic machine, a clock designed by the great watchmaker. Everything could be described in mechanical terms: planets, falling objects, plants, animals and the human body. The task for the natural philosopher was to find the true machinery behind the phenomena. This mechanical worldview suited Polhem the mechanic very well, and it became his philosophy. As such, the modern science that now emerged in Uppsala had an empirical vein, a belief in experiential knowledge and experimentation. The professor of medicine, Andreas Drossander, introduced experimental physics and offered private lectures (which Polhem attended), where he did experiments with air pumps, microscopes, siphons and magnets. Polhem may also have followed Johan Bilberg's lectures on modern mathematics. He probably acquired his basic knowledge of physics and astronomy from the *Synopsis physica* (1678), a widely-used textbook in Uppsala written by the medicine professor, Petrus Hoffwenius. He could learn many of the theoretical explanations for the mechanical worldview from the official lectures at the university. In order to acquire practical knowledge about how to construct useful machines, he had to follow the private lectures that the professors occasionally offered in addition to the standard curriculum. Another professor of medicine, Olaus Rudbeck, taught eight extra technical subjects: surveying, waterworks, artillery, fortification, housebuilding, mechanics, sundials and geography. In addition, he seems to have given classes in shipbuilding, agriculture, forestry, building locks and mills, and the production of iron and copper, as well as gardening. During Polhem's first year, Rudbeck taught a class on building locks and canals. Polhem probably visited on numerous occasions the 'mechanical house' and experimental locks that Rudbeck had constructed.[13] Polhem's later construction of canals, locks, docks, and mining machines suggests that Rudbeck's teaching is likely to have had a significant impact.

The three years at Uppsala University were a hard-working and diligent period. Polhem recounted that he rarely slept more than three hours a night. It was difficult in the beginning, of course, but after a while he felt much more cheerful, and began to understand and learn everything more quickly. After all, he noted, the brain becomes spoiled and inefficient with too much sleep. Theoretical studies may have caused him sleepless nights, but it was probably practical mechanics that occupied most of Polhem's time in Uppsala. His fellow students taunted him and called him 'Spole's smith'. The achievement that attracted most attention during Polhem's stay in Uppsala, however, was his repair of the astronomical clock in Uppsala Cathedral, which he described as having 'promoted all my luck in the world'.[14] The cathedral clock had originally been constructed in c. 1504–1506 by a German monk in Vadstena, Petrus Astronomus.[15] Clockwork mechanics had improved during the seventeenth century, thanks to the pendulum clock, and had become the cutting-edge technology of Polhem's time. Polhem took the cathedral clock apart and examined its different parts and operating mechanisms, in order to find out how it was supposed to work. He learned about new technology and previous solutions by imitating, copying and reconstructing. It took him two years to get the clock running again. After a number of major changes to the mechanism, the clock repair was completed in 1690.

That same year, Polhem built another technically advanced clock, an astronomical clock that showed the sunrise and sunset over a whole year, as well as the phases of the moon. It was one of the first pendulum clocks ever to be constructed in Sweden. As Polhem was aware, the original inventor of the pendulum clock was the Dutch physicist and astronomer Christiaan Huygens.[16] Huygens had constructed a pendulum clock in 1656, described in his *Horologium oscillatorium* (1673), a work that Polhem may have studied. Two of the important regulatory mechanisms in clock-making used by Polhem, the pendulum and the spring mechanism, had been invented by Huygens. With a pendulum, time could simply be divided into equal-sized units. It is likely that Polhem was the only person in Sweden at that time who was able to build a pendulum clock after Huygens's construction.

---

13 Dahl, *Svensk ingenjörskonst*, 289; Eriksson, *Atlantic Vision*.

14 Polhem, *Kort berättelse*, 5.

15 Pipping, 'Christopher Polhems astronomiska ur', 156.

16 Andriesse, *Huygens*.

The main source on Polhem's early education is his own autobiography, which he wrote much later as a member of the Royal Swedish Academy of Sciences.[17] When he tells his life story, he not only tells it for others, but also tells it – and constructs it – for himself. Autobiography, that noble form of art and primary expression of human self-deception, is largely a matter of who one wants to be in one's own eyes and those of others. An autobiography perhaps reveals less about who the author really is, or what they actually achieved, than about what the autobiographer wants the world to observe and recognize. It is all about self-fashioning. In his autobiography, Polhem wanted to create an image of himself as a poor orphan, one who – armed with ingenuity, talent and diligence – worked hard and all alone in a grudging, hostile environment. It is also this story that has repeatedly been told to posterity in the literature on Polhem. Polhem's autobiographical self-fashioning, boasting of his range of skills and knowledge, also made him into a polymath. He was not just a craftsman or carpenter, but also a mathematician; an Archimedean polymath.

Another recurring theme in the history of ingenuity is that geniuses never have any teachers; they teach themselves, directly from nature. Polhem is commonly described as having been self-taught, unprecedented, and independent of the knowledge of others. If we do not immediately accept Polhem's self-construction as a solitary genius, we need to find other contemporary sources that tell us something about how he was regarded by others, who he met, with whom he collaborated, and what the society around him was like. Contrary to the story of the solitary, self-made genius, one distinct trait in Polhem's journey from schoolboy to famous mechanic was his dependence on others: other people's contacts, power, money and knowledge. What one might call his 'network' was a fixed, regulated system of patrons and clients. To be able to progress in one's career, one needs someone to open doors and write recommendation letters. Polhem had Weldt, Dryselius, Spole and Rudbeck; men who not only opened his eyes to new knowledge, but who also opened up new career paths for him.

4 Mechanical Perception

Polhem claimed that he need only see a machine in motion once to be able to build a new one with every one of its construction details, no matter how complicated it was. There were no inventions, he said, that he could not

17 Polhem, *Christopher Polhems efterlämnade skrifter IV*, 387–399.

immediately assess as executable or not.[18] The mathematician Samuel Klingenstierna, who held a memorial speech on Polhem at the Royal Swedish Academy of Sciences in 1753, was deeply impressed by how Polhem had been able to bring so many new machines to Sweden.[19] He had a wonderful ability to conceptualize mechanics. It was remarkable, Klingenstierna exclaimed, that he never used drawings or descriptions, but maintained an abundance of highly complex machines in his memory. Polhem seems to have had a phenomenal spatio-technical ability; an eye for mechanics, motion and transmission. The practical experience of the Archimedean polymath rested on visual and spatial cognition.

To understand how a clock works, one has to take it apart – as Polhem did – and rebuild it. In this way, Polhem learned about technology by seeing and by doing with his own hands, as he encountered the objects around him. As a builder of mechanical devices, Polhem imagined technology visually, both as inner, mental models or representations and as a sort of distributed cognition with external representations, whereby thoughts are placed outside the human body like physical models made of iron and wood. To some extent, technical talent seems to be related to how a person manages to create not only internal, but also external representations of mechanical relations in the outside world. The speculative mind of a mechanic, Polhem claimed, is characterized by 'astute reflection and a good memory that retains the whole structure of the machines in the head until everything is completed'.[20] The concrete external models were not only tools to think with, but can be seen as an integral part of technological thinking. Indeed, models play a number of different roles in technological change. The use of physical scale models had educational purposes; wooden and clay models had already been utilized as teaching aids by Italian engineers and by Stevin in the Netherlands.[21] Polhem himself had a mechanical laboratory and mechanical alphabet, which visualized mechanics and technical ideas in order to understand how machines functioned and how to build them at full scale. Models could be used for experimental purposes, using parameter variations that tested different technological solutions, such as in the case of Polhem's hydrodynamic experimental machine. In the case of Polhem's model chamber, smaller-scale three-dimensional models could be used instead of two-dimensional drawings. Models could also be linked to economic interests and motivations as a kind of advertising or product

18 Polhem, *Kort berättelse*, 76.
19 Klingenstierna, *Åminnelse-tal.*
20 Polhem, *Kort berättelse*, 14.
21 Krüger, 'First Mathematics Curriculum'.

presentation, or simply as pleasure and entertainment, as magnificent technological wonders. The key element here, however, is that models are an externalization of thought; that the inventor thinks with moving objects outside the brain and body.

One important episode in Polhem's technological advancement occurred when he and a friend, Samuel Buschenfelt, won a scholarship from the Royal Board of Mines in January 1694, to undertake a trip to continental Europe and England. This was intended as an educational journey for two promising mechanics to develop their technological skills and learn more about the latest technology. But it also had political and economic motives: new and valuable technology could be brought to Sweden. Polhem's receptive mind was an asset at a time when export bans and tariffs made it difficult to bring original machines and models from England and other countries back to Sweden. By seeing the original machines in operation, he could reconstruct and make copies of them. They were, as one might say, 'industrial spies' sent out on behalf of the Swedish Crown to collect knowledge about new technologies, innovations and manufacturing methods for the benefit of the nation. They studied various industries and mechanical installations, took notes on new inventions and products, and met learned men, technicians and scientists. To a great extent, innovation and technological change depend on carriers of new technologies. As in the case of Polhem and Buschenfelt, skilled craftsmen traveled to other countries or undertook educational journeys, returning home with new ideas and technical solutions.

Technological change was partly a result of the spatial transfer of technological knowledge, but perhaps a more important role was played by the small improvements, step by step, day by day, that rested on acquired experience. Radical transformations and revolutionary innovations were rare. Technological change – inventions and innovations – can be seen as a process whereby someone identifies or experiences a problem that needs a solution, but it can also be a process whereby one recognizes that it is possible to do something that was previously regarded as impossible or simply not perceived at all, as something far beyond the horizon of human thinking and necessities of life. As such, Polhem's tour of Europe was a journey of opportunities, a way of seeing possible solutions to known and unknown problems. His travels, which took place between 1694 and 1696, proved to be of great importance to his technological work. Many of the technological solutions, machines and other things that he saw during the trip would later reappear in his own projects. Immediately after the trip, he launched a couple of projects that would have great significance for technological and proto-industrial change in eighteenth-century Sweden.

Polhem and Buschenfelt traveled through Bremen to the Netherlands, passed Deventer and Amersfoort, and arrived in Amsterdam on 4 December 1694. The Netherlands was known to them as a country with advanced technology and well-developed commerce, trade and manufacturing industry. Buschenfelt wrote that during the summer, Holland looked like a pleasure garden, with beautiful alleys, parterres and canals, and with many hundreds of factories, machines and buildings all around. For good reason, the entire country could be called an 'officina machinarum'.[22] During their stay in Holland, they spent most of their time in the university towns of Utrecht and Leiden, and in The Hague. They observed canals, windmills for grinding grain, sawing boards, paper manufacturing and tobacco cutting, and they also visited a silk factory in Utrecht, a lime kiln and a brick factory. Buschenfelt noted that the windmill near de Heere Port in Leiden had nine floors, was 88 feet high, and had a diameter of 26 feet at the base and 16 feet at the top.

In Scheveningen, they saw an unusual machine – a sailing chariot.[23] Such chariots, Buschenfelt reported, had been sailing on the smooth sandy beach along the coast of Holland. A flying wagon loaded with 26 distinguished people took two hours to cover a distance of 70 kilometers. The building of the wagon, Buschenfelt tells us, had been funded by the Prince of Orange, Maurits of Nassau, and it had been invented by the mathematician Simon Stevin. Due to its age, however, it had been discarded and dismantled by the time Polhem and Buschenfelt visited Scheveningen. A smaller one had been made in its place, Buschenfelt reported, with which the king of England had sailed 'when I was in The Haag'.[24] At the rear, there was a steering rod connected to the wheels, allowing the chariot to be maneuvered like a sailing boat.

During their time in Leiden, they learned to grind lenses – both concave and convex – for microscopes, tubes and telescopes. With such a lens, Buschenfelt recounted, he saw an object eighteen times longer and wider than in reality.[25] The lenses allowed one to see things many hundreds or thousands of times larger; one could clearly see globe-shaped bodies in the blood, the circulation of blood in eels and other transparent fish and insects, worms in vinegar, and other unfamiliar things in nature. They learned about the light refraction of different lenses, whether they were suitable for tubes, and how to assemble tubes with three or four lenses. They witnessed demonstrations

22 Buschenfelt, KB, 7.

23 Buschenfelt, UUB, 46; See also Happel, *Gröste Denkwürdigkeiten der Welt*, 566; Dijksterhuis, *Simon Stevin*, 104ff.; Devreese & Vanden Berghe, '*Magic is No Magic*', 45f.

24 Buschenfelt, KB, 67f.

25 Buschenfelt, KB, 78.

by the physicist Burchard de Volder in Leiden, and met the elderly Christiaan Huygens. As well as studying Antonie van Leeuwenhoek's microscope, they examined Samuel van Musschenbroek's instruments for physics experiments. All of the instruments that they saw enhanced human capacities, strengthened inborn perception, and allowed them to see further into matter and further out into the heavens; as such, they extended the human mind.

The only source for Polhem and Buschenfelt's trip is the latter's travel notes.[26] Polhem himself did not write a travel journal, but after returning home, on 16 October 1696, he sent a brief travel report to the board of mines.[27] He wrote that he had done his very best to obtain knowledge about mechanics. He had made careful observations and memorized all sorts of works and machines that he had seen during his journey through Holland, England, Brabant, France, Germany and Denmark, such as waterworks, manufactures, saw mills, oil mills, paper mills, mills for cement, pigments and plaster, grinding and polishing mills, hammer works, and forges; windmills for flour, tobacco, mustard and so forth; finer sawmills, thimble-making, brick and lime factories, silk weaving, looms for fabric, velvet, ribbons, laces, socks and so forth, and also flatboats, sluices, drawbridges, dams, pilings, dredgers, cranes, glass and mirror factories, astronomical clocks, carillons, mathematical, astronomical and physical instruments, and other artful things in curiosity cabinets 'that I do not have time to mention'. He could reconstruct all of these machines, he claimed, if so required. He had carefully examined all of the machines with his own eyes, but he had never taken any notes during his travels, or made any drawings of the machines; in other words, he did not use any external memorizing techniques. Everything was stored in his memory. To begin with, of course, one could question Polhem's own account. The report was obviously intended to show that he had used the travel grant wisely and efficiently, and that he could now offer his ingenuity and highly-improved technical proficiency to the Swedish state. Nevertheless, the passage indicates a certain ability to recall mechanical connections, partly due to the unambiguous logic of mechanics, his experience of mechanical causalities, and a clear awareness of the objectives. A process based on an unequivocal causality with a limited number of alternatives is easier to remember than a process of inconsequent occurrences with an indefinite number of alternatives.

Polhem constructed machines in his head. Like many other mechanics, he had a spatial ability that allowed him to join together mechanical parts to form

26 Buschenfelt, KB; Buschenfelt, UUB.

27 Polhem, KB, X 265:1.

complex machines. He could see whether a machine would work or not by trying it out in his head. No words were needed, just internal images, whereby mechanical parts were put together in the imagination, in mental representations and models of reality. The imagination – the ability to visualize existent or non-existent things and events in one's head – was an efficient way to anticipate the future. He did not need to perform time-consuming trials and experiments in the physical world. Everything could be perceived with the inner eye, and the movements of the models could be kept in his memory and later constructed on a larger scale. Technology rests partly on non-verbal thinking. The mechanical, technological imagination is the spatial representation of moving objects in three dimensions, not a two-dimensional representation or a linguistic representation in words and characters. When Polhem tried to describe his inventions, language was not sufficient; he gave up after a few lines. It is not possible to describe how a machine works, he admitted; it must be seen in action in order to be understood.

Polhem observed and examined new technology with his own eyes in workshops, mines, factories, mills, canals, in towns and rural areas, and as models in university collections and curiosity cabinets. He registered the technology around him in the physical world, in farm kitchens as well as in palace gardens. His journey through European workshops and factories was a trip through the advances of technology. To a lesser extent, his technological knowledge was the result of reading and theoretical studies. In a few cases, we can make assumptions about the books he may have read. He bought a copy of *Recueil de diverses pieces touchant quelques nouvelles machines* (1695) by the French physicist Denis Papin.[28] He glanced through books on technology with illustrations of fantastic machines, such as Johann Christoph Sturm's *Collegium experimentale* (1676–1685), which describes diving bells, air pumps, siphons, tubes, the Magdeburg hemispheres, the mechanics of muscles and other marvelous machines. A genre of books entitled 'Theatrum machinarum' by Jacques Besson and Vittorio Zonca, for example, excelled in presenting the most strange and imaginative, but often futile machines.[29] Polhem studied the largest technical reference book of the time, *Theatrum machinarum* (1724–1739) by the German experimental physicist and instrument-maker Jacob Leupold.[30] It could be said that the illustrated books on technology were rather intended for communication between engineers, investors and scholars. They were not primarily aimed for those who could actually build the machines, the

28 Papin, *Recueil*; Liljencrantz, 'Polhem', 300f., n. 7.

29 Besson, *Theatrvm instrvmentorvm et machinarum*; Zonca, *Novo teatro di machine.*

30 Leupold, *Theatrum machinarum.*

craftsmen. The costly copper-plate illustrated books on mechanics, Polhem argued, that contained drawings of numerous inventions – each more remarkable than the last – served only to create astonishment among ignorant people.[31] In a German edition of Agostino Ramelli's work on the art of mechanics, *Schatzkammer, Mechanischer Künste* (1620), owned by the eighteenth-century Swedish instrument-maker Daniel Ekström, noted Polhem's assessment of Ramelli: most of the depicted machines were not practicable.[32]

The few works that Polhem published during his lifetime contain almost no illustrations at all, partly due to the lack of skilled engravers and partly because of Polhem's thriftiness. His most important published work, *Kort berättelse om de förnämsta mechaniska inventioner* (1729), contains only one simple illustration (figure 3.3). The full title in translation would be: 'Short account on the chief mechanical inventions that have been invented, time after time, by the councilor of commerce Christopher Polhem and executed for the benefit and service of the public, and about the fate that some of them have had due to the unfavorable changes of time'. As the long title reveals, its purpose was in fact to showcase his genius, and to ensure due attention for his ingenuity. There are lots of drawings by his disciples, however; beautiful sketches in ink, wash drawings and watercolors (figure 3.4). Some of the best ones were made by the students Ehrensvärd and Cronstedt, who stayed at Polhem's manufacturing works the summer of 1729 to learn the art of mechanics. The illustrations depict the machines that they actually saw at Stjärnsund or at Falun mine, in Stockholm and elsewhere, which they had examined in order to acquire knowledge of Polhem's technological innovations. Making illustrations was a pedagogical exercise designed to enhance visual thinking on technology. The visualization of mechanics was one of the major ingredients of Polhem's pedagogy: to see mechanics in action with one's own eyes, to make sketches, and above all, to see, touch and construct models. Another important pedagogical presumption related to the unity of theory and practice; both practical and theoretical education should be provided in Swedish, including a deconstruction of machines into smaller, simpler parts.

The images have a kind of uncertainty to them, however, which often makes them problematic as historical sources. Do the pictures correspond to existing, planned or intended technology, or just imagined technology? Polhem himself never made any detailed plans or drawings of his machines, but his manuscripts include simple sketches that visualize mechanics, the inner structure

31 Polhem, *Christopher Polhems efterlämnade skrifter I*, 364.

32 Ramelli, *Schatzkammer*; Amelin, *Medaljens baksida*, 101f.

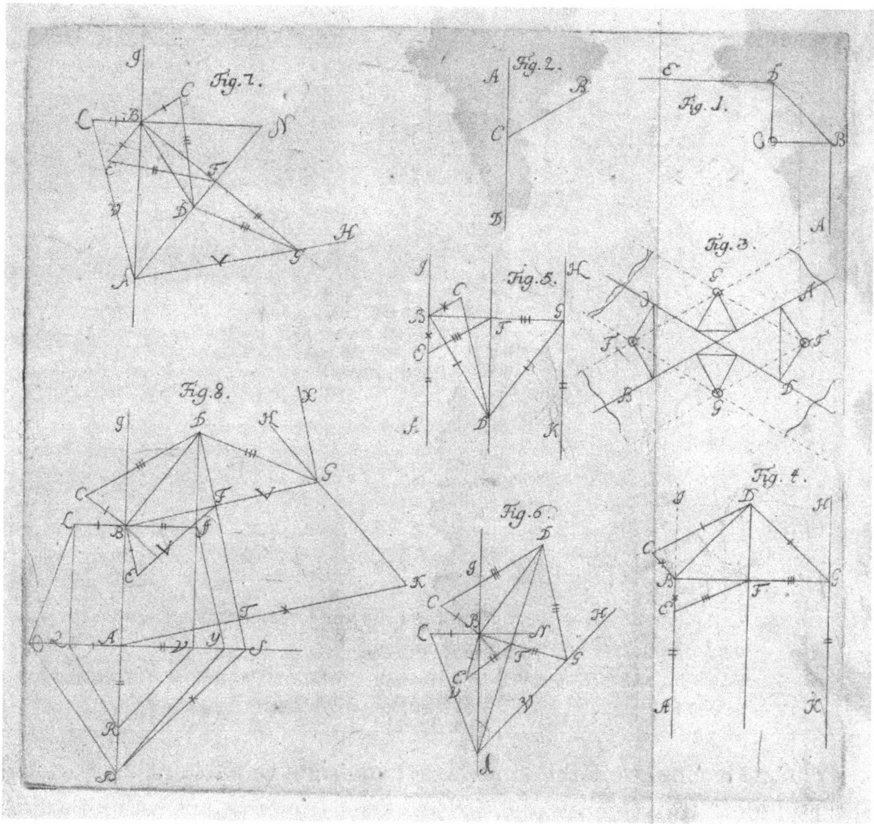

FIGURE 3.3 The geometry of mechanics
SOURCE: POLHEM'S *KORT BERÄTTELSE OM DE FÖRNÄMSTA MECHANISKA INVENTIONER* (1729). PHOTO: AUTHOR

of matter and other phenomena beyond the limits of vision (figure 3.5). There is no evidence that he ever used design drawings; instead, he utilized wooden models. Drawings might be good in general terms, he said, but models are more accurate and precise, especially if one wants to build complicated machines, but lacks experience.[33] Technology could be comprehended more easily through models. Models could also be used to diffuse new technologies. Despite the lack of precise drawings and illustrations, he thought in visual terms; not in two-dimensional, static images on a flat surface, but in three dimensions in motion, in models that one could touch, walk around and see in action.

33 Polhem, *Christopher Polhems efterlämnade skrifter I*, 40.

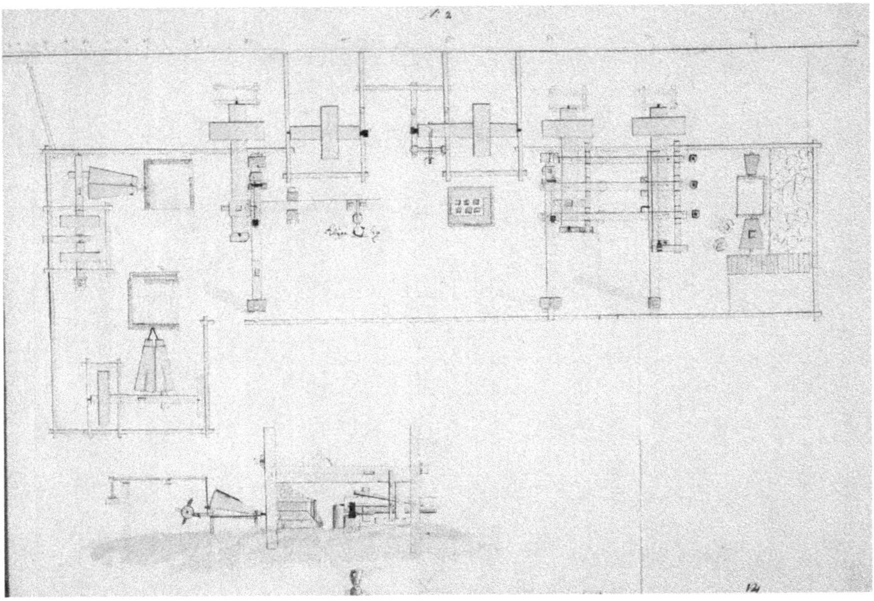

FIGURE 3.4 Carl Johan Cronstedt's ichnography of the iron forge, Stjärnsund, 1729, with ovens, hammers, bellows and water wheels
SOURCE: SKETCH BY CRONSTEDT, 1729. PHOTO: MUSEUM OF TECHNOLOGY, STOCKHOLM

Polhem's technological thinking was a spatial, non-verbal thinking. His technical solutions and scientific ideas existed first as inner visions without words. He then tried to translate the mental image into words, or to transform the internal image into a drawing on paper or a scale model, in order to excite similar mental images in the mind of another person, who would finally construct the mechanical idea at full scale in three dimensions.[34] Polhem's Archimedean polymathy, which put equal emphasis on practical experience, rested on visual thinking in order to visually and spatially experience technology in action.

5 Mechanical Communication

Communication and language – the problem of how to transfer ideas about technology and science – lay at the heart of Polhem's technological pedagogy. He returned many times to the idea of a 'universal language', a perfect language that could be spoken and understood by everybody, irrespective of their

34 Cf. Ferguson, 'The Mind's Eye', 828.

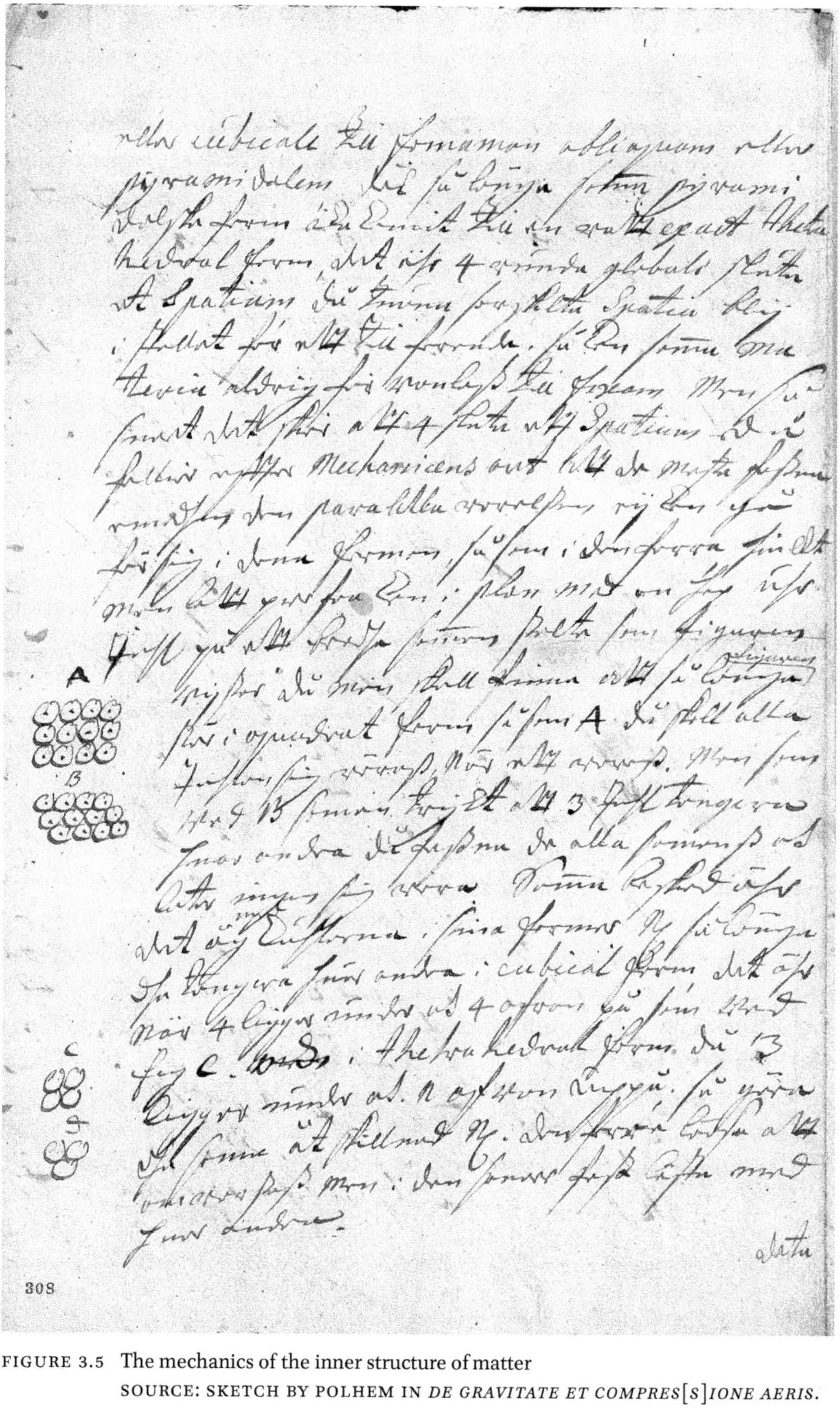

FIGURE 3.5 The mechanics of the inner structure of matter
SOURCE: SKETCH BY POLHEM IN *DE GRAVITATE ET COMPRES[S]IONE AERIS*. PHOTO: ROYAL LIBRARY, STOCKHOLM

education or origin.[35] His preoccupation with a universal language is likely to have been related in part to his own checkered educational history and difficulties with reading books, understanding Latin and writing correctly. Despite these obstacles, they did not stop him from writing an immense number of texts on all kinds of topics, from technology and physics to economics, pedagogics and the philosophy of language. Polhem took as his starting point the contemporary interest in the universally valid, the unambiguous; in logic, order and classification.[36] The fundamental problems of his time concerned logical method, the systematic classification of knowledge and the construction of an encyclopedia of knowledge.

In a sense, a universal language – Latin – already existed, but it was not perfect. Artificial universal languages can be seen as attempts to break the dominance of Latin as the *lingua franca*, its socially exclusive character. Latin had created a broad if not completely insurmountable chasm between elite and popular culture, excluding women, craftsmen, farmers, and a smith and carpenter such as Polhem. His own lack of proficiency in Latin was a clear mark of not fully belonging to the learned category. Polhem was not a polyhistor in the sense of someone who sought both eloquence and erudition; classical rhetorical eloquence and humanistic learning were not ingredients of Polhem's Archimedean polymathy. He often criticized Latin as an obstacle to thinking and the improvement of science. In this respect, he reminds us of Stevin, who advocated the use of Dutch in science, partly in order to make scientific works practically useful for the common people; he introduced new technical words in the vernacular for this purpose. Otherwise, Stevin feared, those who lacked a scholarly education would forever be excluded from scientific activity.[37] For Stevin, language was the key to scientific achievements, and he, like Polhem, stressed the importance of vernacular learning. Learning science in one's mother tongue saved both time and effort. Latin should not be a barrier, Stevin maintained.[38] For Polhem, too, Latin studies sat uncomfortably with his cult of utility and his eagerness to disseminate new findings and inventions to the broader population. Largely because of his own wavering educational path, he had great concern for the teaching of young people. Learning Latin or other subjects by reeling off texts by heart was worth little; it was, Polhem asserted, like giving a book to someone who could not read,

35 Dunér, 'Language of Cosmos'; cf. Yates, *Art of Memory*, Eco, *Search for the Perfect Language*, Rossi, *Logic*.
36 Dunér, 'The Axiomatic-Deductive Ideal'.
37 Stevin, *Principal Works*, I, 7; on language, see also Dijksterhuis, *Simon Stevin*, 126–129.
38 Van der Wal, 'Logic, Linguistics and Simon Stevin', 155.

or spectacles to the blind.[39] Instead, he advocated teaching in Swedish with Swedish books and practical exercises. Unlike Stevin, who viewed Dutch as the ideal language and key to reality, Polhem did not advocate Swedish as a language for science teaching on the grounds that it was a linguistic means of expressing things and ideas. For Polhem, the reason was simply that it was the mother tongue of his young apprentices. It had pedagogical advantages for enhancing understanding. To this was added a modern educational ideal. Polhem did not seek meaning in the classical, traditional wisdom found in books written by the wise men of antiquity. Instead, he sought a particularly empirical form of knowledge appropriated through the senses and with the hands.

Polhem himself never really managed to learn Latin. There is not one single manuscript in Latin in Polhem's writing. Based on what we can conclude from the surviving documents, it seems that Polhem was not a keen reader of books; there are few concrete references to or comments on important passages in other books. His manuscripts show irregular and erratic spelling (more than what one would expect) and inverted word sequences, which might suggest dyslexia. His deficiency in Latin, his cumbersome learning and own shortcomings in reading and writing prompted his endeavors to invent his own language. His first drafts of a universal language were made at a time characterized by a general interest in linguistics and the philosophy of language, in national languages and their sounds and grammar, in the connections between words and objects, in the character and the characterized. The artificial languages of the seventeenth and eighteenth centuries also reflected a desire for cosmopolitanism, to construct a language that could be spoken irrespective of who you were, where you were from and which mother tongue you spoke.

Linguistically – both semantically and phonetically – Polhem's universal language displays a number of advantages. The envisioned universal language would be pedagogical, more efficient and concise than ordinary natural languages, more regular and based on firm logical foundations. The undated manuscript 'Nomina rerum naturalium per philosophiam novam', which despite its title is written in Swedish, provides a fairly good impression of what such a universal language would look like.[40] Polhem covered the speech sounds, vowels and consonants, followed by grammar, syntax, phonology and lexicon. As with other proposed universal languages, great emphasis was placed on nouns and then verbs, after which adjectives were added. The senses and the elements played a central role in Polhem's universal language (figure 3.6).

39 Bring, 'Contribution', 83.

40 Polhem, *Christopher Polhems efterlämnade skrifter* IV, 333–338.

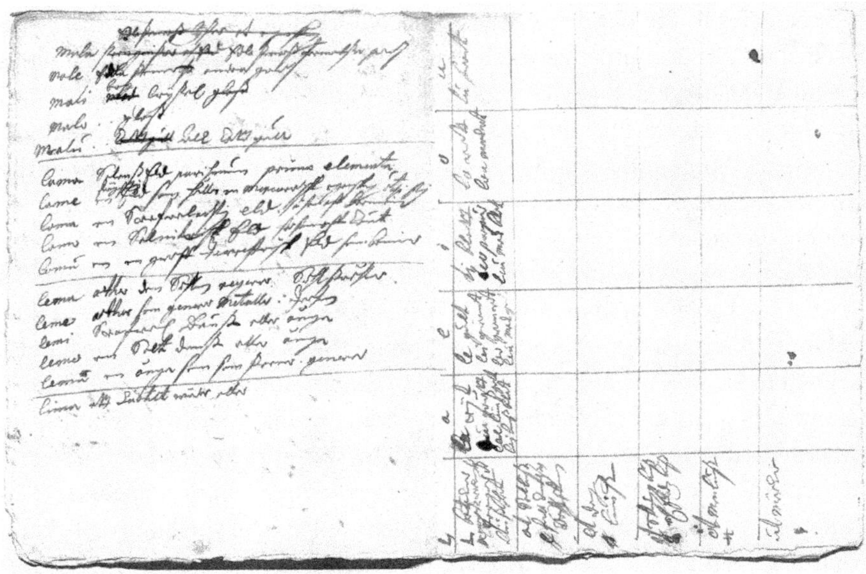

FIGURE 3.6 Polhem's universal language
SOURCE: POLHEM, *ORDA TECKN PÅ NATURENS MATERIALER OCH DESS EGENSKAPER* (C. 1710–1711). PHOTO: ROYAL LIBRARY, STOCKHOLM

The desire to learn different things in the natural sciences tempts many to start reading books, Polhem explains. But we soon tire of the difficulties, and often stop half way. This is because we have to work our way through an innumerable number of irrelevant letters and words, which merely fatigues the eyes and brain and puts a strain on our health. Instead, he speculates, we could invent a new way of writing books with concentrated words and sentences. An entire book could thus be summarized on a sheet or two, 'because then a large book would not so easily frighten many from reading it, as now usually happens, and then people would gain knowledge quicker than otherwise'. 'This I have thought about for a long time', Polhem claimed, but it had always proved difficult and tiring, particularly when other duties and problems had interrupted his studies. But he still considered this 'as a thing that would be no less useful and desirable than finding perpetuum mobile and lapis philosophorum which surely are impossible in themselves, but none the less has led many to spend both time and welfare upon this'. Many learned men had 'put their brains to work on this, but like me, have stopped half way'. But, like all of the gold-makers who had lived and died with the idea that it would be possible some time in the future, 'I also do the same'.

Behind this search for a universal language lies the supposed isomorphism of words and things, between expression and content, and the treatment of language and the world as a kind of combinatorics. We find something like this in Stevin's linguistic ideas concerning monosyllables and compounds, representing non-complex and complex things, respectively.[41] According to Stevin, language could be compared with geometry, where one builds something up from the smallest elements to the more complex. Polhem thought in very much in the same way. The idea of a universal language, Polhem's or others, can be seen as a kind of 'atomization' of reality that rests on the human ability for and obsession with categorizing and classifying things and phenomena in the human lifeworld. In the mechanistic worldview, everything seems to consist of building blocks; mechanical parts that have been put together in a world machine. Thoughts consist of simple ideas, words of letters, music of notes, and nature of numbers. Machines and mechanical movements also have their own, simple parts. Polhem's teaching on technology included 'the mechanical alphabet', which consisted of a large number of simple, educational wooden models showing the fundamental laws of mechanics (figure 3.7).[42] The models represented the simple and indivisible elements of mechanics; quite simply the building blocks of all engineering, such as the steel spring, the cogwheel, the ratchet wheel mechanism, the windlass or other mechanical elements that each represented a 'letter' in the mechanical alphabet. They described different types of mechanical movement, such as the transfer of one type of movement into another, from rotation into straight movement, and other rotating and forwards-backwards movements. Polhem's mechanical alphabet became a pedagogic system, one that was easy to learn, could be observed with one's eyes, and felt with one's hands.

With knowledge of these mechanical letters, a mechanic could build any machine he wished. Just like a poet, Polhem declared, could write the most beautiful poetry with the help of the ordinary alphabet, an engineer could learn the mechanical alphabet and form 'sentences' with mechanical letters; that is to say, construct complicated machines that could carry out useful work. The machines were like words and sentences. It was just as important, Polhem claimed, for a mechanic to know all the cogs, levers and catches in a machine as it was for a scholarly person to know the letters of the alphabet and the meaning of words.[43] There were some particularly important mechanical letters that corresponded to the vowels in ordinary spoken language. In the same way that we could not write words without vowels, it was not possible to build a machine without any of the five mechanical vowels; namely, the lever,

41 Van der Wal, 'Logic, Linguistics and Simon Stevin', 152.

42 Polhem, *Kort berättelse*, 75–77.

43 Cronstedt, TM, 2.

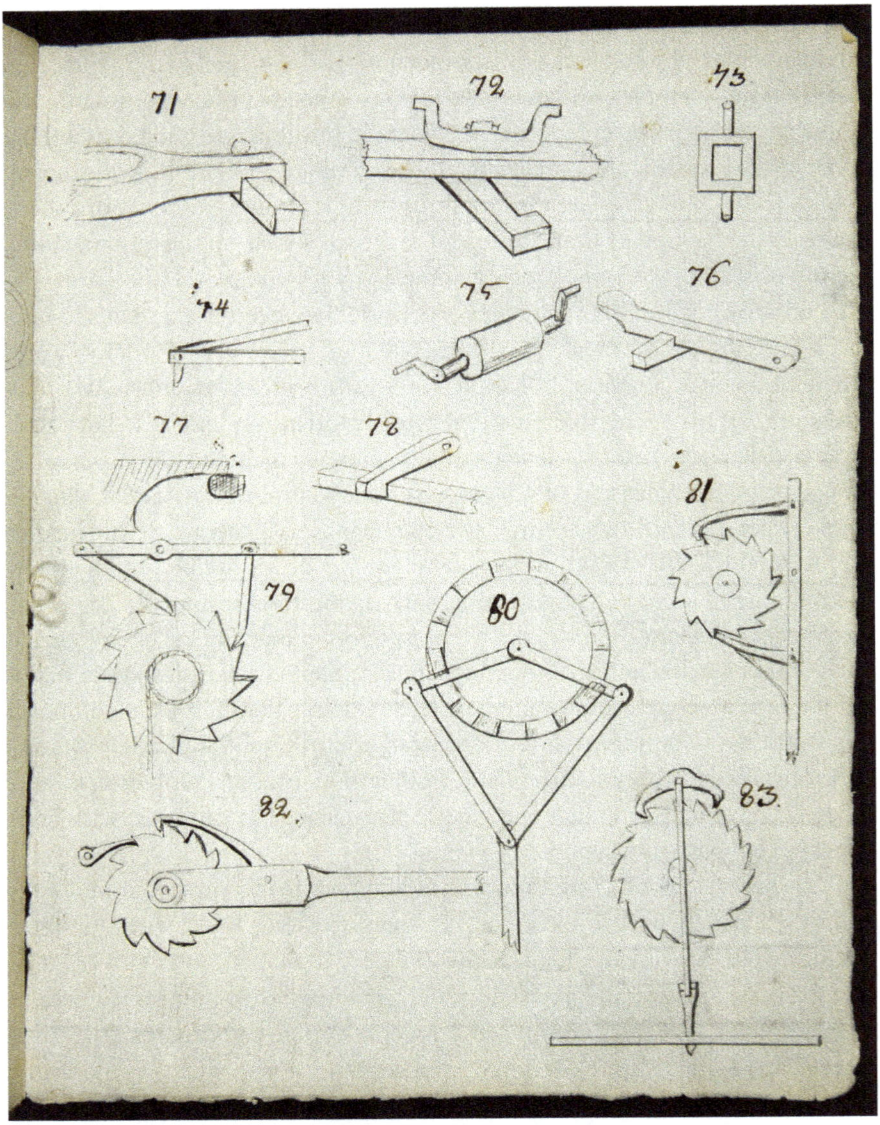

FIGURE 3.7 The mechanical alphabet
SOURCE: SKETCH BY CARL JOHAN CRONSTEDT, 1729. PHOTO: MUSEUM OF TECHNOLOGY, STOCKHOLM

the wheel, the screw, the block and the wedge. The most important was the lever. Cogwheels, chains, bearings, joints and springs were probably regarded as consonants, which did not necessarily have to be included in each machine.

The crucial task was to learn how to write correctly. With a mechanical alphabet, one could make many new, functioning inventions. Polhem's mechanical alphabet was based on the idea of the world as a construction kit, like a character system with an infinite number of possibilities for different combinations. The world consisted of small parts that could be put together into units, small atoms, corpuscles that formed bodies and objects, consonants and vowels that formed words and sentences, digits and numbers, simple geometric figures that described the motions of the universe, small mechanical letters that formed mechanical words and books. Polhem's mechanical alphabet became a world-renowned attraction. The German technologist Johann Beckmann saw this ABC in 1766, as did the future Venezuelan freedom hero, Francisco de Miranda, during his visit to Stockholm in 1787.[44]

6 Conclusion

Many factors shaped the formation of a polymath, of course; they included social, cultural, economic and political factors. Polhem's entrance onto the stage occurred at a time when the Swedish state was pursuing an expansionist mercantilist manufacturing and trade policy. The need to advance the profitable mining industry and protect domestic proto-industry clearly favored entrepreneurs such as Polhem. To some extent, the craftsman-mathematician was an answer to the demand for skilled technological entrepreneurs in technology, individuals who had both insight into the learned world of mathematics and practical experience in the manufacturing industry. We can discern peculiarities in his approach to technology, however, due to his faltering linguistic skills in Latin and reading, partly compensated for by exquisite visual thinking in three dimensions. In addition, his wavering social identity – between craftsman and nobleman, between illiterate and savant – gave a certain nuance to his obsession with the unity of theory and practice. The practical-theoretical combination of polymathy, the craftsman-mathematician, was to some extent an act of self-fashioning to win commissions and transcend the social barriers between craftsmen and the learned. The Archimedean polymath tried to pull these seemingly incompatible – both socially and cognitively – ways of gaining knowledge together.

A fundamental element of the formation of a polymath is learning from others: observing, imitating and copying behaviors, procedures and views through instruction and adjustment. Polhem was dependent on others' instruction, knowledge and experiences, and on his cognitive and intellectual

44 Beckmann, *Schwedische Reise*, 131; Miranda, *Archivo*, 40f.

environment or context. He learned through observing and copying others, through discerning similarities, and from previous experiences and solutions, followed by reconstruction and replication. To achieve this, the craftsman-mathematician needed a number of mental abilities. Two have been highlighted in this chapter: perception and communication.

To a great extent, technological thinking is a non-verbal form of thinking; a way of making visual representations of technology. In Polhem's case, we come across the mental, internal visualization of technology. He seems to have had a certain ability to keep complex mechanical movements in his head, to elaborate an inner representation of external objects. Even external representations of technology played an important role in his technological thinking, not so much in the form of two-dimensional drawings, but rather as three-dimensional, moving wooden models. These models were a kind of externalization of thought.

Another crucial prerequisite for the formation of a polymath was communication; that is, ways of sharing and transferring mental content from one mind to another. For Polhem, language and the communication of technological ideas played a central role in his technological pedagogy. He searched for a universal language; a more efficient, concise and regular way of obtaining information and storing and disseminating knowledge. His mechanical alphabet was a pedagogical tool for understanding fundamental laws as well as finding the smallest components of mechanics from which one could construct complex machines. In this sense, a polymath such as Stevin or Polhem is a thinking subject who perceives connections between seemingly incompatible phenomena, and who combines diverse areas of knowledge in order to come up with new ideas, solutions and explanations. Polymathy is a form of creative thinking.

Bibliography

Unpublished sources

Buschenfelt, Samuel, *Berättelse till Bergskollegium över hans resa till bergverken i Tyskland, Holland, England, Frankrike och Österrike 1694–1696 med tillhörande ritningar*, Royal Library, Stockholm (KB), L 70:54:1.

Buschenfelt, Samuel, *Reseanteckningar 1694–1697*, Uppsala University Library, Uppsala (UUB), X 366.

Cronstedt, Carl Johan, *Machiner, som till största dehlen äro uti wärket stelte* [*av Polhem*] *och af Ehrensverd och mig afritade åhr 1729: tillika med andra tilökningar som iag sielf giort tid effter annan*, Swedish National Museum of Science and Technology, Stockholm (TM), 7405.

Polhem, Christopher, *Afskrift af Chr. Polhammars bref 1696 till Bergs Collegium ang. hans utländska resa och förslag till inrättandet af ett mekaniskt laboratorium*, Royal Library, Stockholm (KB), X 265:1.

Literature

Amelin, Olov, *Medaljens baksida: Instrumentmakaren Daniel Ekström och hans efterföljare i 1700-talets Sverige*, Uppsala: Institutionen för idé- och lärdomshistoria, 1999.

Andriesse, Cornelis Dirk, *Huygens: The Man behind the Principle*, transl. Sally Miedema, Cambridge: Cambridge University Press, 2005.

Beckmann, Johann, *Schwedische Reise nach dem Tagebuch der Jahre 1765–1766*, Lengwil: Libelle, 1995.

Besson, Jacques, *Theatrvm instrvmentorvm et machinarum Iacobi Bessoni Delphinatis, mathematici ingeniosissimi. Cum Franc. Beroaldi figurarum declaratione demonstratiua*, Leiden: Barth. Vincentium, 1578.

Bring, Samuel E., "A Contribution to the Biography of Christopher Polhem", in: *Christopher Polhem: The Father of Swedish Technology*, transl. William A. Johnson, Hartford CT: American Swedish History Museum, 1963, 3–105.

Dahl, Per, *Svensk ingenjörskonst under stormaktstiden: Olof Rudbecks tekniska undervisning och praktiska verksamhet*, Uppsala: Institutionen för idé- och lärdomshistoria, 1995.

Devreese, Jozef T. and Guido Vanden Berghe, *"Magic is No Magic": The Wonderful World of Simon Stevin*, Southampton: WIT, 2008.

Dijksterhuis, Eduard Jan, *Simon Stevin. Science in the Netherlands around 1600*, The Hague: Martinus Nijhoff, 1970.

Dunér, David, "Daedalus of the North: Swedenborg's Mentor Christopher Polhem", *The New Philosophy* 113:3–4 (2010) 1077–1098.

Dunér, David, "Christopher Polhem's Experiments Concerning the Creation", in: *The 18th Century and Europe*, ed. Elizarija Ruskova, Sofia: Sv. Kliment Ohridski, 2013, 114–118.

Dunér, David, *The Natural Philosophy of Emanuel Swedenborg: A Study in the Conceptual Metaphors of the Mechanistic World-View*, Dordrecht: Springer, 2013.

Dunér, David, "The Language of Cosmos: The Cosmopolitan Endeavour of Universal Languages", in: *Sweden in the Eighteenth-Century World: Provincial Cosmopolitans*, ed. Göran Rydén, Farnham: Ashgate, 2013, 41–65.

Dunér, David, "Science: The Structure of Scientific Evolutions", in: *Human Lifeworlds: The Cognitive Semiotics of Cultural Evolution*, eds. David Dunér & Göran Sonesson, Pieterlen and Bern: Peter Lang, 2016, 229–266.

Dunér, David, "The Axiomatic-Deductive Ideal in Early Modern Thinking: A Cognitive History of Human Rationality", in: *Cognitive History: Mind, Space and Time*, eds. David Dunér & Christer Ahlberger, Berlin: De Gruyter, 2019, 99–126.

Dunér, David and Göran Sonesson eds., *Human Lifeworlds: The Cognitive Semiotics of Cultural Evolution*. Pieterlen and Bern: Peter Lang, 2016.

Dunér, David and Christer Ahlberger eds., *Cognitive History: Studies in a Historical Methodology*. Berlin: De Gruyter, 2019.

Eco, Umberto, *The Search for the Perfect Language*, Oxford: Blackwell, 1997 [1995].

Eriksson, Gunnar, *The Atlantic Vision: Olaus Rudbeck and Baroque Science*, Canton MA: Science History Publications, 1994.

Ferguson, Eugene S., "The Mind's Eye: Nonverbal Thought in Technology", *Science* 197:4306 (1977) 827–836.

Gärdenfors, Peter, *How Homo Became Sapiens: On the Evolution of Thinking*, Oxford: Oxford University Press, 2006.

Gärdenfors, Peter, "The Evolution of Semantics: Sharing Conceptual Domains", in: *The Evolutionary Emergence of Language: Evidence and Inference*, ed. Rudolf Botha, Oxford: Oxford University Press, 2013, 139–159.

Happel, Eberhard Werner, *Gröste Denkwürdigkeiten der Welt oder so genannte Relationes Curiosae …*, III, Hamburg: Wiering, 1687.

Huygens, Christiaan, *Horologium oscillatorium, siue, De motu pendulorum ad horologia aptato demonstrationes geometricæ*, Paris: Muguet, 1673.

Kemp, Martin, "Simon Stevin and Pieter Saenredam: A Study of Mathematics and Vision in Dutch Scienceand Art", *The Art Bulletin*, 68(2) (1986) 237–252.

Klingenstierna, Samuel, *Åminnelse-tal öfver Kongl. Vetensk. academiens framledne ledamot, commerce-rådet och commendeuren af Kongl. Nordstjerne-orden, herr Christopher Polhem, på Kongl. Vetenskaps acad. vägnar hållit i Stora Riddarhus-salen, d. 25. junii, år 1753*, Stockholm: Salvius, 1753.

Koetsier, Teun, "Simon Stevin and the Rise of Archimedean Mechanics in the Renaissance", in: *The Genius of Archimedes: 23 Centuries of Influence on Mathematics, Science and Engineering*, eds. S.A. Paipetis & Marco Ceccarelli, Dordrecht: Springer, 2010, 85–111.

Krüger, Jenneke, "A First Mathematics Curriculum: Stevin's Instruction for Engineers (1600)", *International Journal for the History of Mathematics Education*, 10(1) (2015) 79–88.

Lakoff, George and Mark Johnson, *Philosophy in the Flesh: The Embodied Mind and its Challenge to Western Thought*, New York: Basic Books, 1999.

Leupold, Jacob, *Theatrum machinarum …*, Leipzig: Zunkel, 1724–1739.

Liljencrantz, Axel, "Polhem och grundandet av Sveriges första naturvetenskapliga samfund jämte andra anteckningar rörande Collegium Curiosorum", *Lychnos*, 1939, 289–308.

Lindgren, Michael H., *Christopher Polhems testamente: Berättelsen om ingenjören, entreprenören och pedagogen som ville förändra Sverige*, Stockholm: Innovationshistoria förlag, 2011.

Lindroth, Sten., *Christopher Polhem och Stora Kopparberget: Ett bidrag till bergsmekanikens historia*. Uppsala: Almqvist & Wiksell, 1951.

Miranda, Francisco de, *Archivo del general Miranda* III. Caracas: Editorial Sur-America, 1929.

Østergaard, S., "Imitation, Mirror Neurons and Material Culture", in: *Excavating the Mind: Cross-Sections through Culture, Cognition and Materiality*, ed. N. Johannsen, M. Jessen, and H. Juel Jensen, Aarhus: Aarhus University Press, 2012, 25–38.

Papin, Denis, *Recueil de diverses pieces touchant quelques nouvelles machines*, Cassel: Jacob Estienne, 1695.

Pipping, Gunnar, "Christopher Polhems astronomiska ur", *Dædalus*, 121 (1992) 156–163.

Polhem, Christopher, *Kort berättelse om de förnämsta mechaniska inventioner som tid efter annan af commercie-rådet Christopher Polhem blifwit påfundne och til publici goda nytta och tienst inrättade, sampt om det öde, som en del af dem hafft genom tidernas oblida förändringar,* Stockholm: Biörkman, 1729.

Polhem, Christopher, *Christopher Polhems brev*, ed. Axel Liljencrantz. Uppsala: Lychnos-bibliotek, 1941–1946.

Polhem, Christopher, *Christopher Polhems efterlämnade skrifter I: Teknologiska skrifter*, ed. Henrik Sandblad, Uppsala: Lychnos-bibliotek, 1947.

Polhem, Christopher, *Christopher Polhems efterlämnade skrifter IV: Varia.*, ed. Bengt Löw, Uppsala: Lychnos-bibliotek, 1954.

Ramelli, Agostino, *Schatzkammer, mechanischer Künste, …* Leipzig: Liger, 1620.

Rossi, Paolo, *Logic and the Art of Memory: The Quest for a Universal Language*, Chicago: The University of Chicago Press, 2000.

Steele, Brett. D. & Tamara Dorland eds., *The Heirs of Archimedes. Science and the Art of War through the Age of Enlightenment*, Cambridge MA: MIT Press, 2005.

Stevin, Simon, *The Principal Works of Simon Stevin*, I, ed. E.J. Dijksterhuis, Amsterdam: C.V. Swets & Zeitlinger, 1955.

Sturm, Johann Christoph,. *Collegium experimentale, ….* Nürnberg: Endter. 1676–1685.

Wal, Marijke J. van der, "Logic, Linguistics and Simon Stevin in the Context of the Sixteenth and Seventeenth Centuries,", in: *History of Linguistics* 1993, ed. K.R. Jankowski, Amsterdam: John Benjamins, 1993, 147–156.

Yates, Frances A., *The Art of Memory*, London: Pimlico, 2001 [1966].

Zonca, Vittorio, *Novo teatro di machine et edificii per uarie et sicure operationi: co[n] le loro figure tagliate in ramé e la dichiaratione e dimostratione di ciascuna: opera necessaria ad architetti e a quelli ch' di tale studio si ditettano*, Padua: Pietro Bertelli, 1607.

CHAPTER 4

Stevin's Physical Geography

The World as a Chemical Furnace

Rienk Vermij

As a mathematician, Simon Stevin was not only wide-ranging, but also remarkably ambitious. The title of one of his early works, *Dialectike* (1585), states that this work not only teaches the reader how to form a sound judgement, but even opens up 'the path to the most recondite mysteries of nature.' In the dialogue that concludes the work, two characters, Jan and Pieter, discuss the perfection of knowledge. Jan notes the utility of the Dutch language as an instrument and asks rhetorically: '[a]nd also … they [the Dutch] are not lacking in all the other attributes necessary for a more perfect knowledge: what would prevent them from bringing not only the medieval, but also the Greek, Arabic, Egyptian and Chaldean mysteries to light again? Yes, better than before; that is, with knowledge of causes.'[1] Pieter dampens his enthusiasm a little by reminding him of the power of fortune, 'the enemy of reason'.

Similar ideas can also be found in his later works. The perfect nature of the Dutch language is lauded in a treatise included in his *Beghinselen der weeghconst* [Principles of statics, 1586].[2] Stevin further elaborates on this subject in a later work, *Eertclootschrift* ['treatise on the earth'], part of his *Wisconstige ghedachtenissen* [Mathematical memoirs, 1608]. Moreover, in this work he argues for the existence of a pristine period, the 'wijsentijt' [age of the sages], when people's knowledge was superior to our own, or even superior to the classical period. He maintains that the existence of such a time can be proven from certain existing knowledge that can only be explained as a remnant of the knowledge of these sages, 'but without knowing who, where, or when.'[3]

Stevin's considerations on the 'wijsentijt' have not gone unnoticed, but they have generally been considered in isolation.[4] Until now, nobody seems to have

1 Stevin, *Dialectike*, 163.

2 For a discussion of Stevin's ideas on the Dutch language, see Van der Wal, 'Logic, Linguistics, and Simon Stevin'.

3 Simon Stevin, *Eertclootschrift*, 9 (6th edition). *Eertclootschrift* was published as part of Simon Stevin, *Wisconstige ghedachtenissen*, I (Leiden 1608). It is not included in the *Principal Works*.

4 Dijksterhuis, *Simon Stevin*, 317–320; Van den Heuvel, *Huysbou*, 28–30.

 | DOI:10.1163/9789004432918_006

asked why these considerations were published in a work on the earth, or why they were relevant in this context. The fact does not appear to be fortuitous. *Eertclootschrift* is a work in which Stevin engages with major physical problems. Formally, the term 'eertclootschrift' is Stevin's translation of 'geographia'. As such, the work seems to fit well in Stevin's oeuvre as a mathematician, as geography was certainly considered a mathematical subject at the time. Petrus Apianus and Gemma Frisius had popularized a template in which geography and astronomy comprised the two parts of cosmography. Indeed, that is how these fields are organized in *Wisconstige ghedachtenissen*: 'eertclootschrift' and 'hemelloop' (astronomia) are the second and third parts of 'weereltschrift' (cosmographia), respectively. (The first part is on spherical trigonometry.)

On a closer consideration, however, we see that Stevin goes far beyond the content of traditional cosmography in these works. His *Hemelloop* contains not only elementary spherics and astronomical calculations, but also an elaborate discussion of the constitution of the universe, in which he advocates the Copernican system. And whereas contemporary geographies discussed mostly mathematical geography and topography, in his *Eertclootschrift* Stevin included treatises on navigation and the haven-finding art, the height of the atmosphere (closely following the Portuguese mathematician Pedro Nuñes), the tides and the constitution of the earth. In particular, the constitution of the earth would not have been regarded as a mathematical subject at that time. Stevin was thus entering a field that most of his contemporaries would have regarded as philosophical.

Stevin would have felt justified in taking these steps outside what could be regarded as his proper domain, as he explicitly rejected traditional philosophical methods and instead claimed that knowledge of the natural world could only be acquired in an empirical and mathematical way. Despite similarities to some other Renaissance thinkers and practitioners, this approach was quite unusual. It therefore made good sense to include a discussion on the proper way to acquire knowledge of such topics, and on how his methods related to those of the past. The age of the sages is one element in this wider argument on true knowledge; a function that only becomes clear when we consider the whole context. (Interestingly, the past existence of an age of the sages and the superiority of the Dutch language are themselves argued on empirical grounds, as shown in other papers in this volume.)

In this chapter, I hope to shed light on Stevin's wider ambitions by discussing the ideas in *Eertclootschrift* regarding the constitution of the earth. In considering the phenomena on the earth's surface, Stevin was engaging with fundamental questions about the principles of matter and material change. Tackling these with his own methods, he explained why he felt that his results not only went beyond the knowledge of his time, but also beyond the knowledge of classical antiquity. Modern science rejected his theories; they were forgotten and not even included in his *Principal works*. His approach was certainly innovative, however, and can be seen as one of the steps that would ultimately lead to the mechanical philosophy of the seventeenth century.

1 General Principles

Eertclootschrift consists of five books, but only the first two concern us here. The first book is 'on its [the earth's] definitions in general'. Apart from a definition of the earth itself and some geographical concepts, the bulk of this first book consists of Stevin's discussion of the age of the sages and his speculations on the Dutch language. He concludes with a chapter on the relationship between theory and practice ('spiegheling en daet'). This first book offers a set of general principles, not only for his work on geography, but also for his cosmography as a whole. In his astronomy too, he frequently refers to the age of the sages.[5]

Besides the age of the sages, there are some other general points that deserve to be highlighted. Stevin explicitly rejected the opinions of the classical Greek philosophers regarding the four elements: 'Their diligence was laudable, having done what they could, but alas, it was all from hearsay, with few answers, with many errors, and without the knowledge of causes for knowledge of causes in this case was taken in particular from practical chemistry ['dadelicke stofscheyding'], of which they knew nothing'.[6] Stevin's term 'stofscheyding' could also be translated as 'alchemy'. In fact, the Latin equivalent he himself provides is 'Alchimia'; he describes 'Stofscheyders' as 'Chimisten', 'Alchimici' or 'Alchimistae'.[7] However, his chemistry went far beyond recipes for making gold and appears to have been different from medieval alchemy in other respects, too. Thus, to avoid confusion, I prefer the more modern term.[8]

In the words that Stevin used to reject Greek ideas, we find two key concepts that are essential for an understanding of his thought and that frequently recur in his work. In the first place, knowledge of causes ('kennis der oirsaken'). We

5 E.g., Stevin, *Hemelloop*, 157, 187, 214, 293.
6 Stevin, *Eertclootschrift*, 15.
7 Marginal notes in ibidem, 15, 17, 60, 62.
8 It not possible to make a clear distinction between alchemy and chemistry in the early modern period. On the problem of terminology in general, see Principe, *Secrets of Alchemy*, 85. Principe prefers the use of the term *chymistry*.

already came across an instance of this term, which indicates Stevin's ideal of knowledge, in the dialogue in *Dialectike*. Another example can be found in Stevin's dedication of his book on arithmetic to Jan Cornets de Groot: he praises De Groot by stating that he understands things 'not in the vulgar way, but with a solid understanding of the causes'.[9] In *Hemelloop*, Stevin explains that he had described Ptolemy's theories after his own (Stevin's) style, as an instrument for those who wanted to devise better theories; 'but having found thereafter what I deem to come from knowledge of causes ... I have left it unprinted.'[10]

Knowledge of causes was a rather ambitious goal for a mathematician, as mathematics, like history, was regarded as a descriptive art. Causes had always been the province of philosophy. Admittedly, such distinctions were no longer self-evident in the early modern period. In particular, neo-Platonic and Pythagorean philosophies gave mathematicians such as Kepler the confidence to use mathematics as a key to understanding reality. Stevin's confidence was not based on some form of metaphysics, however, but on his knowledge of practice; practical chemistry in this case. The original term rendered here by 'practical' is 'dadelick', which is the second key concept. The word is derived from 'daad' (or 'daet', as Stevin writes it), meaning deed or action. For Stevin, 'dadelick' literally meant 'by deeds'.[11]

As used by Stevin, 'daet' often means the actual practice of an art, as opposed to mere theory. To his work on the principles of statics ('weeghconst', literally the 'art of weighing'), he attached an appendix on the practice of weighing: *De weeghdaet*. *Wisconstighe ghedachtenissen* contains a work on practical geometry entitled *Van de meetdaet* ['on the practice of measuring']. (Earlier, in 1583, he had already published a more theoretical work on geometry: *Problemata geometrica*.) His work on arithmetic of 1585 is accompanied by an annex on the practice of arithmetic.

The relationship between 'spiegeling' (theory) and 'daet' is a central element in Stevin's thinking. Stevin not only considers 'daet' to be an application of theory, but also feels that theory should be based in actual practice. Thus, in some contexts, 'dadelick' can be translated as 'experimental'. Knowledge of the constitution of matter can only be gained by someone who is practicing

9 Stevin, *L'arithmétique* (Leiden 1585), dedication. ('Non pas selon le vulgaire, mais par solide intelligence des causes.'). On Stevin's conception of causes, see also the chapter by Buning in this volume.

10 Stevin, *Hemelloop*, 299.

11 The word has a different meaning in modern Dutch. However, the *Woordenboek der Nederlandse Taal* gives as the first meaning: 'Uit eene daad of uit daden bestaand, zich in daden uitend.'

chemistry, or doing chemical experiments. Stevin's statement that knowledge of causes can only be found by means of practical chemistry derives from a deeper conviction that true knowledge of reality is not obtained by philosophical speculation or humanistic text-interpretation: it can only be found by working actively, 'by deeds'.[12]

According to Stevin, chemistry was highly developed in the age of the sages, as is attested by remnants of knowledge that are preserved in ancient writings, notably by Hermes Trismegistus. (Stevin added that it was not known when and where Hermes had lived, but that he is assumed to have been very ancient.) He expressed regret 'that this worthy art, this inexhaustible source of wisdom, has come by many to be despised.'[13] Chemistry also features on the list of sciences in which Stevin felt knowledge could still be further advanced, by collecting data and undertaking other forms of empirical research. According to him, the scholars of the 'wijsentijt' had obtained their knowledge in this very way. The other fields on this list are medicine – especially anatomy and the knowledge of herbs and remedies 'that are mostly found by chemistry' – and 'Eertcloots stofroersel', literally the 'movement of the matter of the earth'.[14]

2 The Constitution of the Earth

The rest of this chapter will be devoted to 'Eertcloots stofroersel', the second book of *Eertclootschrift*. In this book, Stevin discusses the various changes on the earth's surface. As Stevin explained, his study of this subject had its origin in practical questions: the silting of harbors, the laying of foundations in bad soil, coastal erosion.[15] Indeed, in the book Stevin discusses some topics of practical relevance, such as the courses and the banks of rivers, the shoals at the entrance of harbors and the mouths of rivers, and the welling up of sand. However, these problems are discussed within the framework of a general theory of the constitution of the earth.

Stevin argued that although the earth itself will always maintain its shape, the materials that constitute the earth are constantly moving and changing place. As he emphasized, the various movements are not man-made, but 'naturally ordained in the earth itself'.[16] The cause of these motions lies first in

12 See also Van Berkel, 'Spiegeling en daet'.
13 Stevin, *Eertclootschrift*, 15.
14 Ibidem, 19.
15 Stevin, *Eertclootschrift*, 51, see also 64–69. Cf. Van den Heuvel, *Huysbou*, 79–84.
16 Stevin, *Eertclootschrift*, 51 ('natuerlick int wesen des Eertcloots veroirdent').

the water and second in the air; that is, the wind. The water is moved by the sun, which causes the water to evaporate; by the moon, which causes the tides; and again, by the wind.[17] Stevin does not give an explanation for the action of the wind. It might be added that identifying wind with air contradicted the authoritative theory of Aristotle, who taught that wind was caused by dry exhalations, not by a movement of the air. By Stevin's time, however, this theory was already subject to considerable doubt.[18]

Stevin gives a rather detailed description of how these changes come about. Water from the sea evaporates and falls on land in the form of rain, snow or hail. The waters run downward in the form of streams and rivers, eroding the places they pass; when they slow down in the plains, sand and clay are deposited again. The material that is left in the water is finally deposited on the seafloor. Waters thereby become silted, and finally become dry land. All plains have their origin in this process of deposition of sand in water.[19] Originally sandy, they become dark and fatty when plants start to grow on them. The leaves and fruits that fall down and rot each year are mixed with the sand, producing a dark soil. Taken on its own, this vegetative substance is peat.[20]

As for the origin of mountains, Stevin offers a rather lengthy explanation in which the crucial role is played by the wind, which is able to blow the sand up to heights well above where the water can reach. All mountains therefore have their origin as sand dunes.[21] They become rocky because of the precipitation of stony or metallic substances from rainwater. Stevin argues at some length that water can contain all kind of other materials. 'Since in rain water there is dissolved glass and stone, which one finds in the pot by prolonged boiling [...], it is not strange but natural that this water, when continuously seeping through the sand of the dunes and lands, sets or leaves in it something of its glassy, stony, and clay-like matter, that is stone itself [...].' To the objection that if rocks were formed by rain, they should mostly be found at the surface, whereas in reality they are most often found within the earth, Stevin answers that at the surface they are weathered by the influence of frost, sunshine, and so forth.[22]

In this way, it is obvious that all matter found on the earth consists of mixtures. The water of the ocean is mixed with all kinds of terrestrial matter, partly

---

17 Stevin, *Eertclootschrift*, 54.
18 Borrelli, 'Weather Glass', 67–130.
19 *Eertclootschrift*, 55–57.
20 Ibidem, 59–60.
21 Ibidem, 57–59.
22 Ibidem, 60–63, quote on p. 62.

brought there by the rivers that run down from the mountains, partly originating from the seafloor and shores. The vapors that arise from the ocean when falling on land in the form of precipitation deposit some substances, creating new materials, whilst also carrying some of the older materials back to the sea. The Greeks were right to say that everything in nature is a mixture. They might have inherited that saying from the age of the sages, 'but in doing so, they appear not to have inherited the knowledge of causes, or the way in which this happens in general.'[23]

Eertcloots stofroersel ends with the conclusion that under these various influences, land will be turned into oceans and oceans into land, apparently in some permanent cycle.[24] This may seem like a daring conclusion, but the idea of a cyclical and continuous transformation of the earth was in fact not original. It could already be found in Aristotle, notably in his Meteorology: 'mainland and sea change places and one area does not remain earth, another sea, for all time, but sea replaces what was once dry land, and where there is now sea there is at another time land. This process must, however, be supposed to take place in an orderly cycle.'[25] The topic of meteorology was quite popular in the Renaissance and many authors on natural philosophy, especially in Italy, propagated the view of transformative cycles in the history of the earth, even in vernacular works.[26]

3 Experience and Chemistry

So far, Stevin's ideas are in line with the general philosophical ideas of his age. Speculation on the origins of the various elements of the earth's surface was not entirely new, and many philosophers were looking for alternatives to the classical (in particular Aristotelian) theories. The origin of mountains, for instance, had been discussed by the Italian author Girolamo Cardano in his work *De subtilitate* (1550), one of the most widely-read books on natural knowledge in the sixteenth century. Cardano explained the origin of mountains as lying in earthquakes, or wind or the sea. According to him, mountains formed by wind were especially to be found in Africa; he was probably thinking of the sand dunes in the Sahara Desert.[27]

---

23 Ibidem, 63–64, quote on p. 63.

24 Ibidem, 70.

25 Aristotle, *Meteorologica*, book I, chapter 14. I used the translation by H.D.P. Lee in *Loeb Classical Library*, 107.

26 Dal Prete, 'Being the world eternal …', 296–297.

27 Cardano, *De subtilitate libri XXI*, 151–152.

Things become more interesting when Stevin discusses the question of how we can gain knowledge on these topics. Here, he refers back to the principles laid down in the first book. He is not referring to Aristotle or any other philosopher; as he sees it, the advancement of knowledge is mostly due to practical experience. For example, Stevin refers to the constitution of the soil at certain depths in certain places. In alluvial plains, it often happens that one finds a stony floor, at a depth of about two feet under the earth and about a foot thick, under which one finds soft soil again; 'as we have experienced often in the digging and circumvallation of besieged towns.' He refers in particular to an entrenchment near Hulst in Flanders.[28]

In one of his other works, when it comes to knowledge of the earth, he refers in particular to the new art of digging very deep wells. By digging such wells, 'one can now see in practice ['dadelick'] what different materials are found so deep inside the earth. In this way, a cause ['oirsaeck'] has been revealed to speak thereof with other knowledge than people previously had. I thus reckon this to be among the major discoveries that have come to human knowledge, not only in our own time, but for centuries.'[29] Stevin recorded the observations made when digging a 232-foot well for the charitable home for elderly men in Amsterdam in 1605. His notes on the various types of soil that were encountered at different depths would subsequently be quoted by many authors.[30]

In *Eertclootschrift* itself, Stevin made particular reference to the experiences of chemists. This is a logical consequence of his argument in the first book, where he had claimed that this science (and not the speculations of the Greeks) gives us insight into the true constitution of matter. In his view, the processes in the earth are chemical in nature and are the same as those observed in a chemical laboratory. He even calls the earth an 'eertclootschen scheydoven', a big alchemical furnace.[31] In several instances, he explains the processes in the earth with a reference to laboratory experience. Vapor in the air forms drops 'as one sees similarly in small chemistry'.[32] That rainwater can precipitate ores and metals is argued with reference to the way silver is dissolved and precipitated by nitric acid, 'in the way that is common among

28 *Eertclootschrift*, 61. Hulst had been besieged and conquered by Prince Maurits in 1591.

29 Stevin, *Materiae politicae, Byvough der stedenoirdening*, 85. Quoted from Van den Heuvel, *Huysbou*, 230 (my translation).

30 Published in Van den Heuvel, *Huysbou*, 230–234, 132. See also ibidem, 75–77. The method of drilling was devised by Pieter Pietersz Enten, to whom the States General granted a patent in 1602.

31 Stevin, *Eertclootschrift*, 62.

32 Ibidem, 54.

goldsmiths'.[33] Stevin also reports on his investigations to determine the constitution of soil. By pouring water, he was able to separate black soil into simple sand and a dark substance, similar to peat. He used this experiment to support his claim that all soils consisted of a substratum of sand to which other substances had been added.[34]

For this reason, Stevin argues that knowledge of the transformations in the earth is useful in a more general sense:

> Assuming that the great resolution and consolidation of the terrestrial materials happen as described, as I believe they do, knowledge thereof would be profitable in the investigation of the causes and qualities of the various material substances. [...] In such a way that, investigating by well-founded experiences in a systematic way ['oirdentlick'] what holds true and what is defective, one would be able, as we said, to come to a better knowledge of causes.[35]

We can gain knowledge of processes in the earth because we are able to imitate them in a chemical laboratory, on a smaller scale. It is the practical work of the chemist that teaches us about the constitution of the world. This immediately raises a further question, however: if chemistry is key to understanding natural processes, how should we understand chemical processes themselves?

4 Mechanical Chemistry

With his defense of a chemical interpretation of the processes in the world, Stevin again seems to have been well in line with the dominant tendencies of his age. The use of chemistry as a key to understanding reality and as an alternative to traditional philosophy had become quite common in the sixteenth century. Many philosophers described changes in the world in chemical terms. However, such chemical world-views typically had a strong religious component. Most alchemists regarded chemical processes as analogical to the process of creation described in Genesis. They saw changes in the material world as organic, caused by a life-giving spirit, rather than as a result of mere mechanical causality. The extremely influential philosophical and medical

33 Ibidem, 62.
34 Ibidem, 59–60.
35 Ibidem, 63.

iconoclast Paracelsus taught that Genesis was actually the only source from which one could understand matter and chemical processes.[36]

Against this background, it is striking that Stevin's understanding of chemistry does not appear to have any spiritual dimension. As already noted, at one point he refers to Hermes Trismegistus as an author whose writings contained remnants of past knowledge. At that time, Hermes was generally seen not merely as an authority on chemistry, but also, or even principally, on religion and Divine truth. Thus, it is even more striking that Stevin only mentions him as an authority on the practice of chemistry; there is no indication that he considered him as a source of spiritual knowledge as well.[37] It would be interesting to know which Hermetic writings Stevin had in mind, but he does not identify them. Nor does Stevin's work contain any reference to Genesis, to the process of creation, or to concepts such as clarification or redemption that might be understood in a spiritual sense.

In his explanation of the cyclical motion in the earth, Stevin does refer to a 'first mover' ('eersten Roerder'), but not to a moving spirit: 'It seems that the first mover wants to have everything in a continuous motion, regarding the parts as well as the whole, for the earth not only makes its annual course around the sun and the diurnal on its axis, but the various materials from which it is made have a continuous mixing and motion among each other.'[38] Later, in *Eertclootschrift*, he seems to perceive some final cause here. The sun and the subterranean fire heat the water, so that a visible vapor rises, 'to the end, as has been more broadly explained in *Eertcloots stofroersel*, to wit, that this brings about the great general bringing together and separation of the materials of the earth.'[39] But however one interprets this, it is a far cry from the spiritual interpretation of contemporary alchemists.

This is consistent with Stevin's work as a whole, in which, contrary to the dominant tendency of his age, religious connotations are almost completely absent. For instance, in his long defense of the Copernican system, at no point does he feel compelled to discuss its compatibility with Scripture. He also refuses to see the solar system as the expression of some pre-established or divine harmony, as Kepler or Copernicus had done.[40] He even is unhappy with

36 See among others Debus, *Chemical Philosophy*, I, 84–96; Walton, *Genesis*, 39–41, 51–52.

37 Schilt, *Simon Stevin*, has argued that Stevin's religious thought was inspired by Hermetic ideas. His arguments are not convincing, however.

38 Stevin, *Eertclootschrift*, 54.

39 Ibidem, 72 (the abstract of the third book, on the atmosphere).

40 Something similar is the case for his views on music, as noted in the chapter by Cohen and Kursell in this volume.

the idea of the universe as a well-contrived machine. In his *Hemelloop*, he objected to Copernicus' third motion of the earth, which kept the earth's axis in a constant direction during the earth's course around the sun: 'since I do not like any celestial motions that are so imagined that they move with equal speed, having been adjusted to each other as one makes the wheels of a mechanical clock fitting to each other, as these seem not to exist in nature.'[41]

The closest Stevin comes to recognizing a religious element in nature is when he makes an argument about the state of the world with an appeal to 'the creator of nature'. These instances are rare, however. In his work on geometry of 1585, he explains that the initial point of a half-line corresponds to the number 0, not to 1, adding: 'And may the creator of nature have mercy upon the miserable eyes of he who cannot see this.'[42] Other instances are to be found in his posthumous work on politics, where he refers to the origin of our moral feelings: knowledge of good and evil is not arbitrary, but given to us by 'the creator of nature'.[43] In his (equally posthumous) work on architecture, he advocates a symmetrical building style, because 'our Creator has ordained that we should find pleasure in such like-sidedness and that its opposite should seem unpleasant to us.' Commenting on Vitruvius, he explains: 'just as nature or the creator of animals produces like-sidedness, so the architect should emulate this.'[44] These are all instances in which Stevin referred to some element of divinity in his work. Apparently, he did see the order of nature as being divinely ordained, but it does not appear that he allotted the Divine any other role.

On the contrary, the chemical processes that Stevin identifies as being responsible for the changes on the earth's surface have a character that one is strongly tempted to call 'mechanical'.[45] Whereas traditional alchemists understood their art in qualitative terms and interpreted the changes in the subterranean world as a series of transmutations, Stevin, in discussing the various processes, speaks only of a reallocation of matter, of mixing and separation. At no point in *Eertcloots stofroersel* does he even hint at qualitative change or transmutation. The changes on the earth's surface are due to a 'continuous

41 Stevin, *Hemelloop*, 254; also published in *Principal Works*, III, 128–129. On Stevin's Copernicanism, see Vermij, *Calvinist Copernicans*, 58–68.

42 'Et qui ne le peut veoir, l'auteur de Nature aye pitie de ses infortunez yeulx.' Quoted by Dijksterhuis, *Simon Stevin*, 337, referring to Simon Stevin, *L'Arithmétique* (Leiden 1985) fol 4v.

43 Simon Stevin, *Materiae politicae*, section two, 39.

44 *Materiae politicae*, eerste onderscheid, 13; Van den Heuvel, 'De huysbou', 57; Van den Heuvel, *Huysbou*, 208–209.

45 The concept of 'mechanical' is problematic if one wishes to define it precisely, but the family resemblance is unmistakable.

mingling and moving through each other' of the various materials.[46] During these processes, the different substances remain unchanged. Water can only precipitate materials that have been dissolved in other places, for otherwise 'the water in the end would have to produce ['uytgheven'] something that it did not contain, which would be against order, and goes against all experience.'[47] In addition, referring to the experiences of chemists who claimed that if one evaporated water from a pot while keeping the pot filled, one eventually got 'glass' in the pot, he states that this glass 'would not have come there, had it not been in the water from the beginning.'[48]

These mechanico-chemical processes not only account for all known phenomena on earth, such as the growth and decay of soil and land or the formation of minerals; according to Stevin, these processes also 'appear to happen in a similar way in growing materials, both ensouled and dead ['so versielt als onversielt'].' In other words, living beings are formed by the same processes of displacement of matter as everything else. Stevin even suggested a cosmical understanding of this flux and reflux of matter in growing things: 'And it is possible that because of this, the ancients talked about the common world soul, wherefrom life or the soul enters into the animals and to which it returns, just like the water flows into and out of the sea.'[49] With the 'chemical philosophers' of his age, Stevin shared the all-encompassing ambition to explain everything, but he re-interpreted the spiritual elements in more materialistic terms.[50]

Evidently, Stevin had to admit that the processes by which living beings were formed were as yet far from understood, even less than the formation of ores and minerals within the earth. He did acknowledge a difference between the chemistry that governs the movement of water, sand, clay and peat, and the processes that govern the generation of minerals and living things: 'that the way that everything gathers with its kind is known in the former case by its causes ['deur de oirsaken'], but not in the latter. For how it happens that within the earth gold joins with gold, how a seed, both of animals and herbs, being in the appropriate place, is joined by other materials of the same kind, and how other species each gather with their kind, I have no idea.'[51]

46 Stevin, *Eertclootschrift*, 54.

47 Ibidem, 61.

48 Ibidem, 60.

49 Ibidem, 62. Stevin seems to refer to Stoic ideas here. Stoic ideas were well known at Leiden, especially because of the work of Lipsius.

50 Vermij, 'Putting the Earth' argues that early mechanistic ideas were often a reworking of more esoteric theories; see esp. 135–138, 140–141.

51 Stevin, *Eertclootschrift*, 62.

Stevin hoped that it would be possible to gain some vestiges of causal knowledge of these processes, however, by studying more common chemical experiments: how silver dissolves in aqua fortis and then precipitates on copper, how dissolved copper precipitates on iron, or how iron attaches to a magnet. In that way, it might be possible to understand how these materials were gathered under the earth, and eventually to learn how to make gold – the old alchemical dream. Stevin recognized, however, that the dream of making gold lay beyond reach for the time being. Our knowledge is still too defective. Lacking even an understanding of the causes of visible processes, 'which nevertheless we make happen by experiment ['dadelick']', it is little wonder that we are still far from being able to understand the hidden subterranean processes.[52]

5 Stevin and the Mechanization of Nature

Stevin was clearly influenced by the dominant tendencies of his time. His ideal of the perfection of knowledge is reminiscent of Francis Bacon's, his use of mathematics is similar to Galileo's, his utilitarianism has connections to Ramus, and his anti-Aristotelianism was of course shared by many thinkers of his time. And though dismissing humanist scholarship, Stevin made extensive use of the humanists' ideas on language and a 'prisca sapientia'.

As is clear from the last example, Stevin often gave his own twist to such ideas. The same is true of his ideas on the earth or on chemistry, or on knowledge in general. Although they resembled existing philosophical ideas in many respects, putting them in a new general framework gave them a new and unfamiliar character. That framework was a world-view based on practice and mathematics. Stevin wanted to know the causes of everything in the world, but he was not interested in speculative theories. The world should be explained by the mathematical concepts and processes that were intuitively familiar to the practician.

To students of philosophy, Stevin's 'knowledge of causes' probably looks rather pedestrian. If no other explanations are admitted, little is left of philosophy. Still, it was by a reduction to such basic principles that seventeenth-century natural philosophy would achieve its major successes. Mechanical philosophers would dispose of the majority of philosophical concepts and reduce everything to mechanical laws, and it was mathematicians and engineers such as Stevin and Galileo who paved the way for this approach. In his

52 Ibidem, 62.

deliberate attempt to describe all of reality in terms of the mixing and separating of matter, Stevin appears to have come close to the mechanical philosophy that would emerge later in the seventeenth century as the main alternative to Aristotelianism.

It would be interesting to know the extent to which Stevin's ideas did influence later mechanical philosophers. Isaac Beeckman, in particular, was familiar with Stevin's work and held it in high esteem. In his biography of Beeckman, Klaas van Berkel acknowledges that Stevin made a 'substantive' contribution to Beeckman's natural philosophy. He feels that in the end, however, their approaches were very different: 'Stevin never expressed an interest in natural philosophy. He had never written ... about the structure of matter. ...'[53] Clearly, this characterization should be qualified. Stevin looked down on existing natural philosophy and on philosophical speculation in general, but that was because he felt he could do better. He certainly had an interest in basic natural philosophical questions, including questions about the constitution of matter.

On the other hand, important differences remain between Stevin's original but rather haphazard ideas, and mechanical philosophy as it would later be developed. For one thing, we should acknowledge that Stevin's work is not very systematic. His arguments frequently contain gaps. In the abstract at the beginning of the third book of *Eertclootschrift*, on the atmosphere, Stevin states that the waters of the earth are heated by 'the sun outside the earth, with the heat and the real fire within',[54] but in *Eertcloots stofroersel* there is no mention of any subterranean fire. The chapter on 'Spiegelingh en daet' in the first book of *Eertclootschrift* is mostly on mathematics, with Stevin arguing that mathematical theory is useful even if there are no direct practical applications. This is rather at odds with the argument that knowledge of matter should bè based on practical experience, which is so prominent in *Eertcloots stofroersel*. (Admittedly, the chapter has more relevance for some of the later books, especially the one on the theory of the tides.)

Moreover, even though Stevin is quite emphatic in *Eertclootschrift* about the mechanical nature of chemical processes, he does not pursue this approach consistently throughout all of his work. For one thing, he allows for exceptions. In the dialogue that closes *Dialectike*, already referred to above, Jan argues that in nature, something grows only by means of the addition of something that was not there before; this is consistent with the mechanical ideas in *Eertclootschrift*. In reply to an objection from Pieter, however, he has to admit

53 Van Berkel, *Isaac Beeckman*, 134.

54 Stevin, *Eertclootschrift*, 72. On the theory of subterranean fire, see Vermij, 'Subterranean Fire'.

that gunpowder presents a different case, of 'bodies that change their essential matter into another.'[55]

In this case, Stevin may simply have changed his mind, but such inconsistencies can be found within *Weereltschrift* as well. As is well known, Stevin was deeply attracted to the ideas of William Gilbert, who in his book *De magnete* had described the earth as a big magnet and had stated that the universe is governed by magnetic forces. In his astronomy, Stevin made recourse to these magnetic forces to account for the constant position of the earth's axis and various anomalous planetary motions. Admittedly, Stevin did not adopt Gilbert's idea that magnetism is akin to a soul and that the earth is a living being. As far as is known, he never elaborated on the causes of magnetism, just as he never speculated on the causes of the tides. Still, he appears to have taken for granted that these cosmic forces were in some way non-material. That much can be deduced from his comment, in his work on astronomy (another part of *Weereltschrift*), that the celestial bodies (among which he included the earth) are kept together by a general sympathy ('ghemeene ghenegentheyt'), an idea that he referred to Copernicus.[56]

The most obvious gap, from a modern point of view, is that Stevin never goes into any detail about the basic constituents of matter. It would seem logical to interpret Stevin's conception of chemical processes as the mixing or separation of pre-existing and unchanging matter as implying a belief in atoms or small particles, but Stevin never specifically makes reference to this. The solution of solids in water is described not as a process of dissolving into small particles, but as these solids becoming 'vlietich' (fluid). The water of the sea gets its greenish color from the metals that have been dissolved by the continuous contact of these materials with the seafloor and the shores. 'In such a way that these metallic substances, which formerly were solid and rigid, become fluid in the water. This is caused by the continuous lapping of the sea water ['die gheduerighe strijcking des zeewaters'] over [a long] time, in the way as is seen, in a short time, in *aqua regia* working on metals: it makes them fluid within itself.'[57] Likewise, the phenomenon of concretions in wells or around bridge piles is explained by the process of some 'fluid glass' becoming rigid again.[58]

It is this complete silence on the topic of atoms or particles, above all, that sets Stevin apart from later mechanical philosophers such as Beeckman or

55 Stevin, *Dialectike*, 149.

56 Stevin, *Vanden hemelloop*, 252. Cf. Copernicus, *De revolutionibus*, lib. I cap. 9.

57 Stevin, *Eertclootschrift*, 60.

58 Ibidem ('Dit vlietende glas met steenicheit ghemengt, sietmen hem oock anders tot verstijving begheven …').

Descartes. It is hardly possible that he was unfamiliar with ancient or modern atomistic theories,[59] so this silence must have been deliberate. It may have been that he entertained some alternative theory, but it would seem more probable that he simply thought the issue too uncertain to take a stance. After all, Stevin's aim was not to give a full description of the world or a philosophical system, which would necessarily be speculative. He limited himself to the issues on which he felt he could speak with some degree of certainty, based on mathematical arguments or practical experience. His understanding of processes in the earth or in chemistry was not based on some specific theory of matter, but on what he could see in his laboratory. His emphasizing of processes of addition and separation was not the result of some speculative theory, but was based on the fact that such processes could be understood by doing practical work. For Stevin, there was no real understanding without manipulation.

Bibliography

Sources

Aristotle, *Meteorologica*, translated by H.D.P. Lee (Cambridge (Mass.) & London 1978) (Loeb Classical Library 397).

Cardano, Hieronymus, *De subtilitate libri XXI* (Basel 1611).

Stevin, Simon, *L'arithmétique* (Leiden 1585).

Stevin, Simon, *Dialectike ofte bewysconst. Leerende van allen saecken recht ende constelick oirdeelen; oock openende den wech tot de alderdiepste verborghentheden der Natueren* (Leiden 1585).

Stevin, Simon, *Wisconstige ghedachtenissen*, I (Leiden 1608).

Stevin, Simon, *Materiae politicae. Burgherlicke stoffen*, Hendrick Stevin ed. (Leiden n.d., dedication dated 1649).

Stevin, Simon, *The Principal Works of Simon Stevin*. 5 vols., Amsterdam: C.V. Swets & Zeitlinger, 1955–1966.

Literature

Berkel, Klaas van, "Spiegeling en daet bij Beeckman en Stevin", *Tijdschrift voor de Geschiedenis der Geneeskunde, Natuurwetenschappen, Wiskunde en Techniek*, 2 (1979) 89–100.

59 Atomist theories were defended in the Netherlands by David Gorlaeus. On Gorlaeus, see Lüthy, *David Gorlaeus*.

Berkel, Klaas van, *Isaac Beeckman on Matter and Motion. Mechanical Philosophy in the Making*, Baltimore: Johns Hopkins University Press, 2013.

Borrelli, Arianna, "The Weather Glass and its Observers in the Early Seventeenth Century", in: eds. Claus Zittel, Gisela Long et al., *Philosophies of Technology: Francis Bacon and his Contemporaries*, I, Leiden: Brill, 2008, 67–130.

Dijksterhuis, Eduard Jan, *Simon Stevin*, The Hague: Martinus Nijhoff, 1943.

Heuvel, Charles van den, *"De Huysbou". A Reconstruction of an Unfinished Treatise on Architecture, Town Planning and Civil Engineering by Simon Stevin*, Amsterdam: Edita KNAW, 2005.

Dal Prete, Ivano, "'Being the World Eternal …' The age of the earth in Renaissance Italy", *Isis*, 105 (2014) 296–297.

Debus, Allen G., *The Chemical Philosophy: Paracelsian Science and Medicine in the Sixteenth and Seventeenth Centuries*, New York: Science History Publications, 1977.

Lüthy, Christoph, *David Gorlaeus (1591–1612). An Enigmatic Figure in the History of Philosophy and Science*, Amsterdam: Amsterdam University Press, 2012.

Principe, Lawrence M., *The Secrets of Alchemy*, Chicago and London: The University of Chicago Press, 2013.

Schilt, Kees, *Simon Stevin en het hermetisme. Een cultuurhistorische studie naar zijn methodologie* (unpublished PhD thesis, Vrije Universiteit Amsterdam, 2013).

Vermij, Rienk, "Subterranean Fire. Changing Theories of the Earth during the Renaissance", *Early Science and Medicine*, 3 (1998) 323–347.

Vermij, Rienk, *The Calvinist Copernicans. The Reception of the New Astronomy in the Dutch Republic, 1575–1750*, Amsterdam: Edita KNAW, 2002.

Vermij, Rienk, "Putting the Earth in Heaven. Philips Lansbergen, the Early Dutch Copernicans and the Mechanization of the World Picture", in: eds. Massimo Bucciantini, Michele Camerota and Sophie Roux, *Mechanics and Cosmology in the Medieval and Early Modern Period*, Florence: Leo S. Olschki, 2007, 121–141.

Wal, Marijke J. van der, "Logic, Linguistics, and Simon Stevin in the Context of the Sixteenth and Seventeenth Centuries", in: ed. K.R. Jankowski, *History of Linguistics*, Amsterdam: John Benjamins, 1993, 147–156.

Walton, Michael T., *Genesis and the Chemical Philosophy. True Christian Science in the Sixteenth and Seventeenth Centuries*, New York: AMS Press, 2011.

CHAPTER 5

The Art of Demonstration in the Work of Simon Stevin

Linguistic and Mathematical Innovation

Marius Buning

1 Introduction

Simon Stevin (1549–1620) was one of the most famous mathematicians of the early modern period. He made fundamental contributions to the science of mechanics and hydrostatics and invented, among other things, the decimal system for fractions. This essay examines Stevin's method for studying 'the secrets of nature' by putting the focus on his oft-neglected treatise on logic entitled *Dialectike ofte Bewysconst. Leerende van allen saecken recht ende constelick Oirdeelen; oock openende den wech tot de alderdiepste verborghentheden der Natueren* (Leiden: Plantin, 1585).[1] Stevin's work on dialectics, written in Dutch, has rarely been studied.[2] Dijksterhuis, who devoted one brief but excellent chapter to the *Dialectike* in his 1943 biography on Stevin,[3] concluded that:

---

1 *Dialectics or the Art of Demonstration. Teaching how to Judge all matters skillfully and correctly; also opening the way to the deepest secrets of Nature.* All translations from Dutch are by the author, unless otherwise stated. The terms 'logic' and 'dialectics' are used interchangeably, whereas strictly speaking we are talking about two different methods to evaluate truth claims. For an excellent overview of the trajectory of the tradition of dialectics from the Ancients to the early modern period, see Spranzi, *The Art of Dialectic*.

2 The *Dialectike* was addressed briefly by Heidi de Mare in the context of house-building and (inadequately) by Kees Schilt in a PhD thesis on Simon Stevin and hermetism: De Mare, 'Het huis en de regels van het denken', 228–37; Schilt, 'Simon Stevin en het hermetisme'. Theo Verbeek published a short overview article on the *Dialectike*, concluding that more research was needed: Verbeek, 'La Dialectike'. The *Dialectike* is listed in a thesis on vernacular rationalism in the Low Countries, yet not addressed in any detail: Buys, *De kunst van het weldenken*, 166, 288. Wiep van Bunge did not pay any attention to the *Dialectike* in his Van Bunge, *From Stevin to Spinoza*. In other works, Stevin's ideas on logic are equally overlooked. For example, no mention of the work is made in Devreese and Vanden Berghe, *'Magic is No Magic'.* Paul Dibon, too, completely ignored the work of Stevin in his Dibon, *La philosophie néerlandaise*.

3 Dijksterhuis, *Simon Stevin*, 287–98. The chapter on the *Dialectike* has been omitted from the English translation: Dijksterhuis, *Simon Stevin*, 1970.

 | DOI:10.1163/9789004432918_007

> Though it does not enrich this classical branch of science with any substantial innovation, it is remarkable for two reasons: it is one of the two oldest treatises on logic written in Dutch, and the method of exposition deviates greatly from that followed in the traditional textbooks. By both means Stevin hoped to make logic accessible to all readers, regardless of their previous training. The work is concluded by a *Tsamespraeck*, a dialogue, in which one of the interlocutors defends Stevin's views on the superiority of the Dutch language.[4]

Indeed, the treatise by Stevin has received some attention from linguists.[5] Remarkably enough, however, they have not asked how the use of common language and common sense, so strongly recommended by Stevin, related to his views on dialectics. Moreover, it remains unclear how Stevin's dialectical method relates to his other works. Wilhelm Risse remarked, in his masterly overview of early modern logic, that although Stevin's *Dialectike* appeared mathematical, the intrinsic content had little in common with mathematics.[6] Conversely, it has been noted that a syllogism occasionally cropped up in Stevin's mathematical works; however, the scholarly community has to date largely ignored this, remarking that Stevin merely desired in this way 'to stress the importance' of his innovations.[7] Only Jurgen Naets has examined the further interweaving of dialectics and Stevin's mathematical epistemology, arguing that 'Stevin's concept of "number" was subject to the same epistemological conventions and logical background that also govern the structure and conception of object in general'.[8] However, Naets focused exclusively on Stevin's thinking about number theory. This essay posits that there is a broader relationship between Stevin's linguistic and mathematical innovations, and endeavors to embed Naets' findings in a broader discussion about the status of mathematics in early modern Europe.

As we will see, the *Dialectike* provides important insights into Stevin's views on the problem of universals. Incidentally, the definition of universals was central to contemporary debates on the degree of certainty that could be achieved through mathematics. Given that the *Dialectike* was printed together

4 The citation is taken from Stevin, *Principal Works, Mechanics*, 1:21. The tenor corresponds closely to Dijksterhuis, *Simon Stevin*, 1943, 287.
5 Overdiep, 'Simon Stevin'; Van der Wal, 'Logic, Linguistics, and Simon Stevin'; Van der Wal, 'Simon Stevin'.
6 Risse, *Die Logik*, 1:67.
7 Stevin, *Principal Works, Mechanics*, 1:44.
8 Naets, 'How to Define a Number?', 78.

with Stevin's more well-known mathematical works (including *De Thiende* and *L'Arithmetique*), this gives rise to the following question: to what extent and in what way can Stevin's mathematics and logic be understood as a separate project or as part of a broader design? To answer this question, this essay is divided into three sections. After positioning the *Dialectike* in its social and cultural context and introducing Stevin's dialectical method, I turn to Stevin's treatment of 'empty species' and discuss the importance of vernacular dialectics for Stevin's mathematical outlook.

2 The Cultural and Social Context

Before arriving in the Northern Netherlands, Stevin had been a member of a chamber of rhetoric (*rederijkerskamer*) in Bruges.[9] As far as we know, he was no longer directly involved in the practices of chambers of rhetoric during his Dutch period, yet his work on logic cannot be separated from the broader efforts of rhetoricians to make Dutch the main language of science. The rhetoricians' focus was primarily on grammar, dialectics and rhetoric. The first time that they managed to have the entire *trivium* printed in Dutch was in 1584–1587, when the Amsterdam chamber known as De Egelantier put the following books on the market in sequential order:

1. *Twe-spraack vande Nederduitsche letterkunst* (Leiden: C. Plantin, 1584).
2. *Ruygh-bewerp vande redenkaveling, ofte Nederduytsche dialectike* (Leiden: C. Plantin, 1585), including the *Kort begrip des redenkavelings* (1585).
3. *Rederijck-kunst, in rijm opt kortst vervat* (Leyden: F. Raphelinghen [other title pages: C. Plantin], 1587).[10]

Within the series, the *Ruygh-bewerp* dealt specifically with dialectical method. The book was almost entirely a hotchpotch of the work of others, and in particular of Johannes Rivius' *De dialectica libri sex* (Leipzig 1539), Valerius' *Tabulae, quibus totius dialecticae praecepta etc.* (Antwerp: Plantin, 1582) and Rudolphus Agricola's *De Inventione Dialectica*.[11] In their composition, the members of De Egelantier had opted meticulously to translate every word used in the art

9 Van Dixhoorn, *Lustige geesten*, 125.

10 The series was reprinted in 1614 and 1649. The books have often been attributed to Dirck Volkertszoon Coornhert (1522–1590) or Hendrik Laurensz. Spiegel (1549–1612), yet current research tends to attribute the works to the chamber collectively rather than to an individual. For more background, see Klifman, 'Dutch Language Study'; Peeters, 'Auteurschap'.

11 Kuiper, *Orbis artium*. The authors also indicate Ramus as one of their sources, but there is not a lot of evidence that they made much use of his work.

of dialectics into Dutch, displaying a Latin equivalent of the concept in the margin.

Upon publication, the rhetoricians had decided to dedicate the *Ruygh-bewerp* to the burgomasters of Leiden and the curators of Leiden University. They declared that they did not do so to seek protection from their patrons, but because they hoped that the *Ruygh-bewerp* would help to make Dutch the standard language of the university.[12] Furthermore, the rhetoricians specified that they did not propose that their book be used as a primer; they had simply wanted to show that it was possible to render the *trivium* into Dutch, in the hope that others would follow their example and produce better works (they explicitly referred to Petrus Ramus (1515–1572) in this regard, to whom the French king had given a similar assignment for the French language).[13] There is no further evidence that Leiden University ever seriously considered the use of the vernacular schoolbooks produced by the rhetoricians.[14]

The *Ruygh-bewerp* and the work by Stevin came out in the same year, in 1585, and both were published by the famous Antwerp printer Christoph Plantin (1520–1589). It is unlikely that the printer would have failed to inform the other party about the forthcoming publication of a book on the same topic, yet there are no cross-references to be found. In the introduction to his work, Stevin stated that he had set out to produce a work in Dutch, 'since Dialectics has not yet been described in that [language]'.[15] This suggests that Stevin's *Dialectike* was – despite the uncertainty on this subject that sometimes arises in the secondary literature – the first book on logic ever to be printed in the Dutch language. Stevin further elucidated how he had to invent many of the Dutch words that he used for the first time in order to cover the *Vocabula Artium* (sic!) of dialectics.[16] In doing so, he had explicitly chosen not to translate each and every word, but only to find new words that would correspond to existing Latin concepts when needed.[17] For example, the word 'contrary'

12 Spiegel, *Ruygh-Bewerp*, 68.

13 Ibid.

14 In 1581, Plantin had managed to convince the academic senate exclusively to use the Latin grammar of Valerius; in 1592, the senate was still advising the exclusive use of Valerius' *Rhetorices et Dialectices Tabulae* for teaching. Molhuysen, *Bronnen*, 1:18 (27 April 1581), 232* (16 October 1592).

15 '... overmidts de Dialectike daerin noch niet beschrevenen is ...'. Stevin, *4v. Although Stevin did not mention the *Ruygh-bewerp*, he did refer to the *Twe-spraack* that had 'just appeared' (*onlancx uytghecommen*) in print. Stevin, *Dialectike*, 159–60. This was the book on grammar, authored by De Egelantier.

16 Stevin, *4.

17 Stevin also argues for the use of the nominative and genitive cases only, 'taking after the Dutch habit, because to use these [other grammatical cases] is an unnecessary torment

(*Contrarie*), from the Latin *contrarium*, was already integrated into everyday use to such an extent that there was no point in translating it. Making an explicit reference to Cicero, Stevin silenced his critics in advance by arguing that some words would become incomprehensible if they were 'Dutchified' (*Verduytschen*).[18] This may have been a direct sneer aimed at the members of De Egelantier, who were very strict about finding a Dutch translation for each and every Latin word.

There were also other differences between Stevin's work and that of the rhetoricians. In keeping with tradition, the writers of the *Ruygh-bewerp* had considered dialectics (*redenkaveling*) to consist of two parts: the *inventio* (*vinding*) and the *iudicium* (*oordeel*).[19] They had refused to follow in the tradition of Ramus and Rudolph Agricola (1444–1485), who tackled the invention phase prior to the judgment, with a view to promoting comprehension and students' interests.[20] Stevin clearly wrote with students in mind as well. Yet, as we will see in the next section, he went one step further and dismissed the entire distinction between *inventio* and *iudicium*, proposing instead to use definitions and operations. The entire 'method' of the Bruges mathematician essentially overlapped with books three and four by the rhetoricians; it was not only much shorter than the method of the *Ruygh-bewerp*, but more explicitly oriented towards practical application, too. Stevin's primary aim was not to flaunt his wit or native language proficiency, but to stimulate one and all to use dialectics (*Bewysconst*) in order to investigate the mysteries of nature.

In contrast to the *Ruygh-bewerp*, Stevin's *Dialectike* was published with a printing privilege, issued to Christoph Plantin on 20 December 1584. At the time, Plantin was based in Leiden, where he had taken up the post of university printer and official printer (*hooftdrukker*) of the Dutch State.[21] It was a tumultuous period in the history of the Netherlands. The Act of Abjuration (1581) had rendered the northern provinces of the former Duchy of Burgundy a quasi-independent State, signaling a temporary highpoint in the struggle for independence from the House of the Spanish Habsburgs (even though the

for anyone without any knowledge of Latin language' (*nemende naer de Duytsche ghevvoonte, omdat een onnoodighe quellinghe sijn, voor den gene diese gebruycken vvil, sonder eenighe vvetenschap der Latijnscher talen*). Stevin, fol. **1r.

18 Stevin's argued that many of these words had 'become somewhat Dutch because of longstanding usage' (*door de langhe ghewoonte, eenichsins Duytsch geworden*). Stevin, fol. *7.

19 Spiegel, *Ruygh-Bewerp*, 73.

20 Spiegel, 87.

21 Japikse, *Resolutiën*, 4:717.

situation at the time was highly fragile, and the rebellious provinces were still in search of a new leader as well as a suitable form of state).

The printing privilege issued to Plantin provided him with an exclusive right to produce, sell and distribute Stevin's book in the United Provinces for six years after the first print run (common conditions for printing privileges at the time). Yet the legal document had more than just economic value; it was also a political statement. The printing privilege for the *Dialectike* was issued in the name of Count Maurits of Nassau (1567–1625) and the newly-created Dutch Council of State. Maurits was the son of Prince William of Orange (1533–1584), the leader of the Dutch Revolt, who had been assassinated in Delft a few months earlier. At that time, no clear decision had yet been taken about which institutions were to be responsible for the official printing policy of the United Provinces (the States-General would later take charge of central policy). But it was certainly not by chance that one of the first printing privileges ever to be granted in the Republic was awarded to a set of books produced by the most important printer in the Burgundian domains and written by a rising star who would later personally tutor Maurits of Nassau, by then stadholder of Holland and several other provinces, further in the subtleties of mathematical thinking.[22]

More important for our immediate purposes is the fact that the privilege was given for a set of books: the legal conditions attached to Stevin's *Dialectike* also applied to his other works, *De Thiende* and *L'Arithmetique*, books that would establish his name as a mathematician. The plan was to publish the set of books in Latin, French and Dutch; whereas *De Thiende* appeared both in Dutch and French, we are only aware of a Dutch edition of the *Dialectike* and of a French version of *L'Arithmetique*. In addition to the Dutch privilege, *L'Arithmetique* also mentioned a general printing privilege that was granted to Plantin by Henry III of France (1551–1589) on 5 August 1583 at Fontainebleau, so that the former would not be hindered in his printing activities, 'chose qui seroit de tresgrande consequence & preiudice aux personaiges doctes d'estre priués de la communication de ses oeuures, qui ne leur peuuent apporter que toute emulation de servir au public'.[23] The Dutch printing privilege was in fact an imitation of the French example, except that it was more explicitly aimed at

22 On printing privileges in the early Dutch Republic, see Buning, 'Commentary'. Stevin had just enrolled at Leiden University (1583) and probably also met Maurits there.

23 Stevin, *L'Aritmetique*, sig. Oo3. The French authorities nonetheless added the following provision: 'purueu que lesdictes oeuures & liuvres de ladicte nouuelle composition ne soient en rien contraires à la religion Catholicque, Apostolicque, & Romaine, ni contre nostre estat'. The long 's' has been replaced by its modern equivalent.

a set of books. Indeed, the fact that the books were 'privileged' on the same day provides a first indication that this set of books should be seen as a whole.[24]

In the preface to the *Dialectike*, Stevin declared that he wanted to make dialectical method available in the vernacular for 'beginner students' (*beghinnende Leerlingen*).[25] He had noticed how remarkable it was that those with no knowledge of Latin could master the rules of mathematics by means of self-study, whereas it was almost impossible to learn dialectics, even with an experienced teacher. The reason for this obscurity was that current teaching methods presented dialectics in a disorderly manner; Stevin's primary aim was therefore concisely and comprehensibly to rearrange the existing material. Moreover, there was a clear desire among practical mathematicians to familiarize themselves with the art of dialectics, according to Stevin; these 'great amateurs of the arts' in his surroundings were well aware that the learned made more progress in the field of mathematics because of their understanding of dialectics.[26] That is why Stevin had chosen to prepare a practice manual for independent study without too many difficult examples. The goal was to use language that the 'common man' (*ghemeenen man*) and the 'otherwise unlearned' (*andersins ongheleert*) would understand.[27]

Little is known about the actual readership of Stevin's *Dialectike*. The 172-page octavo booklet must have been priced at a level accessible to larger sections of the population,[28] but we have not come across any versions with readers' notations, and to date there is no evidence of wide distribution.[29] What we do know is that mention was made of the *Dialectike* in Allardus Lodewijk Kok's

24 For a broader elaboration of this argument, see Buning, 'Privileging the Common Good'.

25 Stevin, *Dialectike*, fol. **1v.

26 '... der Consten uytnemende liefhebbers ...'. Stevin, fol. *3v. Stevin recounted how he had encountered 'several persons who were (although otherwise unlearned) well-versed in that Art [Mathematics], having understood it only because of their reason by reading several books, without an oral teacher' (*verscheyden Persoonen dier Consten dapperlick ervaren, nochtans* (*hoe wel andersins ongheleert*) *daer toe gecommen, sonder mondelicke onderwijser, hebbende alleenelick door hare vernuft int lesen verscheydener boucken dat begrepen ...*). Stevin, fol. *5r. They had understood from the learned that dialectics was 'an instrument of science' (*instrument te wesen der wetenschap*). Stevin, fol. *6r.

27 Stevin, fol. *8r and fol. *5r respectively.

28 On book prices in the Dutch Republic, see Van Selm, 'Voor een civielen prijs'.

29 It is worth noting that the *Dialectike* did not appear in the instructions that Stevin prepared, in 1600, for the establishment of the famed Dutch engineering school linked to Leiden University, the *Duytsche Mathematique* (logic is not mentioned at all in the school's curriculum). It confirms the impression that the booklet was not intended for official courses, but primarily for home study purposes. See also Dijksterhuis, 'Duytsche Mathematique and the Building of a New Society'.

(1616–1653) translation of the *Sinopsis logica* (1646), written by the Leiden professor of philosophy Franco Burgersdijk (1590–1635).[30] But the translator's references are only of a linguistic nature, relating to translation problems rather than the content of Stevin's logic. Thus, it is very possible that Stevin's work resulted in little imitation precisely because of its entirely new method of presentation, except perhaps among a group of practical mathematicians, who tended to leave few writings behind.[31]

That there was nonetheless some interest in Stevin's work can be deduced from the fact that a new edition of *Dialectike* appeared on the market in 1621, which was substantively the same as the original version (the typographical errors had been removed, but it was not rewritten).[32] The background to the timing of this new edition is not yet entirely clear, but the death of Stevin in 1620 may have played a role. The publisher, Jan van Waesberghe the Younger from Rotterdam (1556–1626), who republished *De Thiende* around the same time, may have spotted a market for the book. The 1621 edition of the *Dialectike* would eventually be printed by J.C. van Dorp in Leiden.[33]

3 Structure and Content

Before we turn to the details of Stevin's dialectical method, it is worthwhile providing a brief overview of the work. The *Dialectike* is divided into definitions and operations, each covered in a separate book. The definitions are of three kinds: singular, composite and operational. The definition of single issues (*Ynckel saecken*) is composed of dialectics itself and the four causes, of which two are 'in the thing', matter and form, and two come 'from the thing',

30 Burgersdijck, *Institutio logica*, 18, 77, 136.

31 On the cultural environment in which mathematics developed in the seventeenth-century Dutch Republic, see Dijkstra, *Between Academics and Idiots*.

32 I have used this 'corrected edition' for my research, comparing it with the 1585 version available in the Staatsbibliothek zu Berlin. Special mention should also be made of the excellent website on the history of science in the Netherlands, maintained by Ad Davidse, who provides an inexhaustible source of information for researchers on Stevin: Davidse, 'Simon Stevin'.

33 If we are to believe Stevin's son, Hendrik, it would appear that during his lifetime Stevin was working on a revision of his dialectics, which was to be integrated in his masterpiece *Wiskonstige ghedachtnisse*. The new title of this work, similar to 'the one that currently goes out in print, yet corrected and changed by the author' (*die in druck uytgaet, doch is deur den Schryver verbetert en verandert*) was to have been *Nederduytsche Dialectica dats Bewysconst, anders geseyt Redenstryt*. Stevin, *Materiæ politicæ*, 149. However, the work never appeared in print.

i.e., the efficient and the final cause. Together they compose the 'effect' (*Werck* or the *Daet*: in Latin *effectum*). In addition, single issues (i.e., 'unitary' concepts that do not have an indispensable relation to an opposite) are composed of *quantitas* and *qualitas* (two of the ten predicaments of Aristotle), of *auctoritas*, as well as of 'words that mean some something else than they seem'. This last category is differentiated into *fallacia* (the thirteen fallacies addressed in the 'Sophistical refutations') and the *tropoi*, or literary tropes, singling out in particular the notions of metaphor, allegory, metonymy, hyperbole, synecdoche and irony.

The second section of the first book is on the definition of compound things (*samenghevouchde saecken*), which means that it deals with notions that necessarily implicate a relationship between two things (that cannot be thought of independently). These amalgamated things are grouped under eleven headings: substance (translated from *subjectum, substantia, objectum, materia circa quam*: Stevin makes no distinction here) and accident (idem *accidens, circumstantia, antecedentia, consequentia*, also translated as *signa*); genus and species; part and whole; more and less (*majus* and *minor*); reflexivity (*reciprocatio*); relatives; contraries (of things and statements); contradictories; similars (*similia*); and proportion. The final section of Book I deals with definitions of what Stevin calls the 'operational' (*saecken der wecking*), which means that they imply a form of action. This third section is differentiated into three subsections: definitions, differentiations and syllogisms. Syllogisms are either perfect (categorical) or imperfect. The first one has three propositions and three terms (the major, middle and minor term). They exist in four variations, which together have sixteen moods. The imperfect syllogism, in turn, exists in six variations: *indirecte concludens*, sorites, hypothetical, induction, dilemma and enthymeme.

In the second book of the *Dialectike*, Stevin turns to the practice, i.e., how to use the dialectical method. The book looks at operations (*Werckinge*), which are formulated as propositions (six in in total). They deal with analyzing arguments by means of dialectical method, as well as how to construe proper syllogisms, among other things by using the method of conversion, differentiation, by 21 argumentation rules, by *reductio ad absurdum*, and by making use of ten dialectical rules listed separately. The second book is followed by an appendix, in which Stevin explains some of the choices that he made in laying out the structure of his dialectics, as well as his choice of specific definitions. To conclude, Stevin includes a dialogue between two individuals, Peter and John, who discuss various issues using the dialectical method, and by doing so provide an illustration of how dialectics can work in practice.

3.1 *Method and Organization*

Stevin defined dialectics as the 'Art of Proof' (*Conste der Bewijsinghe*) that could be used to test the truth-value of a statement (definition 1).[34] Dialectics was a method that taught one how to:

> ... strip [arguments] of all paraphernalia, deceiving ornaments, and obscuring variety, not counting arithmetically the amount of words; but dialectically weighing the value of simple arguments.[35]

The notion that dialectics opened up the way to learning the other arts was not uncommon, following the works of Peter of Spain. Yet Stevin's presentation of dialectics was radically different from that in any of the other handbooks on logic circulating at the time (see figure 5.1). This is mainly because he divided dialectics into definitions and propositions, making a direct reference to Euclid in this regard. Stevin used the same kind of subdivision in his other works as well (for instance in *Tafelen van Interest* (1582), *De Thiende* (1585) and *L'Arithmetique* (1585), as well as in later works). In his *Geography* (*Eertclootschrift*, 1608), Stevin also explicitly addressed the importance of dealing first with definitions before proceeding to the operations of an issue. He argued that this was the difference par excellence between the 'Mathematico stylo' and 'Historias'.[36] Moreover, the 'mathematical style' formed the basis for the order in which his books came onto the market; practical writings with exercises and applications (*De Weeghdaet, La pratique d'arithmétique, De Meetdaet* and others) were preceded by more theoretical ones (*De Beghinselen der Weeghconst, L'Arithmetique, De Beghinselen des Waterwichts* and so on). Each theoretical component was accompanied by a book on its practical application.

Following the mathematical style, Stevin's first aim in the *Dialectike* was to define the words that he used. He did so in Book I, following a straight line from the most basic single terms to the more complex or compound terms.[37] Differentiation was an essential part of Stevin's method (*methodus* or *oirde*), which he dealt with in proposition 2 of Book II.[38] Very briefly, the right way to

34 Stevin, *Dialectike*, 3.

35 '... blooten van alle haer toerustinghen, als bedriechlicke cieraet, ende verduysterende veelheydt, niet Arithmetelick tellende de menichte der woorden; maer Dialectikelick weghende de cracht der eenvoudige argumenten'. Stevin, 103.

36 In Book I of *Eertclootschrifts*, collected in Stevin, *Wisconstige ghedachtenissen*, 42.

37 *Pace* Dijksterhuis who claimed that 'any particular order could not be discovered' in the presentation of the 47 definitions (*Een bepaalde lijn valt er niet overal in te ontdekkken*). Dijksterhuis, *Simon Stevin*, 1943, 288.

38 Stevin, *Dialectike*, 130.

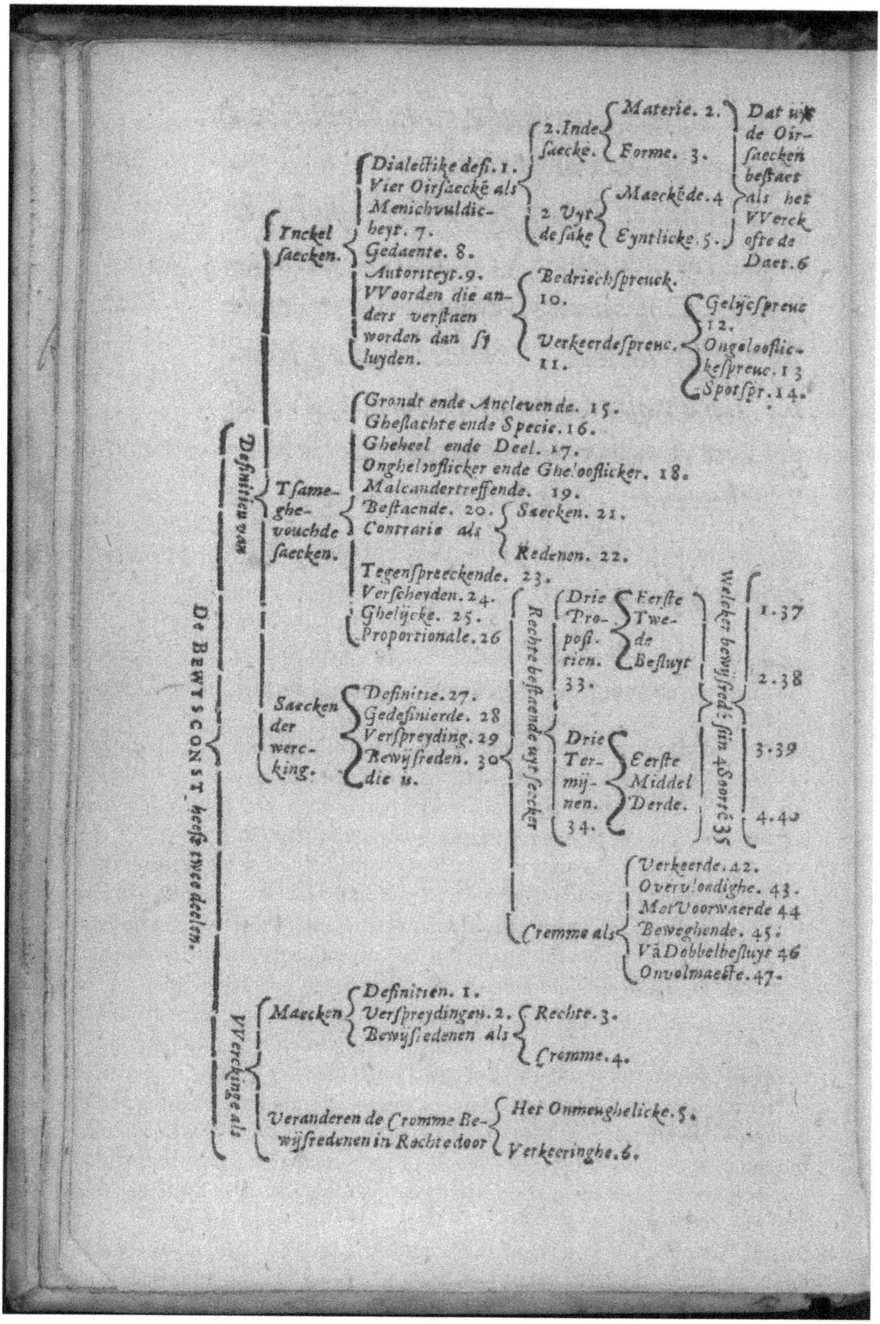

FIGURE 5.1 The *tabula* of Stevin, *Dialectike* (Rotterdam: Jan Van Waesberge the Younger, 1621), f.**4v
COURTESY: GHENT UNIVERSITY LIBRARY

differentiate was to include as few derivations as possible, following the order of nature and always keeping the student in mind.[39]

Stevin's emphasis on clarity and order is reminiscent of some of the Ramist literature of the time.[40] Yet Stevin did not follow Ramus in letting the invention precede the judgment.[41] The writer of the *Dialectike* showed that he was clearly aware of the discussion on the organization of dialectics, but did not wish to elaborate further on the matter because he regarded his own method, with definitions and operations, as much more 'appropriate and adequate' (*eyghender ende bequamer*).[42] In contrast to his French counterpart, Stevin clearly presented himself as an Aristotelian, though a somewhat unusual one, since his presentation of dialectics did not actually fit within the Aristotelian tradition. It would be more accurate to speak of a Euclidian form of dialectics: a dialectics that was based on definitions and propositions.

In this Euclidian dialectics, the definitions were not equal to the classical *iudicium*, but dealt more generally with the theory of knowledge. The

39 Stevin, 59. 'Daerom en ist niet genouch datmen slichtelick Verspreyde, maer men moet noch boven dien altijt inde ghedachtenis hebben, t'gene inder natueren vooren gaedt, ende tot de leeringhe bequaemst is'.

40 Another feature generally associated with Ramist teaching methods is the use of so-called *Tabulae*. However, as Ann Blair and others have noted, one should be careful to interpret this as a confirmation of Ramist tendencies per se. Blair, *The Theater of Nature*, 84–85; Ferguson, 'System and Schema'; Höltgen, 'Synoptische Tabellen'.

41 Aristotle had wanted the study of logic to start with the *iudicium* (judgement). Only after correctly learning the technicalities of a demonstrative syllogism could any student go on to study the ways of presenting the argument (rhetoric), which were to be found in the Topics and Sophistical Refutations. Petrus Ramus (influenced by the humanists Lorenzo Valla (ca. 1405–1475) and Rudolph Agricola (1443–1485), who had written practical compendia for the composing of letters) stressed, however, the importance of *inventio*, the art of finding arguments, rather than the judgment (*iudicium*). Ramus thus added *iudicium* to *inventio*, making it the other part of logic (and thus giving rhetoric more emphasis). He called dialectics the *ars bene disserendi* or the art of correctly discussing and analyzing something. Stevin refers to this definition, but prefers his own definition of dialectics as the 'Art of Proof' (*de Conste der Bewijsinghe*). Stevin, *Dialectike*, 127. On the hybrid relationship between rhetoric and dialectics in the humanist era, see also Traninger, *Disputation, Deklamation, Dialog*.

42 Stevin, *Dialectike*, 108. On Ramist dialectics, see Bruyere, *Methode et dialectique*. On Ramist rhetoric, see Ong, *Ramus*, particularly 171–214. On the influence of Ramus on Stevin, see Verdonk, 'Vom Einfluß des P. de La Ramée'; Van den Heuvel, *'De Huysbou*,' 26–27. On the educational impact of Ramus, see Dibon, 'L'influence de Ramus'; Feingold, Freedman, and Rother, *The Influence of Petrus Ramus*; Hotson, *Commonplace Learning*; Reid and Wilson, *Ramus, Pedagogy and the Liberal Arts*; Schmidt-Biggemann, *Topica Universalis*, 31–67.

second book (containing the propositions) showed how this theory worked in practice: it contained part of what would normally fall under the *inventio* (the 21 rules from the Topics in proposition 4, for instance) but also included many elements that would normally fall under the *iudicium* (notably the finding of a middle term, proposition 3). In the appendix, Stevin elaborated on the fact that 'fallacies, tropes, quantity, and quality belong to dialectics' and not, for example, to rhetoric or grammar.[43] His explanation was that these words, although they 'sound alike' (*ghelijck luyden*), meant something else when used in the various disciplines. Just as the use of Dog, Rabbit, Bull and Lion in astrology was not seen as the improper mixing of the celestial and terrestrial, there was no question here of the improper mixing of branches of science.

Having dealt with definitions, differentiation and the making of syllogisms in Book II (propositions 1 to 4), Stevin adds a list of ten dialectical rules.

1. No rule so general that it does not have an exception
2. One cannot dispute with those who deny the first principles
3. What is similar is not in all aspects the same
4. From truth nothing but truth
5. From what is false, one can conclude the truth
6. Not all is false that cannot be proven
7. Not all is true that appears to be proven
8. The logic from potentiality to actuality does not follow
9. Perhaps can be refuted by perhaps not
10. The unsuited is admitted to understand the truth[44]

Armed with these rules, one could proceed to the technique of how to change imperfect syllogisms into categorical ones. In particular, Stevin singled out the *reductio ad absurdum* (proposition 5), which was a technique much emphasized by Archimedes, as well as the method of conversion (proposition 6). Using the so-called method of conversion, one could render imperfect syllogisms into categorical ones. Stevin made no distinction here between conversion, obversion or contraposition, but instead provided a set of standard rules: A

43 '... BEDRIECHSPRUECX, VERKEERDE-spruecx, Menichvuldicheydts, ende Ghedaenten; inde Dialectike te behoiren'. Stevin, *Dialectike*, 113.

44 Stevin, 85–86: '1. Gheen regel so generael, die niet eenighe uytneming en heeft. / 2. Teghen de ontkenders der beghinselen, en canmen niet disputeren. / 3. Ghelijcke en overcommen niet in alles. / 4. Uyt het ware niet dan waer. / 5. Uyt het valsche can het warachtighe besloten worden. / 6. Ten is niet al valsch, datmen niet bewijsen en can. / 7. Ten is niet al waer, dat schijnt bevvesen te sijne. / 8. Van T'mach, tot Het is, en duecht het vervolgh niet. / 9. Misschien, vvort vveerleyt duer Misschien niet. / 10. Ongheschicte vvorden toeghelaten, op datmen de vvarachtighe daer duer verstae.'

converts to I, E to E and (if needed, but inferior) to O, I to I, whereas O does not allow for conversion (proposition 6, Book II).[45] Following this set of rules, Stevin gave a number of examples on how to convert:
– syllogismus indirecte concludens (*verkeerde bewijsreden*)
– sorites (*overvloedighe bewijsreden*)
– syllogismus hypotheticus (*bewijsreden met voorwaerde*)
– inductio (*beweghende bewijsreden*)
– dilemma (*bewijsreden van dobbelbesluyt*)
– enthymena (*onvolmaecte bewijsreden*)

Content-wise, it seems as though Stevin provides a rather standard – if simplified – overview of Aristotelian logic; yet there are a few elements that Stevin presents quite differently.[46] In definition 35, for instance, Stevin introduces the three standard figures that can be found in Aristotelian logic:

I	II	III
MP	PM	MP
SM	SM	MS
----	----	----
SP	SP	SP

But he adds a fourth figure, which he illustrates with the following example:

First way:
I. JOHN is poor
I. JOHN is a Dutchman
I. Therefore, some Dutchmen are poor.

45 The abbreviations are standard and indicate the type of proposition: A: a universal affirmative proposition – All S is P / E: a universal negative proposition – No S is P / I: a particular affirmative proposition – Some S is P / O: a particular negative proposition – Some S is not P.

In fact, Stevin adds the rule that 'a denial of the indivisible [individual] converts completely' (*Onscheydelicke Ontkenninghe, verkeert volcomentlick*), giving the following example: 'If Peter is not John, then John is not Peter'. Stevin, 89–90.

46 A standard syllogism would go like this: 'Major premise: All humans are mortal / Minor premise: Some animals are humans / The conclusion: Some animals are mortal. There are three terms in this example: 'human', 'mortal' and 'animal'. Mortal is the major term, which functions as the predicate of the conclusion (abbreviated by P); animal is the minor term, which functions as the subject of the conclusion (abbreviated by S). The term that the premises have in common is called the middle term (abbreviated by M), which is 'humans' in our example. The middle term does not return in the conclusion.

Second way:
O. JOHN is not Dutch
I. JOHN is a learned person
O. Therefore, some learned persons are not Dutch.[47]

The fourth figure was characterized by the fact that the middle term was 'indivisible' (*onscheydelick*) as well as the 'grounds' (*grondt, materia circa quam*) for the major and minor premise.[48] One could summarize this in the following figure:

MP
MS

SP

In the appendix to his work, where the author included a glossary of his dialectical terms in Latin translation, Stevin identified the 'fourth way' as the *syllogismus expositorius*. This was a rather common name for this type of syllogism, first introduced in the thirteenth century, and references to it could also be found in other contemporary manuals on logic, such as Cornelius Valerius' *Tabulae totius Dialectices* (Antwerp: Plantin, 1575) and Augustinus Hunaeus' *Dialectica* (Antwerp: Plantin, 1568).[49] As such, the inclusion of a fourth figure tells us little about Stevin's exact position.[50] Yet it does give an indication of how Stevin was silently 'subverting' Aristotle in order to render the work of the great Stagirite more consistent with his own views on the production of knowledge.[51]

47 Stevin, *Dialectike*, 42: 'I. IAN is Aerme, / I. IAN is een Hollander, / I. Eenich Hollander dan is Aerme. / O. IAN en is gheen Hollander, / I. IAN is Gheleert, / O. Eenige Geleerde dan en is geen Hollander.'

48 Stevin, 55.

49 Valerius, *Tabulae Totius Dialectices*, 42v; Hunaeus, *Dialectica*, 212r-215v. On the long life of the fourth figure, see also Rescher, *Galen and the Syllogism*, 22–48. On the logic of individual propositions, see Barth, 'From the History'. Earlier schoolmen had also used the expression 'terminus discretus' and 'propositio Singularis' for a term that was a name or a description of an individual (often subsumed under the third figure).

50 The exact interpretation of the fourth figure was highly controversial and played on the minds of logicians for centuries. On this issue, see Henle, 'On the Fourth Figure of the Syllogism'.

51 Cf. Martin, *Subverting Aristotle*.

In doing so, Stevin stripped Aristotelian dialectics of many of its subtleties. In the appendix to Book II, for example, he provided an overview of the different types of syllogisms as distinguished by Aristotle in the *Topics*:

1. *Syllogismus Demonstrativus,* which he calls the most secure, with the conclusion following from principles or something that is certain (A/E/E: fire is dry, water is not dry, water is not fire).
2. *Syllogismus Dialectius,* which is about probability (A/I/I: pitch is contagious, Peter used pitch, Peter is contaminated).
3. *Syllogismus Contentiosus sue Litigiosus aut Sophisticus,* which seems true but is not (individual premises are true, but not the conclusion: I am not you, I am human, you are not human)
4. *Syllogismus Fallax,* which is a syllogism based on false premises.[52]

Stevin concluded that the distinction between these different syllogisms was pretty useless, because we could usually agree that, when the conclusion of a syllogism appeared to be wrong, something must simply have been wrong with either the matter or the form of the premises. To explain his ideas, Stevin included an extensive 'note' (*Merckt*) at the end of Book II in which he elaborated on how a false conclusion can be due to the 'matter' or 'form', using the example of pepper. He argued that the dialectician would undoubtedly find the deficit in form, but that deficits in terms of matter could not be known by means of dialectics (or by other arts). For example, if someone said 'One pound of pepper is worth 4 *schellingen*, therefore 100 pounds of pepper are worth 20 lb [*Hollandse ponden*]' [1 *Hollandse pond* = 20 *schellingen*], both the mathematician and the dialectician would say that the form of the calculation was correct. But if part of a shipload of pepper was rotten, its market value would not be consistent. This was something that could not be known by mathematics or dialectics, but only by means of 'something else'.[53] Matter was decisive, and eventually common sense took precedence over sophisticated theory when determining the truth value of a premise.

52 Stevin, *Dialectike*, 129.

53 'By exempel, yemandt seght, I. pondt peper is weert 4. sch, daerom sijn 100 ponden peper weert 20 lb. Voorwaer den Arithmeticien can segghen, dat de Forme der rekeninghe goedt is, al en hadde hy gheen verstandt van peper; Maer so hy betoonde datter ghebreck inde Materie ware, als het peper beseeut, ende door wentelinghe in d'asschen (naer de ghebruyck) hem weder een bedriechlick coluer ghegheven te sijne, ende naer martganck soo veel niet weert te wesen; dat en weet hy niet uyt de Arithmetike maer duer wat anders'. Stevin, 101.

4 Common Sense, Common Language and Mathematics

Stevin's ideas about common sense were closely related to his views on language. It was the language spoken by the man in the street that ought to serve as the foundation for any knowledge claim or functional method. Along those lines, Stevin dealt for instance only with four of the ten predicaments (substance, quantity, quality and relations in definition 15, 7, 8 and 20), dismissing the other six (*agere, pati, ubi, quando, situs, habere*) on the basis that these 'words mean what the common man understands them to mean'.[54] It would therefore be superfluous to provide definitions, just as it would be unnecessary to include and describe subcontraries and subalternates (*subcontrariae* and *subalternae*). On this point of parsimony, there might be a link between Stevin and Lorenzo Valla's earlier attack on the ten predicaments of Aristotle – reducing them to three (substance, quality and action) on the basis that the other terms violate what he called 'the common custom of speaking'.[55] In the humanist universe of Stevin, as for Valla, the meaning of language was derived from occasions of use.[56]

For that very same reason, there was not necessarily a right way to define something either.[57] Definitions were dependent on the context in which they were used and it was, so claimed Stevin, more important to be understood, if necessary by using rhetoric, than to define 'something more properly, yet darker and inconvenient for teaching'.[58] This is also precisely the point with which Stevin reproached his predecessors: namely, that dialectics had become so complicated that it was de facto unusable. For how was the student supposed to understand an art in which he 'encountered words that he had to understand differently than what they meant naturally and according to the use of all men'?[59] Such confusion had evolved gradually over time and could only

54 '[t]sijn woorden die beteeckenen, 'tgene den gemeenen mandaer by verstaet'. Stevin, 130.

55 As cited in Waswo, 'Ordinary Language', 257. On Valla's dialectic, see also Cifoletti, 'From Valla to Viète'; Mack, *Renaissance Argument*, 84–88; Nauta, *In Defense of Common Sense*, 211–68. On Valla's discussion of the number one, see Nauta, 65–67.

56 Cf. Copenhaver, 'Translation, Terminology and Style in Philosophical Discourse'.

57 On this issue, see also the exquisite essay by Naets, 'How to Define a Number?'. As Stevin elucidated in the *Dialectike*, Euclid, Plato, and Archimedes had for instance all used different definitions for a straight line 'that are all good' (*die alle goedt zijn*). Stevin, *Dialectike*, 54.

58 '… yet eygender, nochtans duysterder ende tot de leere ongherieviger'. Ibid.

59 'Dit, dat dese valsche propositien inde Dialectike recht het eerste waren, daer den Leerlinck an begonde, alwaer hy woorden bevandt, daer hy al wat anders by verstaen moeste, dan sy natuerlick, ende naer ghebruyck van allen Menschen beteeckenden …'. Stevin, 112.

be sustained by the authority of know-it-all scholars.[60] Generally speaking, the authority of scholars had to be respected; perhaps as a precaution, Stevin had even made this into a rule in his dialectical system, 'from which one is not to depart without a serious reason'.[61] But there was one case that Stevin specifically highlighted for closer scrutiny. It concerned the concept of *universalia*: a problem that cropped up in the section on the construction of species and genera.

> It is common practice, not only among common people, but also among scholars, both when they write and when they speak, to name the indivisible [*Onscheydelick*] a species; for example, someone who sees an unknown animal coming from the sea says: 'It is a kind [*Specie*] of fish; [or] that triangle is a kind of plane or surface, etc'.. Similarly, they will call the container ['*tbevanck*] of the indivisible a genus, like humans [when talking] about Peter and John, or Horses [when talking] about a Gray or a Bayard[62] [= a 'kind' of horse like Pegasus or Bucephalus]. Now, there are two opinions with regard to this way of speaking. One, of the Old [*der Ouden*] who despise it, saying that a fish or triangle thus imagined, being indivisible by itself, cannot be a species; nor can mankind or horse be a genus. Instead, it is an 'empty kind' [*Leechste Specie*] and all who say otherwise (yes, often following customs) err, because they do not speak philosophically, dialectically, or truly. The Newcomers, in contrast, say that it is well put and proper dialectics, and we are also of that opinion: [.... Thus,] the indivisibles Peter and John will be species, and mankind not an empty container but their genus [*Gheslachte*].[63]

60 Stevin, 112.

61 Stevin, 113. Cf. the dialectical rules above, note 44.

62 Stevin spoke here of a *Rosbeyaert* and a *Graeu*. A Ros Beiaard is the name of a magic bay horse in the legends derived from the *chansons de geste*; it became the standard name for any bay-colored horse. Indeed, the color of a horse became the indicator for a specific type of horse. Accordingly, there were many variations on the gray-colored horse (*Graeu*). See Viaene, 'Oude kleurnamen van paarden'.

63 'Het is inde ghebruyck, niet alleene by den Ghemeenen man, maer oock by den Gheleerden, dat sy in hare spreecken ende schrijven, het Onscheydelicke noemen Specie: By exempel, yemandt siende een onbekende beeste, die hem dinct uyt de see te commen, hy seght, *Het is een Specie van Visch: Dien Driehouck is een Specie der Platten ofte Superficien, &c.* Alsoo oock t'bevanck der Onscheydelicken, als Mensche, van Pieter ende Ian; Peert, van Ros beyaert ende Graeu, noemen sy Gheslachte, als Gheslachte der Menschen, Gheslachte der Peerden: Nu sijnder deser manieren van spreecken tweederley opinien, d'eene der Ouden, welcke die verachten, segghende alsulcken voorgestelden Visch, ofte Driehouck, [p.109] alleene Onscheydelick, gheen Specie te wesen; oock Mensche noch Peert, Gheslachte, maer de Leechste Specie te sijne, ende dat alle de

In this citation, we see how Stevin draws the individual unit *into* the species. The individual unit constitutes the species entirely (and he later also calls it 'the most fundamental species').[64] This would be entirely impossible according to the Aristotelian way of thinking, because in order to explain 'what it is' for something to be a species, we would have to rely on a principle that unifies its parts; hence, there could not be a species with just one member. Stevin's approach, entirely based on the use of everyday language, instead made room for the single entity.[65] This radically changed the way in which one could organize and understand the 'deepest secrets of Nature', as the title of Stevin's booklet indeed suggested, if only because all attributes that belong to the species must also belong to the genus (see *Topics*, 4).

As far as Stevin was concerned, access to knowledge concerning the secrets of nature was governed entirely by language. Visual observations lay at the basis of, but at the same time were secondary to, the logic exposed in language. Stevin saw no contradiction here; things were 'true' because they were linguistically true, but also because language represented a true reality that was out there. This applied not only to the consideration of entities such as Peter and John, but also for mathematical entities that were therefore not completely abstract, or an abstraction of tangible objects, but that took their veracity from their place within the discursive notion of thought. Precisely because language was the direct port of access to an understanding of truth, dialectics was the most suitable method for comparing and contrasting different outlooks. For the same reason, it was also perfectly possible to substantiate every conclusive thought with syllogisms, whether these were practical observations or mathematical views.

4.1 *The Certainty of Mathematics*

The relationship between mathematics and dialectics, as well as the place of mathematical entities within Stevin's thinking, is interesting in the light

ghene diese soo noemen (Ia sy selver dickmael de ghewoonte volghende) daer in dwalen, ofte ymmers niet Phylosophykelick, Dialectikelick, ofte Eyghentlick en spreecken: De Naercommers, segghen ter contrarien dat recht ende wel Dialectikelick gheseyt te sijne, van wiens meyninghe wy in desen oock sijn: [...] Twelck billichlick alsoo toeghelaten, seecker den Onscheydelicken Pieter ende Ian, sullen Specien sijn, ende Mensche niet de Leechste Specie, maer haerlieder Gheslachte, [...]'. Stevin, *Dialectike*, 108–9.

64 '[d]e aldereyghenste Specie', Stevin, 109.

65 I am preparing an essay in which I expand on the importance of this issue in the light of Stevin's number theory; there is a direct link between Stevin's view on indivisibles and his use of the fourth figure that can be contrasted with specific remarks in Aristotle's *Metaphysics* Book 10.

of a notorious sixteenth-century debate 'on the certainty of mathematics' (*Quaestio de certitudine mathematicarum*).[66] This debate, which started in the 1550s and continued for centuries, was started by Alessandro Piccolomini (1508–1579), who had claimed that the most powerful logical tool, i.e., what the Aristotelian tradition called 'proof par excellence' (*demonstratio potissima*, explaining both the effect as well as its cause), cannot be applied to mathematics. Thus, Piccolomini argued, mathematics derived its certainty not from the strength of logical proofs, but from the peculiar nature of its objects, which are the products of the imagination and not 'real' (like the objects in natural philosophy). Briefly summarized, mathematics dealt with 'quantity', whereas natural philosophy dealt with moving objects in the physical world. This position implied a sharp distinction between logic and mathematics, which fitted within a longer tradition. Neither Aristotle nor Euclid had ever conceived of a connection between logic and mathematics, and the different disciplines had been taught in different courses of study for centuries (logic in the *trivium* and mathematics, i.e., arithmetic and geometry, in the *quadrivium*).

In response to Piccolomini's attack on the certainty of mathematics, Francesco Barozzi (1537–1604) wrote an *Opusculum*, in which he claimed that mathematics occupied a middle position between the 'divine science' (metaphysics) and 'reality' (the objects of natural philosophy). Based on Proclus and Plato, he then claimed that the sciences were arranged according to the perfection of their entities, and that therefore 'mathematics' was higher on the ladder than 'natural philosophy', mediating between 'matter' (the topic of natural philosophy) and the purely abstract entities of metaphysics. However, the use of mathematics in natural philosophy was not central to Barozzi's argumentation. This would become the pet subject of Pietro Catena (1501–1577), who has been identified by Rivka Feldhay as a third archetype within the discussion on mathematical certainty. Catena introduced 'a view of mathematical universals as predicates of the rational soul that he derived from his Platonic reading of the *Posterior Analytics*'.[67] Based on the idea that any particular participates in a universal mathematical nature, and that the concept of universal science was common both to Aristotle and Euclid, Catena argued

66 My understanding and review of this conflict is based on Ariew, 'Christopher Clavius and the Classification of Sciences'; Feldhay, 'The Use and Abuse'; Giacobbe, 'Francesco Barozzi'; Jardine, 'Epistomology of the Sciences'; Mancosu, 'Aristotelian Logic and Euclidean Mathematics'; Mancosu, *Philosophy of Mathematics*, 24–33; Romano, *La contre-réforme mathématique*, 153–62.

67 Feldhay, 'The Use and Abuse', 87.

that universal intelligibles formed the basis of the principles of science and that therefore the 'causes discovered were the product of rational discourse, not of the senses'.[68]

The debate about the exact status of mathematics had a huge impact, reverberating throughout Europe.[69] It also led to numerous attempts to arrive at a combined method of mathematics and dialectics; after all, in the eyes of some at least, only a syllogistic proof of mathematical theorems could demonstrate that mathematics was worthy of being regarded as a proper science. Of all these attempts to redefine the relationship between dialectics and mathematics, the work by Conrad Dasypodius (ca. 1530–1600) and Christianus Herlinus (1505–1562) was possibly the most extensive. In their *Analysis geometriæ sex librorum Euclidis* (1566), they provided syllogistic proofs for the first six books of Euclid's *Elements*. The attempt was later repeated by Christopher Clavius (1538–1612), albeit only for Euclid's first problem, in the 1574 edition of Euclid's *Elements*.[70]

Although it is still unclear whether Stevin was directly familiar with the work of Dasypodius and Herlinus, he was certainly familiar with the attempt by Clavius.[71] Stevin had used the 1574 edition of the *Elements* for his *Problemeta Geometrica* (1583), and it is also known that Clavius in turn studied Stevin's work.[72] Yet Stevin did not comply with the view advocated by Clavius, namely that mathematics is superior to natural science because of its subject and its certainty (the 'Barozzi point of view'). His position was more in line with that of Catena, although it must be said that Stevin put the emphasis much more

68 Feldhay, 88.

69 Of course, numerous variations were possible between the different positions of Piccolomini, Barozzi, and Catena. The discussion as it arose on the Italian peninsula merely provides a clear picture of the issues at stake; this discussion was soon continued elsewhere in Europe. It should also be noted that attempts were also made from the opposing corner to mathematicize dialectics. On this issue, see Risse, *Die Logik*, 1:555–56. The point here is that there was a coming-together of disciplines that had hitherto been separate.

70 For more background, see Bertato, 'Attempts to Formalize as Syllogisms'; Massa Esteve, 'Symbolic Language in Early Modern Mathematics'.

71 Nevertheless, in my opinion, it is highly likely that Stevin knew the *Analysis geometriæ sex librorum Euclidis* (1566), taking into consideration that, in his *Arithmetique* (1585), he cited from the commentary of Barlaam the monk as published by Conrad Dasypodius in 1564.

72 Gravelaar, *Stevin's Problemeta geometrica*. See also de Wreede, 'Willebrord Snellius', 218, 266. The work of the Jesuit at the Collegio Romano in Rome was, for that matter, widely distributed in the Dutch Republic, where it was not merely read in connection with the 1582 Gregorian calendar reform.

firmly on the importance of language as the essential medium with which to arrive at reliable knowledge.

The emphasis on the importance of the vernacular language points us in the direction of yet another way in which Stevin could have been confronted with the attempt to furnish syllogistic proof for mathematical theorems, namely through the work of Jacques Peletier du Mans (1515–1582), whose 'pagan interpretation of Cicero arrived in sixteenth-century Paris at the same time as the diffusion of interpretations in terms of dialectic, such as those propounded by Lorenzo Valla and Rudolph Agricola'.[73] The French poet-mathematician, whose work was undeniably studied in the circles around Stevin, combined his mathematical innovations with an extensive campaign for a national language, just as Stevin would do later.[74] And Peletier saw in this a clear relationship between mathematical demonstration and the syllogism, writing for example in the introduction to his translation of the first six books of the *Elements* (1557) that:

> The Dialecticians call 'proof' a syllogism which causes knowledge, i.e. one that concludes from proven premises; and this originates from Geometry. Better: every proof which leads us to truth is geometrical in character. As is very truly said, we are not able to distinguish the true from the false if we have not previously been well acquainted with Euclid.[75]

In contrast to his contemporary Petrus Ramus, who had deemed mathematics the higher form of proof and by consequence considered its reduction to syllogisms an impossibility, Peletier regarded mathematics and dialectics as essentially intertwined.[76] Stevin's ideas were clearly moving further in this direction. Much more unequivocally than Peletier, however, he proceeded to bring together dialectics and mathematics, and one can find several well-defined examples of categorical syllogisms in Stevin's work. In *De Beghinselen der Weeghconst* (1586), for instance, after giving three examples of finding the center of gravity of polygons, Stevin comes up with the following proof (see figure 5.2):

73 Cifoletti, 'Mathematics and Rhetoric', 41. See also Jardine, 'Lorenzo Valla'.

74 By way of illustration, Robbert Robbertsz. le Canu (1563–1632), who had met with Stevin on several occasions in the context of patent disputes, cited Peletier's new naming convention for large numbers (from *l'Aritmetique*, 1584). Le Canu, *'tVerscheyden antwoordt*. In this booklet, the Hoorn-based mathematician tried to find an answer to the question of how to pronounce the number 1357328400000761010843278190030030045728345730285927003210 in Dutch. Dijksterhuis claimed that Stevin cited Peletier in his *Aritmétique*; however, I have not yet been able to trace this reference. Dijksterhuis, *Simon Stevin*, 1970, 22.

75 As cited in Mugnai, 'Logic and Mathematics in the Seventeenth Century', 299.

76 Cifoletti, 'Mathematics and Rhetoric', 228.

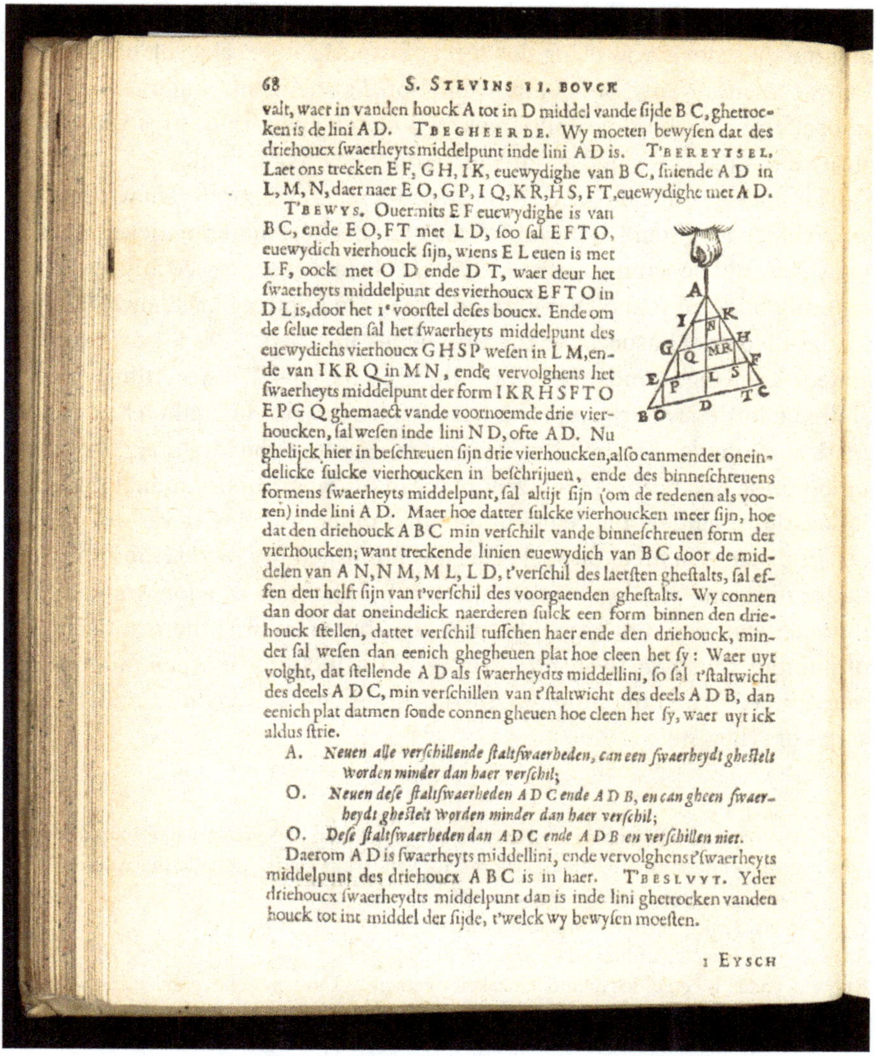

68 S. STEVINS I. BOVCK

valt, waer in vanden houck A tot in D middel vande ſijde B C, ghetrocken is de lini A D. T'BEGHEERDE. Wy moeten bewyſen dat des driehoucx ſwaerheyts middelpunt inde lini A D is. T'BEREYTSEL. Laet ons trecken E F, G H, I K, euewydighe van B C, ſnidende A D in L, M, N, daer naer E O, G P, I Q, K R, H S, F T, euewydighe met A D.

T'BEWYS. Ouermits E F euewydighe is van B C, ende E O, F T met L D, ſoo ſal E F T O, euewydich vierhouck ſijn, wiens E L euen is met L F, oock met O D ende D T, waer deur het ſwaerheyts middelpunt des vierhoucx E F T O in D L is, door het 1e voorſtel deſes boucx. Ende om de ſelue reden ſal het ſwaerheyts middelpunt des euewydichs vierhoucx G H S P weſen in L M, ende van I K R Q in M N, ende vervolghens het ſwaerheyts middelpunt der form I K R H S F T O E P G Q ghemaeckt vande voornoemde drie vierhoucken, ſal weſen inde lini N D, ofte A D. Nu ghelijck hier in beſchreuen ſijn drie vierhoucken, alſo canmender oneindelicke ſulcke vierhoucken in beſchrijuen, ende des binneſchreuens formens ſwaerheyts middelpunt, ſal altijt ſijn (om de redenen als vooren) inde lini A D. Maer hoe datter ſulcke vierhoucken meer ſijn, hoe dat den driehouck A B C min verſchilt vande binneſchreuen form der vierhoucken; want treckende linien euewydich van B C door de middelen van A N, N M, M L, L D, t'verſchil des laetſten gheſtalts, ſal effen den helft ſijn van t'verſchil des voorgaenden gheſtalts. Wy connen dan door dat oneindelick naerderen ſulck een form binnen den driehouck ſtellen, dattet verſchil tuſſchen haer ende den driehouck, minder ſal weſen dan eenich ghegheuen plat hoe cleen het ſy: Waer uyt volght, dat ſtellende A D als ſwaerheydts middellini, ſo ſal t'ſtaltwicht des deels A D C, min verſchillen van t'ſtaltwicht des deels A D B, dan eenich plat datmen ſoude connen gheuen hoe cleen het ſy, waer uyt ick aldus ſtrie.

A. *Neuen alle verſchillende ſtaltſwaerheden, can een ſwaerheydt gheſtelt worden minder dan haer verſchil;*

O. *Neuen deſe ſtaltſwaerheden A D C ende A D B, en can gheen ſwaerheydt gheſtelt worden minder dan haer verſchil;*

O. *Deſe ſtaltſwaerheden dan A D C ende A D B en verſchillen niet.*

Daerom A D is ſwaerheyts middellini, ende vervolghens t'ſwaerheyts middelpunt des driehoucx A B C is in haer. T'BESLVYT. Yder driehoucx ſwaerheydts middelpunt dan is inde lini ghetrocken vanden houck tot int middel der ſijde, t'welck wy bewyſen moeſten.

1 EYSCH

FIGURE 5.2 Stevin, *De Beghinselen der Weeghconst* (Leiden: Plantin, 1586), p. 68
COURTESY: GÖTTINGEN STATE AND UNIVERSITY LIBRARY

A. Beside any different apparent gravities there may be placed a gravity less than their difference;

O. Beside the present apparent gravities ADC and ADB there cannot be placed any gravity less than their difference;

O. Therefore, the present apparent gravities ADC and ADB do not differ.[77]

---

77 Stevin, *Principal Works, Mechanics*, 1:231. *De Beghinselen der Weeghconst* (Leiden: Plantin, 1586) was always published together with *De Weeghdaet* and *De Beghinselen des*

Similar examples can be found throughout the book.[78] There is no reason to assume that Stevin was using this form of exposition because of 'his desire to stress the importance of his innovation', or that he was trying to emancipate himself from 'the burdensome Greek style of demonstration'.[79] Instead, dialectical mathematics formed an integral part of Stevin's method.

The presence of syllogisms in mixed mathematical proofs shows that Stevin's ambitions extended beyond those of a practical mathematician. Accordingly, it would be wrong to assume that Stevin 'never expressed any interest in natural philosophy' and 'was more interested in applying his knowledge than in the theoretical aspects of the disciplines involved'.[80] As Stevin explicitly stated, 'knowing is understanding a thing by its causes'.[81] Accordingly, Stevin shifted effortlessly back and forth between his dialectical method and mathematical methods of proof. This also comes clearly to the fore in the argumentation presented by Stevin to support one of his most important mathematical innovations, namely the redefinition of 'number' as being never discontinuous or discrete (a breakthrough that was published in *L'Arithmetique*, in 1585, under the same privilege as the *Dialectike*).[82] Contrary to a longstanding tradition, Stevin dared to argue that not 'one' but 'zero' was the real beginning of number, thus effectively eliminating all distinctions between number and magnitude.[83] The final proof for Stevin's argument was again formulated in a series of syllogisms:

Waterwichts. N.B. 'Apparent gravity' is the translation of Stevin's neologism 'staltwicht', the effect of both weight and position. See Van Dyck's contribution in this volume.

78 Stevin, 1:207, 253, 263, 267, 269, 272, 278.

79 Stevin, *Principal Works, Mechanics*, 1:44.

80 Berkel, *Isaac Beeckman on Matter and Motion*, 134.

81 '… weten is een dinck deur de oirsaken verstaen …'. Citation to be found at the beginning of the *Driehouckhandel*, bundled together in Stevin, *Wisconstige ghedachtenissen*, fol. 2. Elaborating on the different attempts made by his predecessors to understand the centers of gravities, Stevin also contended at the beginning of *De Beghinselen der Weeghconst* that: 'It is true that with regard to vertical weights it has been observed by experience that two gravities of equal apparent weight are proportional to their arms. But they thought that this proportionality was due to the circles described about the fixed point by the extremities of the arms. From which, as is usual with errors, there followed no knowledge of the causes'. Stevin, *Principal Works, Mechanics*, 1:55. From the dedication letter to Rudolph II (1552–1612). On Stevin's conception of causes, see also the chapter by Vermij in this volume.

82 For more detail on the relationship between Stevin's number theory and dialectics, see Naets, 'How to Define a Number?'.

83 The literature on this is extensive, but see Jones, 'The Concept of 'One' as a Number'; Klein, *Greek Mathematical Thought*; Malet, 'Renaissance Notions of Number and Magnitude'. For a concise overview of Stevin's innovations, see also Corry, *A Brief History of Numbers*, 163–67.

> The part is of the same matter as its whole, Unity is part of a multitude of unities, Hence unity is of the same matter as the multitude of unities; But the matter of a multitude of unities is number, Hence the matter of unity is number.[84]

Just as in the example of the valuation of a shipload of pepper, expanded upon in the *Dialectike*, matter lay at the basis of Stevin's logical considerations. Correct definitions within mathematics were subject to the same rules as any other arguments within logic.

And thus mathematics and dialectics were also closely related within the thematic learning program as it was developed by Stevin. There were several instances in the *Dialectike* where the Flemish mathematician referred to his earlier work, for instance to his *Tafelen van Interest* and his *Pratique d'Arithmetique*.[85] Conversely, he referred directly to the *Dialectike* in his more mathematical works, albeit mainly on the level of linguistic innovation.[86] As demonstrated by other contributions in this volume, Stevin believed in the existence of an 'Age of the Sages' or *Wijsentijd*, when all mankind had spoken Dutch. Stevin's native language was therefore most closely related to pure knowledge and, because of its monosyllabic structure as well as the ease with which one could combine words, the language most suited to scientific activities. It was only through language that mankind had access to unspoiled and genuine knowledge, and it is in that sense that the *Dialectike* must be seen as the basis for Stevin's vernacular epistemology.

5 Conclusion

In his *Dialectike*, Stevin presented a method that was based on the principles of mathematics yet opened up a space for the investigation of nature by means of observation. Although the exact influences on his ideas have only been demonstrated sketchily in this essay, it is clear that Stevin's innovations in this area fitted within a broader discussion that emerged across the whole of Europe in the sixteenth century, about the status of mathematics as a scientific language. It is in that light that one should also see Stevin's endeavors to provide his mathematical innovations with syllogistic proof.

---

84 Stevin, *Principal Works, Vol. 2b*, 2b:495, see also 501.

85 Stevin, *Dialectike*, 58–60.

86 For instance, in his 'Discourse on the Dignity of the Dutch Language' at the beginning of *De Beghinselen der Weeghconst*, Stevin, 1:82. And in *De Thiende* at Stevin, *Principal Works, Vol. 2a*, 2a:446.

Stevin was busy reinventing a new mathematical language, which was related to the notation of decimal fractions and the redefinition of one as a number. At the same time, he was engaged in reinventing the Dutch language, in which a deeper truth was lurking that gave direct access to an unspoiled form of knowledge. The innovations in terms of mathematical and Dutch language cannot be seen in isolation. Not only did Stevin attempt to find new Dutch words for known concepts within dialectics and mathematics, he also used the Dutch language as a portal for developing a new form of dialectics and mathematics. This is, for example, clearly demonstrated in his discussion about 'empty species', where he used everyday Dutch as a point of departure to make claims about the ontological status of mathematical objects. Stevin's vernacular epistemology formed the basis for a kind of rhetorized logic that marked a new turn within the wider network of developments that had materialized in the field of dialectics from the beginning of the sixteenth century.

For Stevin the polymath, it was obvious that different disciplines could not operate in isolation. Dialectics was a 'great liberal Art, [or rather] a pathway of the Arts, leading mankind to all Arts'.[87] This way of thinking fitted perfectly within the Netherlandish *consten*-culture as it had developed in the rhetoricians' chambers (*rederijkerskamers*) from the fifteenth century onwards. As Arjan van Dixhoorn rightly remarked, in this 'pre-modern Dutch arts culture, the knowledge of nature and knowledge of self and the spiritual world were not opposed in the way the two cultures (sciences and humanities) are today'.[88] For Stevin, dialectics did not just relate to the Liberal Arts, but to the cultivation of 'reason' (*de Reden*) that distinguished man from animals.[89] The mastery of dialectics was a form of ingenuity that taught one 'how to Judge all matters skillfully and correctly', as the title of the book indicated: it made no distinction in terms of the objects assessed, but was able to bring together practical crafts and theoretical knowledge. As Stevin specified in the *Dialectike*, when providing examples in his discussion of the predicables genus and species: 'Craft [*Ambacht*] is a genus, the species of which are carpeting, masonry, forging, etc., because each of these is a craft. Similarly, art [*conste*] is the genus of the species craft, rhetoric, music, etc., because each of these is an art'.[90] This

87 '... door dese groote vrije Conste, door desen wech der Consten, leydende den Mensche tot allen Consten ...'. Stevin, *Dialectike*, *2v.

88 Van Dixhoorn, 'Recreating Man's Cunning Virtues', 33.

89 Stevin, fol. *2v.

90 'Alsoo oock is Ambacht Gheslachte, diens Specien sijn Timmeren, Metsen, Smeden, &c. Want Yeder van dien is Ambacht; Wederom Conste, is Gheslachte van Specien, Ambacht, Rhetorike, Musike, &c. want yeder van dien is Conste ...'. Stevin, 22.

broad view of the arts fitted with a way of thinking that was based on a linguistic understanding of the world. It lay at the core of a new attitude regarding the value of real-world experiences in philosophy and was essential for Stevin's innovations in mathematics.

Bibliography

Ariew, Roger. "Christopher Clavius and the Classification of Sciences." *Synthese* 83, no. 2 (1990): 293–300.

Ashworth, E.J. "Some Notes on Syllogistic in the Sixteenth and Seventeenth Centuries." *Notre Dame Journal of Formal Logic* 11, no. 1 (January 1970): 17–33.

Ashworth, E.J. "Developments in the 15th and 16th Centuries." In *Mediaeval and Renaissance Logic*, edited by Dov M. Gabbay and John Woods, 2:609–43. Amsterdam: Elsevier, 2008.

Ashworth, E.J. *Language and Logic in the Post-Medieval Period.* Dordrecht: Springer, 1974.

Barth, E.M. "From the History of the Logic of Individual Propositions." In *The Logic of the Articles in Traditional Philosophy: A Contribution to the Study of Conceptual Structures*, edited by E. M. Barth, 141–79. Dordrecht: Springer, 1974.

Berkel, Klaas van. *Isaac Beeckman on Matter and Motion: Mechanical Philosophy in the Making*. Baltimore: JHU Press, 2013.

Bertato, Fabio. "Attempts to Formalize as Syllogisms the Propositions of the Elements of Euclid Made by Herlinus, Dasypodius, Clavius and Hérigone." Forthcoming. Accessed October 1, 2018. https://www.academia.edu/37271146.

Blair, Ann. *The Theater of Nature: Jean Bodin and Renaissance Science.* Princeton, NJ: Princeton University Press, 1997.

Bruyere, Nelly. *Methode et dialectique dans l'oeuvre de La Ramee: Renaissance et Age classique*. Paris: Librairie Philosophique, 1984.

Bunge, Wiep van. *From Stevin to Spinoza: An Essay on Philosophy in the Seventeenth-Century Dutch Republic*. Leiden: Brill, 2001.

Buning, Marius. "Commentary on the privilege for the Deux-Aes Bible and the privilege for the Squaring of the Circle." In *Primary Sources on Copyright (1450–1900)*, edited by Martin Kretschmer and Lionel Bently. 2018.

Buning, Marius. "Privileging the Common Good. The Moral Economy of Printing Privileges In the Seventeenth-Century Dutch Republic." In *Buying and Selling*: The Business of Books in Early Modern Europe, Andrew Pettegree and Shanti Grahell eds., Leiden: Brill, 2019.

Burgersdijck, Franco. *Institutio logica, dat is, reden-konstigh onder-wys, be-ghreepen in twe boeken*. Translated by Allardus Lodewijk Kok. 2nd ed. Amsterdam: Gerrit Willemsz., 1658.

Buys, Ruben. *De kunst van het weldenken: lekenfilosofie en volkstalig rationalisme in de Nederlanden (1550–1600)*. Amsterdam: Amsterdam University Press, 2009.

Cifoletti, Giovanna Cleonice. "From Valla to Viète: The Rhetorical Reform of Logic and Its Use in Early Modern Algebra." *Early Science and Medicine* 11 (2006): 390–423.

Cifoletti, Giovanna Cleonice. "Mathematics and Rhetoric: Peletier and Gosselin and the Making of the French Algebraic Tradition." PhD thesis, Princeton University, 1992.

Copenhaver, Brian P. "Translation, Terminology and Style in Philosophical Discourse." In *Cambridge History of Renaissance Philosophy*, edited by Eckhard Kessler, Quentin Skinner, and Charles B. Schmitt, 77–110. Cambridge: Cambridge University Press, 1992.

Corry, Leo. *A Brief History of Numbers*. Oxford: Oxford University Press, 2015.

Davidse, Ad. "Simon Stevin," 2018 2003. https://adcs.home.xs4all.nl/stevin/index.html.

De Rijk, L.M. *Logica Modernum: A Contribution to the History of Early Terminist Logic*. 2 vols. Assen: Van Gorcum, 1962.

Devreese, Jozef T., and Guido Vanden Berghe. *'Magic is No Magic': The Wonderful World of Simon Stevin*. Southampton: WIT, 2008.

Dibon, P. "L'influence de Ramus aux universités néerlandaises du XVIIe siècle." *Proceedings of the XIth International Congress of Philosophy* 14 (1953): 307–11.

Dibon, Paul Auguste Georges. *La philosophie néerlandaise au Siècle d'or*. Amsterdam: Elsevier, 1954.

Dixhoorn, Arjan van. *Lustige geesten: rederijkers in de Noordelijke Nederlanden (1480–1650)*. Amsterdam: Amsterdam University Press, 2009.

Dixhoorn, Arjan van. "Recreating Man's Cunning Virtues. The Philosophical Project of Netherlandish Arts Culture." *Renaissance Studies* 32, (2018): 23–42.

Dijksterhuis, Eduard Jan. *Simon Stevin*. The Hague: Nijhoff, 1943.

Dijksterhuis, Eduard Jan. *Simon Stevin: Science in the Netherlands around 1600*. The Hague: Martinus Nijhoff, 1970.

Dijksterhuis, Fokko Jan. "Duytsche Mathematique and the Building of a New Society: Pursuits of Mathematics in the Seventeenth-Century Dutch Republic." In *Mathematical Practitioners and the Transformation of Natural Knowledge in Early Modern Europe*, 167–81. Cham: Springer, 2017.

Dijkstra, Arjen. "Between Academics and Idiots: A Cultural History of Mathematics in the Dutch Province of Friesland (1600–1700)." PhD thesis, University of Twente, 2012.

Edwards, D.W. "Randall on the Development of Scientific Method in the School of Padua: A Continuing Reappraisal." In *Naturalism and Historical Understanding. Essays on the Philosophy of J.H. Randall, Jr.*, edited by J.P. Anton, 42–45. Albany, NY: State University of New York Press, 1967.

Feingold, Mordechai, Joseph S. Freedman, and Wolfgang Rother, eds. *The Influence of Petrus Ramus: Studies in Sixteenth and Seventeenth Century Philosophy and Sciences.* Basel: Schwabe, 2001.

Feldhay, Rita. "The Use and Abuse of Mathematical Entities: Galileo and the Jesuits Revisited." In *The Cambridge Companion to Galileo*, edited by Peter Machamer, 80–145. Cambridge: Cambridge University Press, 1998.

Ferguson, Stephen. "System and Schema: Tabulae of the Fifteenth to the Eighteenth Centuries." *Princeton University Library Chronicle* 49 (1987): 9–30.

Giacobbe, Giulio Cesare. "Francesco Barozzi e la Quaestio de certitudine mathematicarum." *Physis* 14, no. 4 (1972): 357–74.

Goulding, Robert. *Defending Hypatia: Ramus, Savile and the Renaissance Rediscovery of Mathematical History*. Dordrecht: Springer, 2010.

Henle, Paul. "On the Fourth Figure of the Syllogism." *Philosophy of Science* 16, no. 2 (1949): 94–104.

Heuvel, Charles van den, *"De Huysbou": A Reconstruction of an Unfinished Treatise on Architecture, Town Planning and Civil Engineering by Simon Stevin*. Amsterdam: KNAW-Edita, 2005.

Höltgen, K.J. "Synoptische Tabellen in der Medizinischen Literatur und die Logik Agricolas und Ramus." *Sudhoffs Archiv* 49 (1965): 371–90.

Hotson, Howard. *Commonplace Learning: Ramism and Its German Ramifications, 1543–1630*. Oxford: Oxford University Press, 2007.

Hunaeus, Augustinus. *Augustini Hunnaei dialectica, seu generalia logices praecepta omnia, quaecunque ex toto Aristotelis organo in academia Lovaniensi et Coloniensi philosophia tyronibus ad ediscendum proponi consueverunt*. Antwerp: Plantin, 1568.

Japikse, Nicolaas, ed. *Resolutiën der Staten-Generaal van 1576 Tot 1609*. Vol. 4. 14 vols. The Hague: Nijhoff, 1919.

Jardine, Lisa. "Lorenzo Valla: Academic Skepticism and the New Humanist Dialectic." In *The Skeptical Tradition*, edited by Myles Burnyeat, 253–286. Berkeley, CA: University of California Press, 1983.

Jardine, Nicholas. "Epistemology of the Sciences." In *The Cambridge History of Renaissance Philosophy*, edited by Charles B. Schmitt, 675–711. Cambridge: Cambridge University Press, 1988.

Jones, Charles Virgil. "The Concept of 'One' as a Number." PhD thesis, University of Toronto, 1979.

Klifman, Harm. "Dutch Language Study and the Trivium: Motives and Elaborations." In *The History of Linguistics in the Low Countries*, edited by Jan Noordegraaf, Kees Versteegh, and E.F.K. Koerner, 63–84. Amsterdam: Benjamins, 1992.

Le Canu, Robbert Robbertsz. *'tVerscheyden antwoordt uyt vele steden in Hollant, op de vraghe van numeratio, het eerste A. B. C. der talkonst*. Hoorn: Willem Andriesz., 1612.

Mack, Peter. *Renaissance Argument: Valla and Agricola in the Traditions of Rhetoric and Dialectic*. Leiden: Brill, 1993.

Mancosu, Paolo. "Aristotelian Logic and Euclidean Mathematics: Seventeenth-Century Developments of the Quaestio de Certitudine Mathematicarum." *Studies in History and Philosophy of Science Part A* 23, no. 2 (1991): 241–265.

Mancosu, Paolo. *Philosophy of Mathematics and Mathematical Practice in the Seventeenth Century*. New York: Oxford University Press, 1996.

Mare, H. de. "Het huis en de regels van het denken: Een cultuurhistorisch onderzoek naar het werk van Simon Stevin, Jacob Cats en Pieter de Hoogh." Amsterdam: Vrije Universiteit, 2003.

Massa Esteve, M. Rosa. "Symbolic Language in Early Modern Mathematics: The Algebra of Pierre Hérigone (1580–1643)." *Historia Mathematica* 35, no. 4 (2008): 285–301.

Molhuysen, Philip Christiaan, ed. *Bronnen Tot de Geschiedenis Der Leidsche Universiteit. 1574-[1811]*. Vol. 1. 7 vols. Rijks Geschiedkundige Publicatiën 20. The Hague: M. Nijhoff, 1913.

Mugnai, Massimo. "Logic and Mathematics in the Seventeenth Century." *History and Philosophy of Logic* 31, no. 4 (November 1, 2010): 297–314.

Naets, Jurgen. "How to Define a Number? A General Epistemological Account of Simon Stevin's Art of Defining." *Topoi* 29, no. 1 (2010): 77–86.

Nauta, Lodi. *In Defense of Common Sense: Lorenzo Valla's Humanist Critique of Scholastic Philosophy*. Cambridge, MA [u.a.]: Harvard University Press, 2009.

Ong, Walter J. *Ramus, Method, and the Decay of Dialogue: From the Art of Discourse to the Art of Reason*. University of Chicago Press, 2004.

Overdiep, G.S. "Simon Stevin en de Nederlandsche taal." In *Simon Stevin, Het Burgherlick Leven*, edited by Annie Romein-Verschoor and G.S. Overdiep. Amsterdam: De Wereldbibliotheek, 1939.

Peeters, L. "Auteurschap en tekst van 'Spiegels' 'Twe-Spraack' (1584)." *Tijdschrift Voor Nederlandse Taal- En Letterkunde* 98 (1982): 117–29.

Randall, John Herman. *The School of Padua and the Emergence of Modern Science*. Padova: Antenore, 1961.

Reid, Steven J., and Emma Annette Wilson. *Ramus, Pedagogy and the Liberal Arts: Ramism in Britain and the Wider World*. Burlington, VT: Ashgate Publishing, Ltd., 2013.

Rescher, Nicholas. *Galen and the Syllogism*. Pittsburgh: University of Pittsburgh, 1966.

Risse, Wilhelm. *Die Logik der Neuzeit. Bd. 1. 1500 - 1640*. Vol. 1. 2 vols. Stuttgart-Bad-Cannstatt: Frommann, 1964.

Romano, Antonella. *La contre-Réforme mathématique. Constitution et diffusion d'un culture mathématique Jésuite à la Renaissance (1540–1640)*. Rome: Bibliothèque des Ecoles françaises d'Athènes et Rome, 1999.

Schilt, Kees. *Simon Stevin en het hermetisme: een cultuurhistorische studie naar zijn methodologie*. Unpublished PhD thesis, Vrije Universiteit Amsterdam, 2013.

Schmidt-Biggemann, Wilhelm. *Topica Universalis: Eine Modellgeschichte humanistischer und barocker Wissenschaft*. Hamburg: Meiner, 1983.

Sellberg, E. "The Usefulness of Ramism." In *The Influence of Petrus Ramus: Studies in Sixteenth and Seventeenth Century Philosophy and Sciences*, edited by Mordechai Feingold, Joseph S. Freedman and Wolfgang Rother, 107–26. Basel: Schwabe, 2001.

Selm, B. van, " '... Te bekomen voor een civielen prijs'. De Nederlandse boekprijs in de zeventiende eeuw als onbekende grootheid." *De zeventiende eeuw* 6 (1990): 98–116.

Spiegel, H.L. *Twe-Spraack. Ruygh-Bewerp. Kort Begrip. Rederijck-Kunst*. Edited by W.J.H. Caron. Groningen: Wolters, 1962.

Stevin, Simon. *Dialectike ofte Bevvysconst. Leerende van allen saecken recht ende constelick Oirdeelen; oock openende den wech tot de alderdiepste verborghentheden der Natueren. Beschreven int Neerduytsch door Simon Stevin van Brugghe*. Rotterdam: Janssonius van Waesberge, 1621.

Stevin, Simon. *L'Aritmetique*. Leiden: Plantin, 1585.

Stevin, Simon. *Materiæ politicæ. Bvrgherlicke stoffen: Vervanghende ghedachtenissen der oeffeninghen des doorluchtichsten Prince Maurits van Orangie*. Edited by Hendrik Stevin. Leiden: Adryaen Rosenboom, 1649.

Stevin, Simon. *The Principal Works of Simon Stevin, Mathematics*. Edited by Dirk Jan Struik. Vol. 2a. 5 vols. Amsterdam: Swets & Zeitlinger, 1958.

Stevin, Simon. *The Principal Works of Simon Stevin. Vol. I, Mechanics*. Edited by Eduard Jan Dijksterhuis. Vol. 1. 5 vols. Amsterdam: Swets & Zeitlinger, 1955.

Stevin, Simon. *The Principal Works of Simon Stevin. Vol. IIb, Mathematics*. Edited by Dirk Jan Struik. Vol. 2b. 5 vols. Amsterdam: Swets & Zeitlinger, 1958.

Stevin, Simon. *Wisconstige ghedachtenissen*. Leiden: Jan Bouwensz, 1608.

Traninger, Anita. *Disputation, Deklamation, Dialog: Medien und Gattungen europäischer Wissensverhandlungen zwischen Scholastik und Humanismus*. Stuttgart: Steiner, 2012.

Valerius, Cornelius. *Tabulae Totius Dialectices*. Antwerp: Plantin, 1575.

Wal, Marijke van der, "Logic, Linguistics, and Simon Stevin in the Context of the Sixteenth and Seventeenth Centuries." In *History of Linguistics 1993 : Papers from the Sixth International Conference on the History of the Language Sciences (ICHoLS VI), Washington DC, 9–14 August 1993*, edited by K.R. Jankowski, 147–56. Amsterdam: John Benjamins, 1995.

Wal, Marijke van der. "Simon Stevin: Taalbeschouwer en taalgebruiker." In *Simon Stevin 1548–1620. De geboorte van de nieuwe wetenschap*, ed. Wouter Bracke, Leuven: Brepols, 2004, 170–77.

Verbeek, Théo. "La Dialectike (1585) de Simon Stevin." In *Simon Stevin. De la vie civile, 1590*, edited by Catherine Secretan and Pim den Boer, 181–90. Lyon: ENS Éditions, 2014.

Verdonk, J.J. "Vom Einfluβ des P. de La Ramée auf Simon Stevin." *Centaurus* 13, no. 3 (1969): 251–62.

Verdonk, Johannes Jacobus. *Petrus Ramus en de wiskunde*. Assen: Van Gorcum, 1966.

Viaene, Antoon. "Oude kleurnamen van paarden." *Biekorf* 63 (1962): 151–52.

Waswo, Richard. "The 'Ordinary Language Philosophy' of Lorenzo Valla." *Bibliothèque d'Humanisme et Renaissance* 41, no. 2 (1979): 255–71.

Wreede, Liesbeth C. de. "Willebrord Snellius. A Humanist Reshaping the Mathematical Sciences." PhD thesis, Utrecht University, 2007.

Zabarella, Giacomo. *Opera logica*. Venice: Paulus Meietus, 1586.

CHAPTER 6

Causality and the Reduction to Art of Simon Stevin's Mechanics

Maarten Van Dyck

1 Bringing Mechanics to the Form of Art

Simon Stevin is a seminal author in the so-called 'renaissance of mathematics'.[1] His vernacular books on statics and hydrostatics (published together in 1586) not only helped readers to become familiar with Archimedean mechanics, but they also contained strikingly innovative additions, such as his proof of the law of the inclined plane and the hydrostatic paradox. They also stand out for their self-confident character: Stevin was well aware of the novelty of his approach, and he claimed in the introduction to his text that he was the first author to allow statics, which he called the 'art of weighing' (*weeghconst*), to 'come to the form of Art'.[2]

The idea that a field of practical operations could be subject to what was usually called 'reduction to art' was very common in the period.[3] It was an idea with ancient Ciceronian precedents that had become imbued with the Renaissance notion of method. The ideal was to provide a systematic, textual presentation that would bring as many actions as possible under the purview of a small number of principles that allowed one quickly and efficiently to decide on how to proceed to attain one's goals. Stevin's books from 1586 did not provide many explicit statements on how he understood this ideal, but we can read his discussion on the 'restoration of the Age of the Sages', included in his *Wisconstighe Ghedachtenissen* (published in 1605–1608), as an articulation of the method behind his attempted 'reduction to the form of art'.

1 For the influential use of the expression and a detailed analysis that is limited to the Italian context, see Rose, *Italian Renaissance*.

2 Stevin, *Principal Works*, I, 93. Here and in some of the following quotations, I adapt the translation given in the *Principal Works*, but I refer to the English translation for ease of reference.

3 See Vérin, 'Rédiger', for an essential introduction.

 | DOI:10.1163/9789004432918_008

Stevin introduced five components that characterize the order to be followed in the written presentation of any art.[4] One should start the presentation by laying out the definitions of all the *constwoorden* ('terms of art'). One should then strive always to express similar things with similar words throughout the text (using what is known in rhetoric as *anaphora*), so that their relation is immediately clear. Each proposition in the treatise should be presented according to the model provided by Euclid's *Elements*, clearly separating the different parts, such as the givens, proof and conclusion. This not only transparently exhibits the deductive structure responsible for the proposition's truth, but it should also allow each reader immediately to identify the part he is interested in on different occasions. Using properly constructed dichotomous divisions (visually presented at the opening of a text in the typical tables using curly brackets) allows one immediately to see that the text as a whole offers a complete treatment of its subject, without leaving out any parts, and that its parts will be presented in the right order, where later parts presuppose earlier parts but not the other way around. Interestingly, these tables will also help rationalize the inverse translation back into practice, by functioning as a kind of decision tree that allows all the relevant options to be seen at a glance.[5] And lastly, one should present the theoretical treatment separately from the practical applications that form its ultimate goal, so that the specifically systematic character of the former can be grasped by the reader. As stressed by Stevin, this will again aid the decision process by allowing one easily to locate the theoretical proposition relevant for the particular purpose at hand, without having first to separate it from any other, extraneous practical operations that have been included unnecessarily in the presentation.

In what follows, I will offer an analysis of how Stevin's books on *The Principles of the Art of Weighing* (*Weeghconst*) and *The Practice of Weighing* (*Weeghdaet*) can be seen as exemplifying this typical Renaissance ideal of reduction to art.[6] There was yet another prominent discourse concerning the arts, though, and I will also try to see how this relates to Stevin's endeavor. According to the Aristotelian conception, the arts are characterized by two features: that they aim at useful ends, and that they do so by investigating the causes of things (see, e.g., the first chapter of the first book of the *Metaphysics*). Any reflection

4 Stevin, *Wisconstige Ghedachtenissen, 1 Vant Weereltschrift*, 41–48. I present the components in a different order from Stevin, to streamline further discussion. *The Principal Works* only contains a partial translation of this text, cf. Stevin, *Principal Works*, III, 591–623.

5 See Vérin, *Gloire*, 271–277, for further comments.

6 See Descotes, 'Rhétorique' for a complementary approach that focuses on what the author calls Stevin's 'mathematical rhetoric'.

on the notion of causality is conspicuously missing from Stevin's characterization of the proper order to be used in the arts, but this did not stop him from using causal terminology throughout his texts. Thus, how is his use of causal language related to the ideal of a reduction to art? As we will see, Stevin's text on dialectics, or what he calls *Bewysconst* ('the art of demonstrating'), which he published a year before his books on statics, offers a number of important clues. This text also bears many hallmarks of the ideal of reduction to art, but it has the added interest that because of the traditional nature of its subject matter, it also engages explicitly with the Aristotelian discourse.[7]

2 Delineating a Field

When Stevin stressed the importance of first gathering definitions in his discussion of the 'restoration of the Age of the Sages', he gave a seemingly trivial example: he explained how, when interacting with practitioners such as masons or carpenters, he would first ask them the meaning of the specific words they used in their practice, after which he would write these down and learn them by heart. Having learned them, he was then able to interact fluently with these practitioners.[8] This example seems trivial, because it suggests that defining is merely about learning the meaning of words.[9] But when we approach the example from the perspective opened up by the ideal of a reduction to art, there is a more illuminating way to understand its importance. By introducing specific terms of art, these definitions drew attention to the relevant characteristics of the practical situation under consideration. By learning the meaning of these words, Stevin also learned to see which aspects of their practice were of special importance for these practitioners. Good definitions should thus in the first place communicate the relevant characteristics of the situation under consideration – where relevance is always relative to what one wants to be able to achieve.

In his *Bewysconst*, Stevin had already explained that he translated *definitio* with the Dutch *bepaling* because this has the primary meaning of demarcating a piece of land by surrounding it with poles or *palen*.[10] Accordingly, definitions

7 See the chapter by Bunning in this volume.

8 Stevin, *Wisconstighe Ghedachtenissen*, 1, 43.

9 This is exactly the criticism that Arnauld and Nicole would later level against Stevin's definition of number in their *Logique ou l'art de penser* of 1662; as we shall see, Stevin had anticipated and dismissed this kind of criticism in his *Bewysconst*. See Naets, 'How to Define a Number', for further discussion with respect to Stevin's arithmetic.

10 Stevin, *Dialectike*, 31, 121–122.

must primarily succeed in properly delineating the field of study, and it does not matter how they achieve this. In this context, Stevin explicitly stated that he did not distinguish between definitions that are descriptive (by giving predicates that can be used to describe the subject corresponding to the term that is defined), explanatory (by merely indicating the meaning of words), or essential (by giving the essence of the subject). He justified this position with two revealing reasons: he appealed to the authority of mathematical authors such as Archimedes, and he strongly doubted the actual possibility of giving essential definitions privileged by philosophical authors such as Aristotle.[11]

Let us now consider Stevin's *The Principles of the Art of Weighing* with this in mind. The definitions introduce the general subject matter of the work, the crucial properties of the bodies' weight, and the kind of configurations that will be addressed. The latter configurations single out geometrically significant relationships within the instruments used for weighing bodies, by 'explanatorily' introducing specific terms for them, such as 'beam', 'handle' and 'fixed point'. The most important properties of weight (which is 'descriptively' defined as the known value of a body's 'power of descent in a given place')[12] have to do with the bodies' center of gravity, which Stevin 'descriptively' defined as 'the point such that if the body is conceived to be suspended from it, it remains at rest in any position given to it'.[13] The subject matter of the book is defined as 'the ratios, proportions, and properties of the weight or gravities of bodies'.[14]

There is one further definition that merits comment, as it shows the crucial importance of Stevin's use of *anaphora* in his text. In definition XI, Stevin introduced the neologism *evenstaltwichtigh* to denote two gravities that keep each other in equilibrium. The need for a specific term is clear from the fact that two such gravities will in general not be of equal weight (and thus *evenwichtigh* – or equiponderant). *Evenstaltwichtigh* bodies only have the appearance of equal weight, but the real equality is one according to their position (*'na den ghestalt'*).[15] The requirements of textual presentation prompted Stevin to introduce this term, which was needed to avoid ambiguous expressions – but in the same move, it allows the reader to adopt a more abstract perspective on the situations described. The possibility of using the same term to describe a whole array of situations shows that there is a way in which they are all crucially

11 See also Stevin, *Dialectike*, 53.
12 Stevin, *Principal Works*, I, 97.
13 Stevin, *Principal Works*, I, 99. Stevin took this definition from Pappus, and learned it from Federico Commandino's writings.
14 Stevin, *Principal Works*, I, 97.
15 Stevin, *Principal Works*, I, 104.

similar. Bringing out the mathematical characterization of this similarity was the main goal of Stevin's treatise. Knowing the general conditions under which bodies are *evenstaltwichtigh* will allow any practitioner to reach his goals efficiently in all particular circumstances, once these are properly described, and will thus successfully reduce the field of operations involving heavy bodies to the form of art. As we will see, this knowledge is grounded in two fundamental propositions: the law of the lever and the law of the inclined plane.

3 Demonstrating the Law of the Lever and the Law of the Inclined Plane[16]

The first proposition of *The Principles of the Art of Weighing* establishes the law of the lever. By assuming that equal bodies are in equilibrium when hanging from equal distances, Stevin was able to prove that if a heavier body is *evenstaltwichtigh* with a lighter one, the distances from which they are hanging have the inverse ratio of their weights.[17] The proof is based on the similarity of both situations which can be almost directly visually ascertained in Stevin's first diagram (see figure 6.1). The whole body *ABCD* is in equilibrium when suspended from its center *T* (which can be considered its center of gravity), since we can imagine it to be two equal bodies hanging from equal distances – and everyone agrees that these will be in equilibrium. If we now mentally divide the body into two unequal bodies rather than two equal bodies, they will also be in equilibrium, as long as they are suspended from the same point. For example, we can take the heavier body to be *LMDA* and the lighter *LMCB*, which have their own respective centers of gravity located in *S* and *X* (since these points divide the bodies again into equal parts). It is now easy to see that the distances of the centers of gravity of these supposedly *evenstaltwichtigh* bodies to their point of suspension *T* are inversely proportional to the weights of the bodies (as represented by their magnitudes).

The validity of the demonstration rests on two related, crucial steps: the assumption that the unequal bodies will indeed remain in equilibrium when we actually hang them from their respective centers of gravity, and the further assumption that the shape of the bodies does not matter (such that the inverse

16 This section contains a more elaborate discussion of some claims introduced in Van Dyck, 'Motion'.

17 This is actually the converse of Archimedes' original proof, where it is shown that given the inverse ratio, equilibrium will follow. Stevin assumed without proof that the converse also holds.

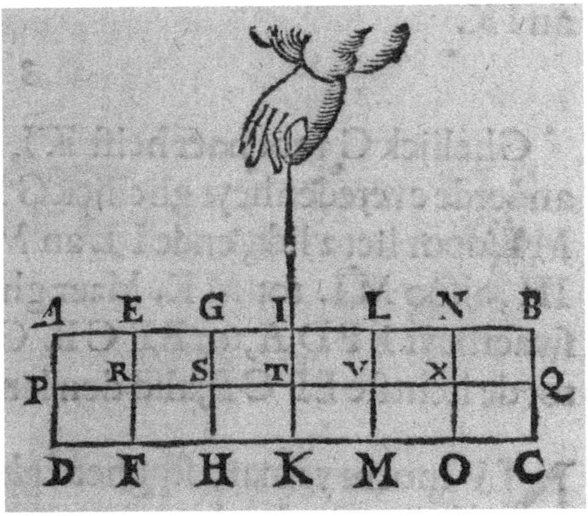

FIGURE 6.1 Demonstrating the law of the lever
SOURCE: GHENT UNIVERSITY LIBRARY, BIB.ACC.056517

proportionality established can be interpreted as proving the stated proposition in full generality, as shown in figure 6.2, where Stevin replaced the regular bodies *LMDA* and *LMCB* with irregular bodies of the same weight). What is of interest here are the judgements of similarity that lie at the basis of both steps: that the situation with the irregular unequal bodies is similar to the one with the regular unequal bodies, which in turn is crucially similar to the one with the solid body hanging from its middle, so that the initial equilibrium will not be disturbed under these imaginary replacements. Crucially, these similarities can only be noticed once one has identified a body's center of gravity as being of special interest. Its definition draws attention to the existence of a uniquely privileged point within all bodies: if we want to neutralize the effect of a body's gravity, we have to suspend the body either from its center of gravity, or from somewhere in the vertical going through it (this is implicitly assumed throughout Stevin's treatment, and made explicit in a comment on proposition 6, where it was called a 'common rule in the Art of Weighing')[18]; and if we suspend it from somewhere else, the body's non-balanced gravity will cause a motion that will stop when the body has reached its preferred position as

18 Stevin, *Principal Works*, I, 139. The translation in the *Principal Works* renders 'ghemeenen Reghel' as 'general rule'; but Stevin had already introduced a number of postulates as known by 'ghemeene wetenschap', i.e., common (or maybe more precisely: commonly-shared) knowledge (*Principal Works*, I, 111). This rule can thus be taken to have the status of a postulate, a shared element of the practice being reduced to art.

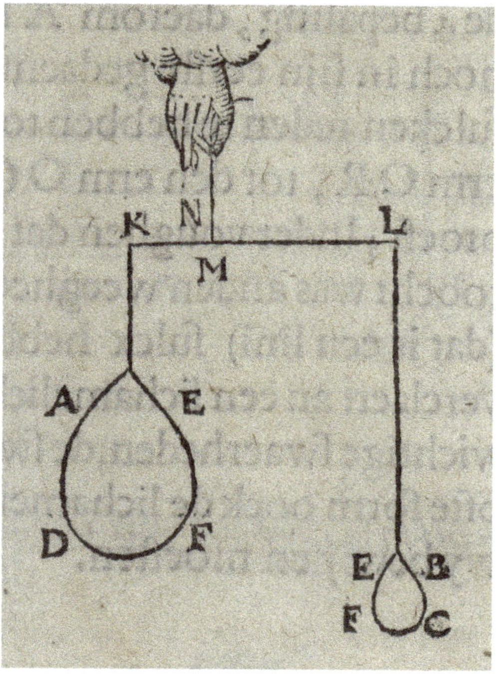

FIGURE 6.2 Generalizing to arbitrarily shaped bodies
SOURCE: GHENT UNIVERSITY LIBRARY, BIB.ACC.056517

determined by the center of gravity (this is further elaborated upon in propositions 6, 7 and 8 of Stevin's treatise). In other words: the center of gravity is the unique point at which we can imagine all of the body's weight to be concentrated – and it is the latter idea that underwrites both steps of the demonstration.

The demonstration of the law of the lever can be usefully understood along the following lines.[19] It is assumed (1) that all bodies possess a unique center of gravity (in which, for the reasons stated, their weight can be thought to be concentrated), and (2) that we can all agree that for symmetrical bodies this point is located in the geometrical center. The mathematical proof then shows how to determine the location of the center of gravity of composed bodies: it is the place from which the distances will be inversely proportional to the weights of the composing bodies. The proof thus brings us from an almost purely qualitative characterization of the center of gravity to a precise quantitative determination.

In the appendix to the volume containing *The Principles of the Art of Weighing*, Stevin criticized an alternative demonstration of the law of the

19 See Van Dyck, 'Epistemological Foundations', for more discussion.

lever.[20] Taking inspiration from the Aristotelian *Mechanical Problems*, some authors had claimed that the cause of unequal bodies being *evenstaltwichtigh* could be found in the circles they would describe under the conditions stated: the smaller body would describe a proportionally greater circle, and the greater body a proportionally smaller circle. Stevin merely pointed out that since bodies in equilibrium do not move, these postulated motions cannot be part of the cause of the equilibrium. Starting with Duhem, he has often been criticized for not having seen the potential fruitfulness of this approach, which can be supposed to have led to the principle of virtual velocities.[21] For our purposes it is more interesting to point out that Stevin also emphatically stated that the true cause of *evenstaltwicht* was to be found 'in that which has been proved about it mathematically' in the demonstration of the law of the lever. Crucially, he added that these other authors' inability to devise the causes was related to the fact that they could not 'reduce to the form of an Art of Weighing' (*sonder te crijghen form van Weeghconst*) – thus explicitly relating the goal of reduction to art to knowledge of causes.

The preceding analysis makes clear how, according to Stevin, the mathematical demonstration would have been able to uncover a cause. Two bodies are *evenstaltwichtigh* because their configuration is one that can be mathematically transformed into one where their common weight is distributed evenly around the point of suspension; and the possibility of such a transformation rests on the fact that all bodies do have a unique center of gravity. In this way, the mathematical law is nothing but the encoding of the similarity between these different configurations, and it is this similarity that provides the appropriate causal grounding. We will further analyze how this idea relates to the possibility of a successful reduction to art in the next two sections, but first we will consider the second foundational proposition underlying this reduction, the law of the inclined plane.

The cause that grounds *evenstaltwicht* on an inclined plane can be understood along the same lines, although in this case the underlying similarity is easier to miss. Stevin famously demonstrated that when two connected bodies are lying on two inclined planes that form an upright prism, equilibrium will necessarily follow when their weights have the same proportion as the lengths of the planes. The demonstration consists in showing that if this were not the case, a perpetual motion would follow for a wreath lying around this prism consisting of connected equal spheres lying at equally spaced distances (since each postulated motion of the wreath would reproduce the original situation,

20 Stevin, *Principal Works*, I, 507–509.

21 Duhem, *Origines*, 267.

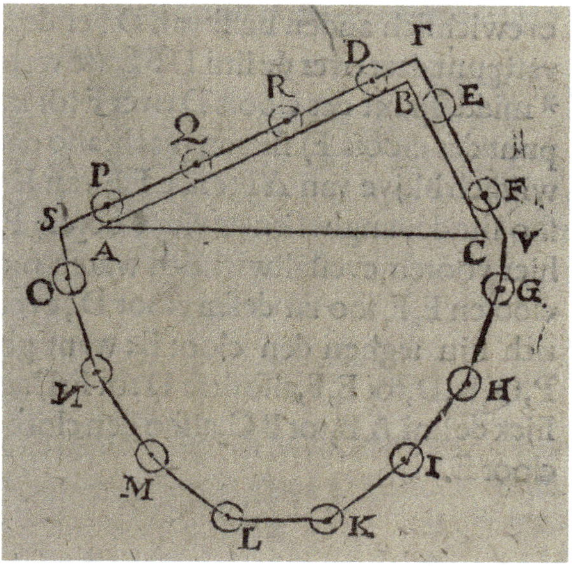

FIGURE 6.3 Demonstrating the law of the inclined plane
SOURCE: GHENT UNIVERSITY LIBRARY, BIB.ACC.056517

with the number of spheres lying on each of the inclined planes determining the relevant weights stated in the condition; see figure 6.3, which again allows one immediately to see that this is the case). Crucially, the perpetual motion that is excluded by Stevin is one that starts spontaneously, a fact all too often passed over in silence by commentators, but explicitly stated by Stevin. This is important, because we are dealing with heavy bodies, which by definition have an intrinsic tendency to descend; and as we saw, if their resulting motion is limited by the presence of fixed constraints, they also have a preferred position where they will come to rest, as determined by the relative position of their center of gravity and the constraints. The wreath of spheres lying around the prism can be considered a composite, heavy body: claiming that it could spontaneously start its motion by its weight, but that it would not reach a position of equilibrium, would amount to claiming that this body was completely dissimilar to bodies characterized by a center of gravity. It is the latter fact that Stevin judged to be absurd. It is thus again the underlying similarity of this situation to that of a body suspended in or above its center of gravity that gives this proportion its causal grounding.[22]

22 In 1644, Evangista Torricelli would be the first to render this similarity completely transparent by reducing the proof of the law of the inclined plane to a direct application of the

4 Disclosing the Causes of the Properties of the Balance

Stevin's complete reduction to the form of art rested on a preliminary division of the field of operations, as shown in the dichotomous table at the very beginning of *The Principles of the Art of Weighing*. It announces that the text will deal with four kinds of basic situations: vertical lowering weights, vertical lifting weights, oblique lifting weights and oblique lowering weights. The proportions stated by the different propositions will characterize the forces exerted by these weights in terms of the weight of the body to be held in equilibrium by them and the geometry of the situation, where the main division is between the forces exerted in a vertical direction and those exerted in an oblique direction. These propositions are based on the two fundamental propositions just analyzed: the law of the lever for vertical forces, and the law of the inclined plane for oblique forces, respectively.

As we have seen, the demonstrations of these two laws were implicitly based on physical facts involving bodies' centers of gravity, and more specifically on considering the body composed of any two bodies as also having a single center of gravity that remains invariant through the replacements (for the law of the lever) or through the counterfactual motion (for the law of the inclined plane). The conclusions were anaphorically formulated in terms of each composing body's *staltwicht*, though, which has the important advantage that it allows one to focus on the effect that each of the bodies has on the other; i.e., it allows one to interpret the mathematically demonstrated proportions as characterizing forces. In a further move, Stevin abstracted away from the inclined plane as a physical device, until it only represented the triangle of forces implicit in the law of the inclined plane that gives a general recipe for determining forces exerted in an oblique direction (see figure 6.4).[23]

Based on his two fundamental propositions, Stevin was in a position to 'disclose the causes of the properties of the balance from their grounds'.[24] This disclosure further extends the realm of relevant similarities, by exploiting the power of proportional reasoning. Manipulating mathematical proportions, it becomes easy to translate the basic cases into other cases with different geometrical configurations of forces and weights – but still characterized by

---

general principle that the center of gravity of any composite body always takes the lowest position allowed by the constraints of the system: when the stated proportion holds, the center of gravity is already located in the lowest position possible. See Torricelli, *Opera geometrica*.

23 See Dijksterhuis, *Simon Stevin*, 53–57 for further comments and analysis.

24 Stevin, *Principal Works*, I, 151.

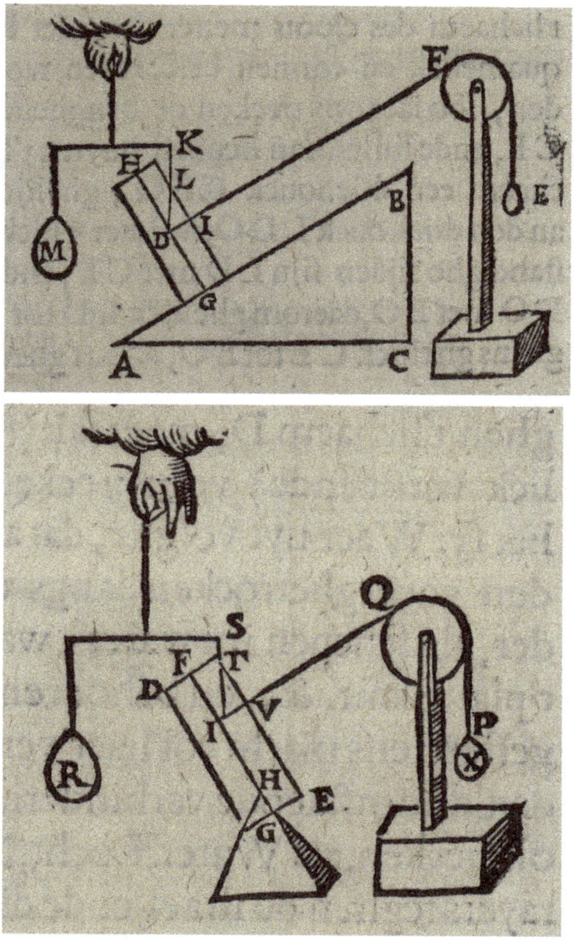

FIGURE 6.4 According to the law of the inclined plane, there is equilibrium if the weight *M* of the body is to the weight of *E* as *AB* is to *BC*, which is in the same ratio as *LD* to *DI*. This implies that the physical plane can be neglected in further considerations, as is done in the figure below, where triangle *ITV* encodes the relevant ratios
SOURCE: GHENT UNIVERSITY LIBRARY, BIB. ACC.056517

equilibrium. Perhaps the most relevant instance is given by proposition 20, which shows how to relate vertical lifting weights (grounded in the law of the lever) to oblique ones (grounded in the law of the inclined plane):

> If there is in the axis of the prism a fixed point and a movable point, at which it is kept in a certain position by a vertical lifting weight and an

> oblique lifting weight: as the vertical lifting line is to the oblique lifting line, so is the vertical lifting weight to the oblique lifting weight.[25]

The accompanying diagrams show how to recognize the relevant elements of the proposition in different situations, situations that are all essentially similar to the mathematical perspective opened up by Stevin's new art of weighing (see figure 6.5). Looking at the diagrams, the reader is taught immediately how to isolate the crucial ratios that determine equilibrium – thus disclosing the causes.

Proposition 20 brings a further, related aspect of Stevin's reduction to our attention. In the preface to his work, where he extolled the superior qualities of the Dutch language, he had referred to the formulation of this proposition as a crucial example. Stevin's defense of Dutch as a scientific language was obviously connected both to a broader reassessment of the status of the vernacular in the late Renaissance, and to the specific political context of the newly-emerging nation in which he wrote his books. But it is important not to lose sight of the fact that his insistence on the cognitive reasons for this superiority had an intrinsic connection to his ideal of reduction to art. Dutch was the most perfect language, according to Stevin, because of the combined effect of the high prevalence of monosyllabic words that correspond to simple things in nature and the possibility of combining these words to form new words, the meaning of which is perfectly clear from that of their composing parts.[26] This made it eminently suitable for teaching the arts, as shown by the ease with which complex propositions such as the one presently under consideration could be perspicuously expressed. The compositional nature of Dutch allowed the terms to be permuted in the right order: 'Ghelijck rechtheflini tot scheefheflini, alsoo rechthefwicht tot scheefhefwicht'.[27] As in the proofs of the two basic propositions, the printed diagrams and the words of the accompanying written text jointly rendered the relevant structures transparent for the reader.

It is no coincidence that Stevin chose the formulation of a proportion as his crucial example. Proportional reasoning was the central engine of his reduction to art, by which one could show that a whole field of operations (in this case those involving heavy bodies) could be covered by the same principles. Moreover, Stevin had already introduced the term 'proportion' itself as a general example of the superiority of Dutch in the *Bewysconst*. The Dutch term *evenredenheyt* expressed its content directly: 'Evenredenheyt is de Evenheyt

25 Stevin, *Principal Works*, I, 195.

26 Stevin, *Principal Works*, I, 65. See Van der Wal, 'Logic, Linguistics, and Simons Stevin' for further discussion.

27 Stevin, *Principal Works*, I, 91.

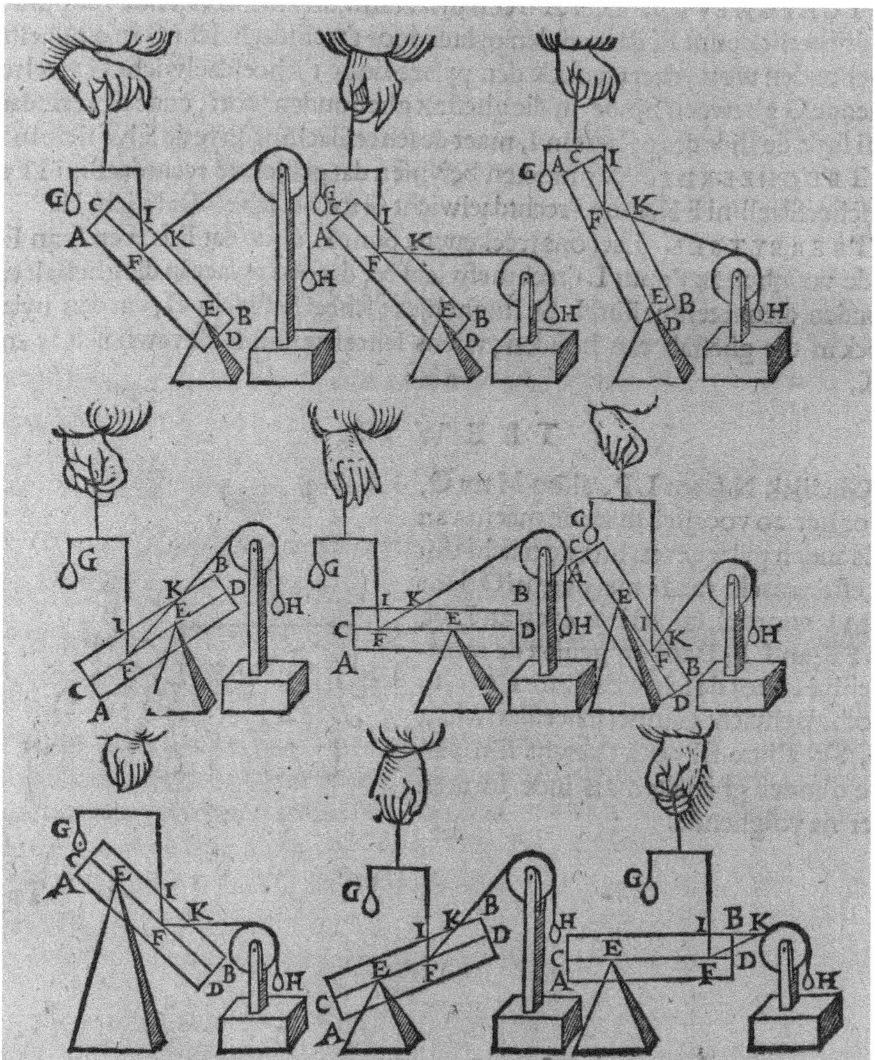

FIGURE 6.5 Systematically relating vertical forces to oblique forces
SOURCE: GHENT UNIVERSITY LIBRARY, BIB.ACC.056517

der Redenen' (in contrast to, e.g., its abstruse Latin rendering: '*Proportio est Rationum similitudo*'). As a result, the term could be used directly to guide correct applications.[28] Formulating the goal as a search for relevant *evenredenheden*

28 Stevin, *Dialectike*, 156. On *evenredenheyt*, see also the chapter by Cohen and Kursell in this volume.

immediately made it clear that one was looking for ratios that remain invariant throughout different configurations. Significantly, Stevin often used *reden* (which primarily means 'reason' or 'ratio') as almost directly synonymous with *oirsaeck* or cause (as is possible in many languages).[29] An *evenredenheyt* could thus be understood as simultaneously expressing a mathematical relationship and a causal equivalence.

5 Formal Causality and the Arts

Stevin's way of using of causal language to describe his mathematical art of weighing was perfectly consistent with his denial of real definitions expressing underlying essences in the *Bewysconst*. Indeed, in that same work he also approached causal structures from the perspective of human agency rather than that of essential powers inhering in natural things.[30]

Stevin opened the *Bewysconst* with definitions of the four traditional causes: material, formal, efficient and final; but he used them exclusively to refer to artificial constructions where the causes jointly result in an effect that he called the *werck* or *daet*. Stevin silently transformed the fourfold distinction from an ontological description of the structure of all natural events into a pragmatic scheme that could be used to characterize everything that is done by art.[31] To give an example, similar to Stevin's own of the beer barrel: a mechanical practitioner (efficient, or 'making' cause) constructing a balance (effect or *daet*) by imposing a certain geometrical configuration (formal cause) on some pieces of iron (material cause) to be able to weigh bodies accurately (final cause).

According to this scheme, the theoretical art of weighing is completely devoted to investigating the formal cause.[32] Again, the explicit definition that

29 In the preface to his book, for example, Stevin contrasts (in the translation given in the *Principal Works*) 'things we understand through their causes [*oirsaeken*] and those whose causes [*redenen*] are unknown to us', Stevin, *Principal Works*, I, 59.

30 Stevin, *Dialectike*, 4–9.

31 Of course, Aristotle also used the four causes to characterize artificial objects, but only as a way of motivating the fourfold division, which he then extrapolated to what were the central cases for him: the causes of natural events. This latter step is completely absent from all of Stevin's writings.

32 Machamer, 'Galileo and the Causes' already suggested relating the new mathematical sciences of the sixteenth and seventeenth century to formal causality, but without the level of detail offered here. Biener and Schliesser, 'Certainty' and Schliesser, 'Newtonian Emanation' further explore this in relation to the case of Newton, an author who was much more of a metaphysician than Stevin. Closer to Stevin, Mancosu, *Philosophy of Mathematics*, chapter 1, has discussed how appeals were regularly made to formal

Stevin gave of a formal cause superficially appears as though it focuses on purely verbal questions: 'Form is the cause out of which the matter receives the name of some species'.[33] In this case, the resulting material object will be called a balance, because the geometrical structure imposed upon the matter makes it suitable for performing a *specific* kind of operation, i.e., weighing. It is the latter suitability that is guaranteed by the theoretical analysis of 'the causes of the properties of the balance'. All 'formally' possible configurations of the balance are determined by the realm of relevant similarities underlying the propositions proven in the art of weighing that allow these configurations to play their causal role. Or to put it differently: these similarities provide appropriate *causal* grounding because they determine the extension of the species 'configurations in equilibrium' to which all balances belong. This does imply that the causal status of claims ultimately depends on our classificatory interests. Discovering causal factors comes down to identifying the structural features of systems that allow them to be classified in specific ways.

Stevin's causal scheme from the *Bewysconst* also gives us more insight into the relation between *Weeghconst* and *Weeghdaet*. Whereas the theoretical book on the art of weighing was limited to an abstract study of the formal cause, the other book, which was published simultaneously (its title can be translated as the 'The Practice of Weighing', following Stevin's own Latin annotation), dealt with the effects that could be achieved when the formal configurations uncovered by the theoretical investigation were imposed on matter by practitioners in order to achieve their goals.

6 Coordinating Theory and Practice, Cause and Effect

Stevin's book on the *Weeghdaet* was devoted to teaching its readers to recognize the theoretically established proportions in material constructions. It shows how our basic ways of dealing with heavy objects in practice exemplify certain configurations of forces, helping us to achieve our goals. By translating practical challenges in terms of the *staltwicht* of objects involved, it becomes possible to make efficient decisions regarding the proper actions to be taken or instruments to be used in diverse circumstances (see, e.g., figure 6.6). By reducing mechanical knowledge to the form of art, Stevin thus rationalized the structure of our practical actions in a way that also allows us to extend our

causality in an attempt to give mathematical demonstrations a causal status in the debates surrounding the *Quaestio de Certitudine Mathematicarum*.

33 Stevin, *Dialectike*, 6.

FIGURE 6.6 Translating a practical challenge into theoretical terms
SOURCE: GHENT UNIVERSITY LIBRARY, BIB.ACC.056517

initial know-how to ever-new situations that can be brought under our control. In several places, Stevin commented that he only gave a few examples, and that these should suffice to clarify how to effect the coordination between theory and practice. After this, one should be able to translate other situations at will by drawing the appropriate diagrams.[34]

As a reduction to art, Stevin's theoretical treatment started with a practical field that was already partly articulated by know-how. For example, he commented on how people had noticed the effect of a longer lever arm in practice,

34 E.g., Stevin, *Principal Works*, I, 333. Van den Heuvel. 'Architectuurmodellen', provides further analysis of Stevin's visual way of reasoning, with a focus on his work in architecture.

before gaining any theoretical insight into the underlying mathematical proportions.[35] This prior experience of heavy bodies helped delineate the field by drawing attention to crucial aspects that should be incorporated into the theoretical treatment. This is why it was so crucial to Stevin to appeal to very simple, qualitative experiences that could help ground his theory. After having stated propositions 6, 7 and 8 in the *Weeghconst* (the crucial importance of which was already stressed above, since they describe the different kinds of equilibrium that result from the relative position of a body's center of gravity with respect to its fixed point), Stevin added that 'If anyone should wish to see the contents of the preceding three propositions by some experience, he might take a ruler of wood'. By making some holes in the ruler and making it freely pivot around a needle that had been put through these holes, one would immediately see the qualitative truth of these propositions for this simple object.[36]

Thus, it is no coincidence that the first proposition of the *Weeghdaet* explained in detail how to practically (*werckelick*) find any body's center of gravity through a combination of simple operations of suspension. Given that it is the most important notion in the theoretical elaboration, it was crucial that everybody could acknowledge its practical existence. It is revealing to compare this first proposition to the first proposition of Stevin's text on practical geometry, the *Meetdaet*, which explained how to construct a straight line using a number of different methods.[37] We can see this proposition as providing instructions on how materially to perform the crucial elementary operation that was already given purely theoretically by Euclid's first postulate. This shows that more complicated, theoretically-established propositions can 'in principle' also be applied to empirical situations, since their validity depends on this postulated possibility. Exactly the same role is played by the first proposition of the *Weeghdaet*. If you can constructively determine any given body's center of gravity, you can 'in principle' construct all configurations of material bodies in equilibrium. As in the case of geometry, reducing a practice to the form of art involved the identification of elementary operations that allowed one to ground the coordination of the practice with an elaborated theoretical structure.

Note how Stevin defined theory in the later text on the 'restoration of the Age of the Sages': 'Theory [*Spiegheling*] is an imaginary operation [*verdochten handel*] without natural matter'.[38] He had likewise opened *The Principles of the*

35 Stevin, *Principal Works*, I, 321–323.
36 Stevin, *Principal Works*, I, 143.
37 Stevin, *Principal Works*, II, 771.
38 Stevin, *Principal Works*, III, 619.

Art of Weighing with the characterization of its subject matter as 'gravity, dissociated in thought [*duer t'ghedacht ... gheweert*] from physical matter',[39] and defined the center of gravity as 'the point such that if the body is conceived to be suspended [*door ons ghedacht hanghende*] from it, it remains at rest in any position given to it'. He commented on the latter definition by stressing its imaginary character: it is a suspension that is only effected in one's mind.[40] Following Stevin's use of the Dutch term *spiegheling* (speculation or reflection) for theory, we can understand these imaginary operations as ideally 'mirroring' the structure of practical activities: theoretical structures are in the first place linked to possibilities for acting – and we can see the stylized hands suspending the geometrical bodies in Stevin's theoretical treatise (see the figures above) as striking visual reminders of this crucial link. It is only through this speculative reduction to art that the structure of our practical activities can for the first time be fully and transparently articulated.

The second proposition of the *Weeghdaet* played a similar role to the first one, but it also brought a further kind of consideration into play. This proposition described how to construct 'a most perfect balance', one that would remain at rest in any position it was given, thus directly exhibiting the defining property of a body's center of gravity for bodies of more complex shape than the simple wooden ruler already alluded to in the *Weeghconst*. But this object also had its own practical use as a measurement instrument for establishing precise quantitative relationships. The conditions for such precise measurements were further analyzed in propositions 2–6. In the last of these propositions, Stevin described a small device that he called an oblique balance, with which 'to make us see, examine, and understand through visual experience the truth of the propositions on the proportionality of oblique weights'. He thus introduced an instrument for the measurement of oblique forces that could play the same role as the 'most perfect balance' did for the vertical forces.[41]

These propositions were absolutely crucial to establish in full the coordination between theory and practice. The conditions required for high precision instruments were theoretically established, but could never be perfectly verified in practice due to the limitations of working with material objects. Stevin explicitly allowed that practitioners had to decide how closely they needed to approximate these conditions for the purpose at hand;[42] but his detailed description, with an eye to the empirical challenges encountered, clearly

39 Stevin, *Principal Works*, I, 95.

40 Stevin, *Principal Works*, I, 101.

41 Stevin, *Principal Works*, I, 321.

42 Stevin, *Principal Works*, I, 306, 318.

revealed that he supposed that these approximations could in principle always be improved upon. In his later reflection on reduction to art in the text on the 'restoration of the Age of the Sages', he formulated the relationship between theory and practice as follows: 'The property and the end of theory is that it furnishes a sure foundation for the method of practical operation, in which by closer and more painstaking care one may get as near to the perfection of the theory as the purpose of the matter requires for the benefit of man'.[43] The possibility of this ever-closer approximation was thus an essential part of the successful reduction to art, as it guaranteed the relevance of the theoretically proven, mathematically exact proportions.

After the first six propositions, Stevin considered situations for which such approximation could nevertheless not be expected. In the preface, he had already warned his reader not to mistake this for a defect of the theory.[44] When machines are put into motion to realize useful effects, this inevitably brings material impediments to this motion that cannot be eliminated in any way.[45] But given the successful establishment of the first six propositions, this should no longer worry us unduly. We are now in a position legitimately to ascribe the differences between theoretically proven proportions and practical results to the effect of material disturbances on the 'ideal' phenomenon.

The theoretical proportions remain relevant for the latter kind of situations, since we can trust them to articulate correctly the structure of the situation we are analyzing. Consider, for instance, the situation in figure 6.7: if the force applied by the man is to the weight of the load *H* as *KG* is to *EG* (the radius of the axle around which the rope pulling the load is being coiled), this will in theory keep the load in equilibrium. Again, in theory, the smallest additional force would suffice to start raising the load – but in practice this will of course not be the case, because of the material impediments that need to be overcome. We do not know in advance how much more force we will actually have to exert to move the load, but we do know that the force determined by the theory is the minimum force: it puts a theoretical limit on the operation of the machine. The same theoretical consideration also tells us how to identify the relevant causal factors that might be taken into consideration if we want to adapt the machine to new circumstances (e.g., if we wanted to move a heavier load or if less force were available).

43 Stevin, *Principal Works*, III, 619.

44 Stevin, *Principal Works*, I, 297–299.

45 See Van Dyck, 'Motion', and Van Dyck, 'Idealization', for further analysis of this claim and some of its consequences.

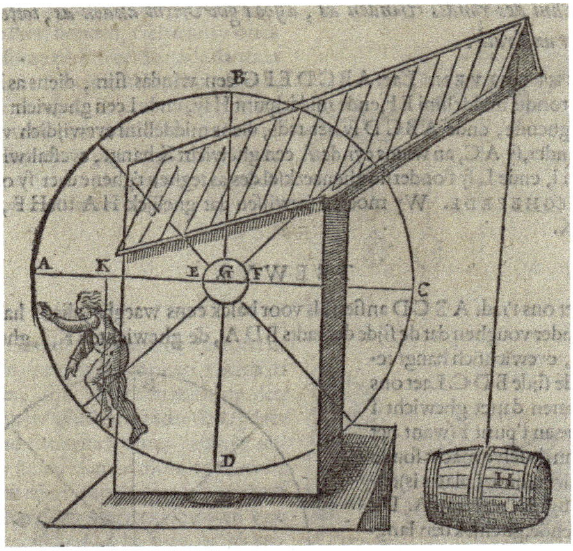

FIGURE 6.7 Putting machines in motion
SOURCE: GHENT UNIVERSITY LIBRARY, BIB.ACC.056517

The last proposition of the *Weeghdaet* provides the most spectacular example of this way of structurally linking theory to practical challenges. It introduced a machine that Stevin daringly called *het Almachtich* ('the Almighty').[46] This machine was proposed to actualize the theoretical possibility of producing a multiplication of power that could in principle be indefinitely extended, such that any conceivable practical challenge could be met. The machine basically consists of a system of geared cogwheels that multiply an initial force delivered with a crank and transmit the output force using a windlass. Stevin stressed the practical advantages of his design over alternatives, due to the sturdiness of its construction, the continuous nature of its output motion, and the ease with which parts of the system could be disengaged, depending on the amount of multiplication that was needed for the task at hand.[47] Yet it is not simply a practical machine, as is made clear by its provocative name. Having illustrated

46 Jan de Groot made clear that the connotations of the name were not accidental, by rendering it as *Pantocrator* in his laudatory Latin poem that opened Stevin's treatise. Snellius's translation of Stevin's treatise in the Latin version of the *Wisconstighe Ghedachtenissen* published in 1605–1608 opted for the more cautious neologism of *Pancratium*; see Stevin, *Mathematicorum Hypomnematum*. Girard's French translation published in 1634 dropped all potentially problematic references altogether and simply referred to it as a *cricq* ('crank'); see Stevin, *Oeuvres mathématiques*, 480–481.

47 The States of Holland had already granted Stevin a patent for a device with which to haul ships over a dam in 1584, plausibly a version of the *Almighty*; see Dijksterhuis, *Simon*

its use to haul a ship over a dam, Stevin went through elaborate calculations to show the effects of adding more and more gears to the system, until he finally concluded that this would allow a mere child to move the earth, thus making true on Archimedes' famous boast. But, he added, continually operating the machine over a period of ten years would have only lifted the earth a distance of 1/2,400,000,000,000,000 feet.

The clearly fantastic nature of this last example raises a question about its place in Stevin's *Weeghdaet*. The general focus of the treatise is on the challenges posed by concrete materiality, but it is obvious that this postulated motion would never take place, due to all the imperfections that would arise in the transmission of the motion and the stresses that would be put on the whole device. The discussion seems to have been included with a dual purpose.[48] On the one hand, the play with immense numbers (all printed in full, some with more than thirty digits) was no doubt intended to dazzle the reader. On the other hand, and more importantly, the comments on the time spent and distance moved that accompanied these numbers simultaneously pointed in another direction: they warned the reader that multiplication of force must always be balanced against other factors. And this last point does have important practical implications. It is no coincidence that having gone through these spectacular extrapolations, Stevin ended his book by returning to some of the *Almighty*'s practicable applications, by giving some examples of tasks for which it would be suitable – but added, tellingly, that he would not enter into any further discussion of these, since the reader who had followed his discussion would know best how to adapt the machine 'according to the occasion'.[49] This adaptation to particular occasions crucially includes considerations involving the balance between multiplication of available force, distance moved and time spent: depending on the challenge, a different balance will need to be struck. Theory shows that such choices have to be made, but it does not dictate which one is right. At this point, the practical intelligence of the practitioner necessarily takes over.

Stevin, 91. Stevin's dedication of the *Weeghconst* to Rudolph II refers to a patent that he would request for an 'unheard-of' contrivance, for the understanding of which the book contained the necessary theory (see Stevin, *Principal Works*, I, 65), which can again be interpreted as referring to the *Almighty*; and a patent granted to him a few months after the publication of the books on mechanics included the machine in a new device for dredging canals, with an explicit reference to the proposition from the *Weeghdaet* (see Stevin, *Principal Works*, V, 35).

48 Stevin's discussion would be taken up by later authors such as Marin Mersenne and John Wilkins, who would again stress both sides, but with somewhat different purposes (see Van Dyck and Vermeir, 'Varieties of Wonder', for a discussion of Wilkins).

49 Stevin, *Principal Works*, I, 373.

This last point highlights the importance of 'the occasion' for correctly understanding the relationship between theory and practice.[50] It is never a matter of simply 'applying' theory, but rather one of guiding the process of situationally determined, practical decision-making by mathematically explicating the structure of a range of possible actions and their effects.

All the foregoing should also allow us to better gauge the place of mathematical reasoning in Stevin's proposed reduction to art. We start from a set of qualitative observations that draw our attention to a number of significant phenomena. Theoretical speculation shows that these qualitative observations in principle allow the establishment of precise quantitative relationships. These relationships can indeed be exemplified to a very good approximation in well-determined material circumstances. Mathematical reasoning furthermore allows the identification of a much larger class of situations that are – *in principle* – also similar to these cases. Even if precise quantitative relations can no longer be empirically exhibited for this larger class, mathematics has still played a crucial role. It has allowed the identification of a causal structure underlying the messy empirical situations that puts theoretically-determined constraints on what we can and cannot put into effect.

There is one last aspect that needs to be commented upon. We saw how the existence of a unique center of gravity was absolutely crucial for the theoretical articulation of the causal structures in question. Stevin was well aware that this could only be assumed under specific circumstances.[51] If we were to take into consideration that all material bodies' 'power of descent' is actually directed towards the center of the world, then we would have to conclude that only perfectly spherical bodies have a unique center of gravity; for all other bodies, there would be no longer a 'point such that if the body is conceived to be suspended from it, it remains at rest in any position given to it' (see figure 6.8, where the balance would need to be supported at a different point to be kept in equilibrium when no longer oriented horizontally). We can safely ignore this complication, though: the scale of our instruments is so insignificant with respect to the size of the earth that any convergence of directions must remain imperceptible. But this does imply that the causal structures uncovered by Stevin's art of weighing must again be relativized to a specific domain of reality that is singled out by the demands of our practical and classificatory goals. If we were to consider bodies on a different scale, it would no longer be possible to establish the relevant configurations described in the treatise as similar.

50 See Vérin, *Gloire*, for a classic study of this aspect of engineering practice in the early modern period.

51 Stevin, *Principal Works*, I, 113–115.

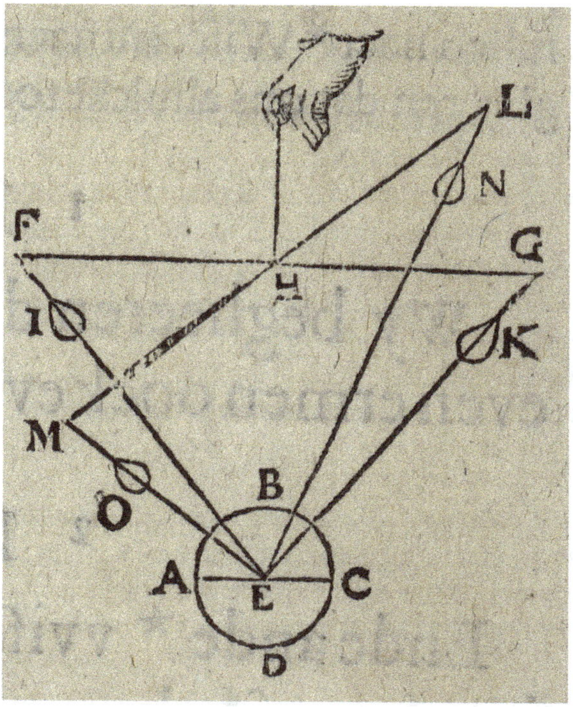

FIGURE 6.8 Non-uniqueness of center of gravity for weights along converging directions
SOURCE: GHENT UNIVERSITY LIBRARY, BIB.ACC.056517

7 Conclusion: 'This Long-Hidden Matter'

Stevin was rightly proud of having successfully reduced practical dealings with mechanical instruments to the form of an art. Archimedes' highly abstract mathematical treatise on the equilibrium of plane figures had been well known since its publication in 1544, and some important contributions to the medieval *scientia de ponderibus* had also been published by Tartaglia and had found their way into Cardano's popular writings, for example. Sixteenth-century Europe was also characterized by a rich material culture where skillful production and use of diverse mechanical contrivances played a central role. What was lacking was a way of clearly linking the theoretical treatment to this practical knowledge, which is exactly what Stevin excelled in doing.[52]

52 The only work comparable in this respect to Stevin's was Guidobaldo del Monte's *Liber Mechanicorum*, which had been published in 1577, but which probably was unknown to Stevin (and which was conceptually less accomplished, as it lacked the resources to deal with forces exerted in any other direction than the vertical). See Van Dyck, 'Motion', for

As I hope to have shown, this articulation of mathematically-structured theory on material practice also determined Stevin's perspective on how to uncover causal factors in the world. The main goal of his reduction to art was to show how a whole array of seemingly diverse operations could be seen as characterized by an underlying systematic unity – and this unity could also be seen as expressing a causal structure. In this way, the ideal of reduction to art implicitly introduced a strikingly un-Aristotelian epistemology of causal knowledge, in which causes could be identified through a combination of mathematical reasoning that isolates invariant ratios, and material construction that shows how to exemplify these mathematical structures.

Most of the themes we have been discussing came together in the opening paragraph of Stevin's dedication of his *Principles of the Art of Weighing* to Emperor Rudolph II:

> That number, magnitude, and weight are in all existing things inseparable attributes, full of profound and useful properties, is attested not only by several scholars, but it is also known to all by experience in all things. It is also known that the first two, to the great profit of man, have reached the form of written arts, viz. arithmetic and geometry; but not so weight, because its fundamental properties have remained hidden from our predecessors. It is true that with regard to vertical weights it has been observed by experience that two *evenstaltwichtigh* gravities are proportional to their arms. But they thought that this proportionality was due to the circles described about the fixed point by the extremities of the arms. From which, as is usual with errors, there followed no knowledge of the causes. As to the oblique weights, these were not known at all, with the result that this subject matter could not receive the form of an art like the others. But when the situation changed, and this long-hidden matter was revealed through its fundamental principles, it at last reached this status, in the formal shape in which it is here being dedicated to Your Imperial Majesty.[53]

Stevin's recurring focus on the proper formal features of the 'written' presentation of any art parallels the role of formal causality as the *Weeghonst*'s main subject matter, as explained in Section 5. In both cases, the central issue is how to dispose given matter in a structure that allows one to achieve certain effects. The discussion on the 'restoration of the Age of the Sages', summarized

some further comments on the similarity between Stevin and Guidobaldo, and Bertoloni Meli, *Thinking with Objects*, for Stevin's place in the history of mechanics.

53 Stevin, *Principal Works*, I, 55.

in Section 1 above, can thus be seen as intended as a description of the formal cause of all theoretical arts.

This resolutely structural perspective is in line with Stevin's translation of *theoria* as *spiegheling* (reflection or speculation), as noted above in Section 6. As Stevin (who also wrote a treatise on perspective and captoptrics) was perfectly well aware, one can never reflect the whole situation in which one finds oneself: by orienting one's mirror, one delineates that part of the situation which is of prime interest; as we saw in Section 2, this was the role of Stevin's theoretical definitions. The translation into the reflected form is then brought about by the propagation of light. The reflection shows a projection of the spatial relationships on a two-dimensional surface, where the relationships are presented purely in the light, abstracted from their original material *relata*. It is the light that writes the 'theoretical' or 'speculative' proportional relations, leaving formal traces that can be translated into action by the person who has learned to use the mirror.[54] Abstracting away from matter to 'reveal' the formal relations that can be exploited in further practice is exactly what mathematics did for Stevin, as we saw in Sections 3 and 4: it 'brought to light' what had remained 'hidden' in the obscurity of non-reflected forms of action.

Bibliography

Bertoloni Meli, Domenico, *Thinking with Objects: The Transformation of Mechanics in the Seventeenth Century*, Baltimore: Johns Hopkins University Press, 2006.

Biener, Zvi and Schliesser, Eric, "The Certainty, Modality, and Grounding of Newton's Laws", *The Monist* 100, 3 (2017) 311–325.

De Pace, Anna, *Le matematiche e il mondo. Ricerche su un dibattito in Italia nella seconda metà del Cinquecento*, Milan: FrancoAngeli, 1993.

Descotes, Dominique, "La rhétorique mathématique dans la Statique de Simon Stevin", in: eds. Catherine Secretan and Pim den Boer, *Simon Stevin. De la vie civile. 1590*, Lyon: ENS Editions, 2005, 131–180.

Dijksterhuis, Eduard Jan, *Simon Stevin*, The Hague: Martinus Nijhoff, 1943.

Dijksterhuis, Eduard Jan, *Simon Stevin: Science in the Netherlands around 1600*, The Hague: Martinus Nijhoff, 1970.

Duhem, Pierre, *Les Origines de la Statique*, Tome Premier, Paris: Hermann, 1905.

54 Stevin explicitly comments on the link between theoretically-established propositions and mirror-images to justify his use of the term *spiegheling* in the posthumously published *Materiae Politicae*, 180–181.

Heuvel, Charles van den, "Architectuurmodellen en voorstellingen van een logische orde: Stevin en Ramus", *Feit & Fictie* II, 1 (1994), pp. 95–111.

Machamer, Peter, "Galileo and the Causes", in: eds. Robert E. Butts and Joseph C. Pitt, *New Perspectives on Galileo*, Dordrecht: Springer, 1978, 161–180.

Mancosu, Paolo, *Philosophy of Mathematics and Mathematical Practice in the Seventeenth Century*, Oxford: Oxford University Press, 1996.

Naets, Jurgen, "How to Define a Number? A General Epistemological Account of Simon Stevin's Art of Defining", *Topoi* 29 (2010) 77–86.

Rose, Paul Lawrence, *The Italian Renaissance of Mathematics: Studies on Humanists and Mathematicians from Petrarch to Galileo*, Geneva: Droz, 1976.

Schliesser, Eric, "Newtonian Emanation, Spinozism, Measurement and the Baconian origins of the Laws of Nature", *Foundations of Science* 18, 3 (2013) 449–466.

Stevin, Simon, *Dialectike ofte Bewysconst*, Leiden: Christoffel Plantijn, 1585.

Stevin, Simon, *Mathematicorum Hypomnematum ...: Tomus Quartus: De Statica*, Leiden: Ioannis Patius, 1605.

Stevin, Simon, *Wisconstige Ghedachtenissen. 1 Vant Weereltschrift*, Leiden: Ian Bouvvensz., 1608.

Stevin, Simon, *Les Oeuvres Mathematiques de Simon Stevin de Bruges*, Leiden: Bonaventure & Abraham Elsevier, 1634.

Stevin, Simon, *Materiae Politicae. Burgherlicke Stoffen*, Leiden: Iustus Livius, 1648.

Stevin, Simon, *The Principal Works of Simon Stevin. Vol 1: General Introduction. Mechanics*, Amsterdam: Swets & Zeitlinger, 1955.

Stevin, Simon, *The Principal Works of Simon Stevin. Vol 2: Mathematics*, Amsterdam: Swets & Zeitlinger, 1958.

Stevin, Simon, *The Principal Works of Simon Stevin. Vol 3: Astronomy. Navigation*, Amsterdam: Swets & Zeitlinger, 1961.

Stevin, Simon, *The Principal Works of Simon Stevin. Vol. 5: Engineering. Music. Civic Life*, Amsterdam: Swets & Zeitlinger, 1966.

Torricelli, Evangelista, *Opera geometrica*, Florence: typis Amatoris Masse & Laurentii de Landis, 1644.

Van Dyck, Maarten, "On the Epistemological Foundations of the Law of the Lever", *Studies in History and Philosophy of Science Part A* 40, 3 (2009), pp. 315–318.

Van Dyck, Maarten, "Motion and Proportion in Simon Stevin's Mechanics", in: eds. Marcus Adams, Zvi Biener, Uljana Feest and Jacqueline A. Sullivan, *Eppur si muove: Doing History and Philosophy of Science with Peter Machamer*, Dordrecht: Springer, 2017, 21–37.

Van Dyck, Maarten, "Idealization and Galileo's Proto-Inertial Principle", *Philosophy of Science* 85, 5 (2018) 1–11.

Van Dyck, Maarten and Vermeir, Koen, "Varieties of Wonder: John Wilkins' Mathematical Magic and the Perpetuity of Invention", *Historia Mathematica* 41, 4 (2014) 463–489.

Wal, Marijke J. van der, "Logic, Linguistics, and Simon Stevin in the Context of the Sixteenth and Seventeenth Centuries", in: ed. K.R. Jankowski, *History of Linguistics*, Amsterdam: John Benjamins, 1993, 147–156.

Vérin, Hélène, *La gloire des ingénieurs. L'intelligence technique du XVIe au XVIIIe siècle*, Paris: Albin Michel, 1993.

Vérin, Hélène, "Rédiger et réduire en art: un projet de rationalisation des pratiques", in: eds. Pascal Dubourg Glatigny, and Hélène Vérin, *Réduire en art. La technologie de la Renaissance aux Lumières*, Paris: Editions de la Maison des sciences de l'homme, 2008, 17–58.

CHAPTER 7

The Wise Origins of the *Consten*

Stevin and Sixteenth-Century Debates on Arts, Mathematics and Language

Fokko Jan Dijksterhuis

The first volume from 1608 of the *Wisconstighe ghedachtenissen* contains Stevin's reflection on the 'Wijsentijt':

> The Age of the Sages is what we call that time in which exceptional learning was to be found among men, something that we perceive with certainty from certain signs, but without knowing among whom, where or when.

However hazy the traces of this pristine wisdom, it was Stevin's intention to

> ... explicate how in my view we might set out to attain once more such great learning, both in astronomy and in other subjects, as we perceive was formerly found among men, it appeared necessary to me to define this time as above.[1]

The 'Wijsentijt' addresses the theme of the origin of wisdom. Stevin argued that original wisdom had existed in a pre-classical age, and that the current revival of learning should thus look back beyond the Greeks. Stevin combined this with a reflection on the role and value of language in learning, both linguistically and epistemically. He came to the conclusion that original wisdom required a pure language and that Dutch was most fitting for this end: the Sages

1 The two quotes are from the following passage, with the middle part paraphrased: Stevin, *Eertclootschrift* (*Wisconstighe ghedachtenissen* volume 1, part 2), 9. 'Wysentyt noemen wy die, waer in by de menschen een seltsaem wetenschap gheweest heeft, t'welck wy deur seker teyckens ghewisselick mercken, doch sonder te weten by wie, waer, of wanneer. [Anghesien het int volghende dickwils te pas sal commen, t'onghewoonlick woort Wysentijt te noemen, en dat mijn voornemen is t'sijnder plaets te weten inde navolghende vernieuwing des Wysentijts,] verclaring te doen, hoemen mijns bedunckens de saeck an soude meughen legghen, om weerom te gheraken tot alsulcken groote wetenschappen soo in Hemelloop als ander stoffen, ghelijck wy mercken by de menschen eertijts gheweest te hebben, so heeft my noodich ghedocht den selven tijt te bepalen alsvooren'.

 | DOI:10.1163/9789004432918_009

had spoken Dutch. The 'Wijsentijt' takes the typical Renaissance approach of discussing the nature of knowledge by reflecting on its origins. It outlines an epistemic framework based on knowledge of the arts and natural logic.

Historians have treated Stevin's exposition on the Age of the Sages in a rather stepmotherly way, if at all. It is often presented as a bit of a curiosity and detached from its textual and historical context. As a result, it is absent from the relevant historiography, which has the effect of obscuring Stevin's connections to sixteenth-century debates on mathematics and wisdom and vice versa. In fact, with the 'Wijsentijt', Stevin's ideas were linked to a contemporary debate on the origins of mathematics that was instigated by Petrus Ramus. This debate has been discussed at length by Robert Goulding, but Stevin's contribution does not feature in his account. My aim in this paper is to re-read the 'Wijsentijt' in both the context of Stevin's biography and sixteenth-century debates on language, the arts and knowledge. My account is intended to supplement that by Van den Heuvel, who has developed an insightful new interpretation by placing it in the context of Stevin's oeuvre.

1 The 'Wijsentijt' in Stevin's Oeuvre

It is not easy to make sense of the 'Wijsentijt'. It is a complex and impenetrable text, quite hermetic in both language and content. The sixteenth-century Dutch is hard to grasp in itself, but in addition Stevin used many idiosyncratic terms that he introduced for epistemic topics. Whilst they are wonderfully inventive words, they are difficult to comprehend, because Stevin does not explain the purport and underlying contemporary debates. This makes it hard to read, even for a native speaker. The content matter and argument are rather obscure, with elliptic and oblique reasoning. The structure and position of the text complicates interpretation yet further. Its place in the *Wisconstighe ghedachtenissen* is not obvious, and raises the question of when and how Stevin conceived the discourse on 'Wijsentijt' and its revival.[2]

The 'Wijsentijt' is one of the definitions of the *Eertclootschrift*, which in turn is the second book of the *Weereltschrift*. 'Weereltschrift' – discourse on the world – can be translated as 'cosmology'. It consists of three books: *Driehouckhandel* – treatment of triangles, trigonometry; *Eertclootschrift* – discourse on

2 'Wijsentijt' is in Stevin, *Eertclootschrift* (*Wisconstighe Ghedachtenissen* volume 1, part 2), 9–48. Ad Davidse provides a valuable guide (in Dutch): https://adcs.home.xs4all.nl/stevin/weerelt/wijsentijt.html (accessed February 2018). See also Van den Heuvel elsewhere in this volume.

the terrestrial globe; and *Hemelloop* – the course of the heavens, astronomy. *Eertclootschrift* can be translated as 'geography', but it covers geology and the theory of matter as well as navigation (see Vermij in this volume). Besides the definitions, the *Eertclootschrift* itself consists of a book on the constitution of the Earth (a bit of chemistry), the measuring of the atmosphere, and three books on hydrographia (*Zeeschrift*): compass directions, navigation and a theory of tides. The discourse on the 'Wijsentijt' and its revival is the sixth and last definition of the *Eertclootschrift*. The first five definitions concern basic geographical concepts: the Earth, its axis and equator, longitude and date. They are concise and fill just four pages. The definition of the 'Wijsentijt', on the other hand, runs up to forty pages, including a fourteen-page comparative list of Dutch words. It is a multi-layered section that evolves into an ever-further bifurcating exposition of the nature and origins of the sciences and their methods. The introduction to the 'Wijsentijt' is subdivided into four 'signs' and followed by an exposition on the revival of the 'Wijsentijt'. This is in turn divided into four clauses that are first stated and then explained. The fourth clause is then subdivided into five 'oirden' (orders) of the 'Wijsentijt'. The typography of the section does not make entirely clear whether the discussion of the revival is a subsection of the definition or a new section.

All in all, the discourse on the 'Wijsentijt' is more or less an independent section of the first chapter of the *Eertclootschrift*. The place of the discourse in the whole of the *Wisconstighe ghedachtenissen* is curious. It seems a rather excessive conclusion to an introduction to a theory of geography and navigation. The purport is much wider than geography as such; it refers to geometry and astronomy and is intended to provide an exposition of the nature, aims and methods of the mathematical sciences. One would expect the 'Wijsentijt' to be some sort of programmatic statement at the beginning of the *Weereltschrift* or the *Wisconstighe ghedachtenissen* as a whole. The five-volume *Ghedachtenissen* Stevin published at the end of his life is in itself a complex work that is difficult to interpret. It is a compendium of old and new, published and unpublished works, reordered, appended, edited and rather pragmatic in places. It is a mathematical memoir published within the courtly context of Stevin's advisory role to the Stadholder, presented as a discourse *for* as well as *with* the prince. Stevin had a clear sense of the order of knowledge, and it seems that the discourse on the 'Wijsentijt' has not found its definitive place.[3]

3 See Van den Heuvel in this volume, as well as: Van den Heuvel, 'Wisconstighe Ghedachtenissen', 107–110; 117–119. Stevin, *Wisconstighe ghedachtenissen* volume I consists of the *Weereltschrift*. The books are individually page-numbered. Davidse believes that the 'Wijsentijt' was intended as part of the *Hemelloop*.

The *Principal Works* do not really help here. They select from the *Wisconstighe ghedachtenissen* in accordance with twentieth-century disciplinary distinctions and prejudices, reordering Stevin's structure beyond recognition. The *Eertclootschrift* is completely torn apart: part six on the tides is combined with astronomy in the astronomical section; parts four and five as navigation; and the geographical parts are absent altogether. Selections from the 'Wijsentijt' are appended after the section on navigation, as a sort of freestanding afterthought. It has been read as an isolated exposition and a mere curiosity; a witty but 'unscientific' excursion.[4]

The 'Wijsentijt' was probably a late text in Stevin's oeuvre.[5] Nevertheless, it is highly consistent with his ideas on knowledge and the arts, and it contains elements that appeared in earlier publications. This is most clear in the third clause on the restoration of the 'Wijsentijt'. This argues that Dutch is the optimal language for the mathematical arts, as Stevin had argued previously in the *Weeghconst* of 1585. The introduction to this work contained an 'Uytspraeck vande Weerdicheyt der Duytsche Tael' – a eulogy to the dignity of the Dutch tongue. Stevin returned to this in the 'Wijsentijt'. Although he condensed the argument, Stevin included the full comparative lists of Dutch, French, Latin and Greek words. The fourth clause also relates to 1585 with an exposition on theory and praxis, as well as the elements of logic. This brings us from the *Wisconstighe Ghedachtenissen* of 1608 to Stevin's early works of 1585.

2 *Stevin's Program for the* Consten

The succession of books that Stevin published with Plantin's Leiden office in 1585 and 1586 express a coherent vision on learning and the mathematical sciences: *L'Arithmetique, Dialectike, De Thiende,* the treatises on *Weeghconst* and *Waterwicht.* If we discard modern disciplinary divisions and approach these publications collectively, we see a coherent program on knowledge and the arts. This coherence is substantiated by the fact that the *Weeghconst* and the *Dialectike* shared the same printing privilege.[6] Again, the *Principal Works*

---

4 In Stevin, *Principal Works*, the following parts of *Weereltschrift* are present: in volume IIB (Mathematics) selections from the *Driehouckhandel*; in volume III (Astronomy and Navigation) selections from *Hemelloop* and treatise on tides as 'astronomical', *Havenvinding* and *Zeylstreken* as 'nautical', without an indication that all are in fact part of the *Hemelloop*. The 'Wijsentijt' is also in volume III, with a note (p. 609) suggesting it was originally intended to precede the *Hemelloop*.

5 See Van den Heuvel's chapter in this volume.

6 See Buning's chapter in this volume.

are not very helpful here; in this edition of his work, the 1585/6 publications are reordered in accordance with modern disciplinary divisions, and the *Dialectike* is not included at all. Stevin was a newcomer in Holland, but he was ambitious and already quite well-connected. In a dispute on the quadrature of the circle in 1584 that involved Stadholder Willem's chief engineer, he was referred to as an expert. He had enrolled at Leiden University and may already have become acquainted with Willem's son and the future stadholder, Maurits. It was at this stage that Stevin began elaborating his ideas on knowledge and the arts.[7]

The best-known feature of Stevin's epistemics is his conviction that theory and praxis are mutually dependent. From his very first documented activities, Stevin propagated the idea that the two should be blended, and this is evident from the 1585 publications. The works on weighing are a carefully ordered ensemble that can be regarded as a programmatic statement on the mathematical sciences. *De Beghinselen der Weeghconst* and *De Beghinselen des Waterwichts* addressed the principles of the art of weighing. They were accompanied by *De Weeghdeat* – the act of weighing – and a planned 'waterwichtdeat'. Stevin explained:

> As it would be useless to lay a large and strong foundation that can bear a heavy weight without intending to erect any building thereafter, in the same way the reflection on the principles of the arts is a lost effort if it does not serve action in the end. As by natural order that foundation goes before the upper building, in the same way this reflection goes before its action.[8]

The 'theory' of statics was linked to practice, and vice versa; they were complementary. Stevin's idea of 'blending' separated theory and praxis, while at the same time combining them. *De Thiende* can be seen as an example of this; it unveiled the use of decimals by practitioners and codified the notation and method. Stevin's views on theory and practice are not as straightforward as

7 Stevin, *L'Arithmetique, Dialectike, De Thiende, Weeghconst, Weeghdaet, Waterwichts* (the treatises on weighing have separate title pages but are usually bound together). On Stevin's environment, see for example Dijksterhuis, 'Stevin, Huygens'.

8 Stevin, *Weeghdaet*, 3. 'GHELIICT onnutte cost waer, een groote stercke GRONDT te legghen, die een swaer ghesticht draghen can, sonder eintlick eenich GEBAU daerop te willen brenghen; Alsoo is de SPIEGHELING {*Theoria.*} inde beghinselen der consten verloren arbeydt, daer t'einde totte DAET {*Effectum.*} niet en strect. Ghelijck oock na de natuerlicke oirden, dien grondt voor t'oppergebau gaedt, alsoo dese spiegheling voor huer daet'.

they seem, however, and below I will discuss their further elaboration in *Wisconstighe ghedachtenissen*.

Although it has played a less prominent role in the historiography on Stevin, the *Dialectike* may be the most ambitious of the 1585 treatises. *Dialectics* or the art of proof, learning to be the judge of all matters right and able; it was an exposition on logic and, as Maarten van Dyck has pointed out, it directly informed the concepts of quantity, matter and abstraction that lay at the basis of the *Weeghconst* and were discussed in *L'Arithmetique*. The *Dialectike* is written in Dutch. In the case of the treatises on weighing and *De Thiende,* the vernacular more or less follows automatically from its subject matter and its audience. *De Thiende* can be read as a plea for the use of vernacular arithmetic (i.e., Arabic) in the sciences, not just the arts. The *Dialectike,* however, went a step further, being a treatise in the vernacular on as academic a subject as the trivium.

Stevin was aware of his audience, anticipating questions such as 'what drove him to discuss such profound matters in the Dutch tongue'.[9] Together with the treatises on statics, the *Dialectike* offered a sustained argument and a proof of concept of the suitability and permissibility of using Dutch in learning. In the first place, Stevin was reaching out to readers who had not mastered Latin, but were well-versed and interested in the 'consten' – so-called 'idiots' yet virtuosi.[10] This was the main reason to publish in Dutch on the mathematical sciences. In the case of the *Dialectike,* there was a second, more profound, reason in the subject matter itself. Dialectics is the science of reasoning; that is to say, it provides concepts to explicate and judge inferences. However, reasoning as such was not an academic subject at that time. On the contrary: every right-minded person has mastered the art of reasoning. (For that matter, the same was true of arts such as reckoning, geometry and algebra, as well as social conduct.) Furthermore, it was not the case that the topic was incomprehensible or that practitioners lacked sufficient intellect or interest. They were perfectly capable of reflecting on reasoning; they only lacked the vocabulary to talk *about* it. Stevin's intention was to provide a good introduction to the principles of reasoning.

The *Dialectike* was based on the conviction that logic is a natural capability and that the science of logic should reflect this natural reasoning, by explicating the rules of reasoning. As Stevin added, this meant that the science of logic should also speak natural language, rather than the language of the schools.

9 Stevin, *Dialectike*, dedication fol 3. 'Aengaende my yemandt segghen mochte, wat gaet u doch over, soodanighen diepsinnighen Materie te willen beschrijven inde Neerduytsche tale?'

10 Dijkstra, *Between Academics and Idiots*.

To this end, he developed a whole new vocabulary in Dutch, one that is highly original but also quite impenetrable. According to Stevin, scholarly Dutch should contain true Dutch words rather than Dutchified Latin terms: *menichvuldicheyt* rather than *kwantiteit*; *ghedaente* rather than *kwaliteit* (and *evenredigheid* rather than *proportie*).[11] Stevin had outspoken ideas about the learned application of Dutch as such. He elaborated these in the eulogy to the dignity of the Dutch language that preceded his exposition of mechanics in the *Weeghconst*: 'Uytspraeck vande Weerdicheyt der Duytsche Tael'. This makes the *Weeghconst* a rather odd treatise: the mathematical theory is preceded by a 24-page argument justifying the linguistic form of the theory, including a thirteen-page list of words.[12] One is tempted to say that the goal of the *Weeghconst* is linguistic rather than mathematical: to provide an example of how learning can be done in Dutch, the mathematics illuminating the argument of the eulogy. At any rate, Stevin's ideas on language are searching, systematic, precise and outspoken.

The 'Uytspraeck' opens with Stevin's adage 'wonder en is gheen wonder', or 'A miracle is no miracle': we call phenomena wondrous because we do not know their causes, not because they are actual miracles. Knowledge is a human faculty and language is the key to reality, and the right language provides the right key. The right words disclose the causes of phenomena and thus reveal wonders. When Man developed his language, the workings of nature were known, but this original wisdom was blinded when fortune (*'tschicsel*) obscured the reason of many. Stevin's exposition on language was an epistemic position as much as a linguistic one. Following well-trodden lines of Renaissance thinking, Stevin explained how language and understanding coincided. This concerned not only the form of the language used, as elaborated in the communicative measures and conceptual dichotomies he used in his accounts; Stevin also argued that a specific language existed for true knowledge: Dutch. In other words, Dutch was not only a functional language for the *Weeghconst*, but it was also the only possible language with which one could acquire an understanding of the art.

In order to substantiate his position, Stevin first had to explain that Dutch was the original language. His main argument was that in prehistoric times from which no writings exist – that is, before the age of barbarism – the

11 See also the chapters by Van Dyck and Cohen and Kursell in this volume, as well as the Introduction.

12 I have not systematically examined how it is bound with the treatises. In the books I have seen, it is printed on separate sheets – together with the *cortbegryp* and errata – and bound between the preliminary matters and the text.

Dutch had been the most powerful and honorable people. Envious and impressed, other peoples, such as the French and the Spanish, adopted Dutch words. Dutch in this regard meant 'Germanic': Stevin counted both the 'Netherlanders' and the 'Upperlanders' among the Dutch, the 'Dutch/German'-speaking world extending well towards the Alps and Slavic regions.[13] When the Dutch lost their prominence in the age of barbarism, the languages of other peoples diverged and became diluted. (In particular, the French of the Gauls, which from the viewpoint of Dutch became some sort of a speech impediment.) Stevin based his account on a phonetic analysis of a set of basic words. In this way, he argued that Dutch was not just a vernacular but *the* vernacular; the language of reason and accordingly the key to understanding.

Stevin went on to explain what made Dutch the most comprehensible and lucid language. It was the most concise language because it was the founded upon the widest collection of monosyllabic words. Furthermore, these monosyllables could be combined in a logical and straightforward way, without loss of meaning or clarity. The order of a compound univocally determined its meaning: *waterput* and *putwater* (water well, well water). He referred to the *Dialectike* for a further discussion of the compounding of concepts. In this way, Stevin gave a more or less atomistic account of language: simple elements that could be compounded into more complex structures. He had considered mathematics in *L'Arithmetique* in the same way, identifying the basic elements of quantity and the effects of combination.

Next, he explained that the example of the *Weeghconst* showed that Dutch was the only language truly to teach the arts. Only Dutch could express the basic concepts of mechanics by combining the simple terms of weight and proportion: *evenstaltwichtich, rechthefwicht, scheefdaellini.* 'Nature has fitted Dutch most exquisitely for that'.[14] Accordingly, Dutch words were preferred. Finally, Stevin wrote, Dutch had the greatest power of expression, referring to the way the Protestant sermons in the vernacular incited true faith. To round off his eulogy, Stevin reflected that the proof of the superiority of Dutch lay in practice, and he challenged other languages to translate, without loss of meaning or clarity, proposition XX of the *Weeghconst*: 'Ghelijck rechtheflini tot scheefheflini, also rechthefwicht tot scheefhefwicht'. This promotion of Dutch as the language of true learning had a lasting result: Dutch is the only language

13 The ties within the family of Germanic languages are close: in Dutch, the word for 'Dutch' is 'Nederlands', for 'German' it is 'Duits'.

14 Stevin, *Weeghconst*, XXVIII: 'de Natuer heeft daer toe aldereyghentlicxt het Duytsch veroirdent'.

with its own word for mathematics, *wisconst* – the art of certitude (*wis* has the same root as wise, wisdom).

The eulogy to Dutch is not an oddity within the famed mathematics of the *Weeghconst*. Together with the *Dialectike* and the other publications of 1585, the mechanical works provide a coherent epistemic program for the *consten*. Perhaps we should read the mathematical treatises the other way around: as essays on this discourse on knowledge in the arts. By means of this discourse, Stevin related his ideas to sixteenth-century debates on knowledge, the arts and the schools. It was quite ambitious; after all, he was a relative newcomer to Holland's scholarly scene. He did not shy away from confronting the learned establishment with a Dutch treatise on logic. In 1590, he would respond to the prominent Leiden professor Lipsius' *Politicorum* with *Vita politica. Het Burgherlick leven*, a Dutch exposition on politics in the new Republic. In 1584, Lipsius had said he was 'just a mathematician without any knowledgeability, yes almost without any linguistic competence'.[15] In the 'weerdigheit', Stevin left a final message for the scholars who had done nothing but defend Latin and get no further than linguistic discourse, rather than uncover the truths of nature.

3 A Grammar of the Arts

In the 'Wijsentijt', Stevin returned to the epistemic program of 1585. He elaborated some elements and added some new ones. In 1585, the emphasis had been on language and the arts; now he broadened the scope to the arts as such and the matter of the foundations of knowledge. The 'Wijsentijt' opened with the common Renaissance notion that following the obscuring of reason in the *Barbarum sæculum*, man was now recovering the learning of the classics. Stevin explicitly referred to the Reformation and linked religious freedom to the general freedom to pursue learning. However, Stevin continued, the origins of wisdom did not lie in classical times; they stretched back far beyond the age of the Greeks. An Age of Sages had existed when pure wisdom had been present, wisdom that had been eroded or lost to the Greeks. No records had been passed down, but there were indisputable signs of its existence. These signs consisted of knowledge that the classical authors did not have. It came through other channels or could be surmised from particular circumstances. Stevin listed six signs of this prehistoric wisdom: astronomy, algebra, geometry,

15 Lipsius, *Epistolae*, 110. 'De** Mathematico scripseram olim ad Ortelium, sed nescio reverà sítne futurus ad salivam vestram & gustum. Mathematicus enim merus est, sine aliâ arte, imò pænè linguâ: ...'.

atmosphere, alchemy and magic. Astronomy and algebra were by far the most elaborate lemmas, the other four consisted of brief remarks.

In the section on astronomy (*hemelloop*), Stevin argued that previous concepts and data had already been lost in classical times. Through Arabic writings, as Scaliger had uncovered, it was clear that constellations had been known that were not mentioned by classical authors. The clearest sign was the repudiation of the motion of the Earth by Ptolemy and the Greek philosophers. Previous astronomers who had learned the 'roersel des Eertcloots' were Pythagoras and Aristarchos, as Archimedes explained in the *Sand Reckoner*. Stevin then embarked on a lengthy discussion of notable problems with the system of Hyparchos and Ptolemy.

The signs given by arithmetic were connected to astronomy, showing that original numerical systems were more suitable for trigonometry and the calculation of sine tables. For Stevin, the ideal numerical system was, of course, the decimal system, referring to the arguments of *De Thiende*. Once again, he mentioned Scaliger's studies to explain the eastern origins of the decimals as well as the concept and notation of zero as a number, and he recalled his argument in *L'Arithmetique*. Besides trigonometric calculations, decimals were ideal for 'the measure of the circle', calculating the proportion between diameter and circumference. In addition to decimals, Stevin mentioned 'reckoning with letters', algebra, which had even remained unknown to the Hebrews and Chaldeans. Stevin contrasted Arabic methods for solutions with the number theory of the Greeks, in particular Euclid's discourses on proportions.

For his views on geometry, Stevin referred to the final section of the 'Wijsentijt' and with regard to Arabic accounts of the atmosphere, he referred to the geography (see Vermij). He concluded by emphasizing the age and importance of alchemy (*Stofscheyding* – the separation of substances/matter) and magic (*Gheesthandel* – the operation of ghosts). Stevin's account of the six signs of the Age of Sages thus placed mathematics alongside alchemy and natural magic. This may defy modern disciplinary categories, but – without professing any general conclusions – it was not uncommon in the sixteenth century.[16] Reading the 'Wijsentijt' as a reflection upon the nature and scope of the *consten*, we can say that Stevin emphasized the operative character of the mathematical sciences alongside alchemy and natural magic.

In a short intermezzo, Stevin related the testimony of Hugo Grotius on traces of lost philosophy. The son of an earlier collaborator in Delft, Jan Cornets de Groot, Grotius was a close acquaintance and developing into a notable scholar.

16 Dijksterhuis, 'Magi'; Keller, *Drebbel*; Long, *Artisan/Practioners*.

He had studied ancient men of letters and found all kinds of signs that in philosophy, too, wisdom had existed before classical times. In this way, Stevin utilized the philological investigations of academics such as Grotius and Scaliger to enrich and reinforce his argument.

Finally, Stevin explained how the 'Wijsentijt' could be revived: 'how to arrive again gradually to such great sciences as have existed in the Age of Sages'. He was convinced that man's power had not declined – in the barbaric times, it had only been obscured by papal powers – and in four clauses, he explained the requirements for true wisdom. What Stevin was setting out to describe was a 'science of arts' reflecting the program of 1585.

First of all, true wisdom was founded on experiential knowledge (*dadelicke ervaringen*) of the arts, which in turn was acquired in collaboration. He argued that a revival had begun in the initiatives of Tycho and Moritz of Hessen-Kassel. He then made the switch to arguing that knowledge was ultimately something civic rather than princely, and thus the people formed the foundation of (arts) knowledge. Second, collective efforts were only fruitful if people used their own language (*angeboren tael*). This language ought to be particularly suited to the arts, just as Greek had been suitable for geometry. Latin is external to the community; it is suitable for academic subjects such as law, theology and medicine, but not for the arts and the mathematical arts in particular. This explained why there had not been notable mathematicians amongst the Romans. The French had progressed in mathematics, but had been hampered by their use of artificial, Latin terms. Pure words were necessary to avoid confusion and misunderstanding. Taking an example from the science of music (*Singconst*, the art of singing), he argued that the word 'evenredigheid' conveyed the proper understanding of the harmonic interval ('equi-ratio' rather than proportion). This brought Stevin to the third clause, the reprise of his discourse on the dignity of the Dutch tongue of 1585. He explained the qualities that make a language fitting for the arts, the 'atomistic model' illustrated by the table of words showing the power of Dutch. These qualities can be discerned in the liberal arts and principally in mathematical practice. He contrasted the mathematical arts with literary and historical activities. Linking back to the first clause and the importance of experiential knowledge, Stevin argued that the Dutch word 'anschauwen' captured the nature of observation perfectly. In seeming contrast to scholasticism, the arts show that we do not experience the essence of things, but their appearance. In other words, the sages must have spoken Dutch.[17]

---

17 See Van der Wal, 'Logic, Linguistics, and Simon Stevin' and 'Verrotens taalbeschouwing'.

In the fourth clause, Stevin's argument once again branched out into subsections. Consisting of five parts, it provided a 'grammar of the arts', explaining the proper order conducive to the presentation and teaching of the arts. It might seem, Stevin began, that the knowledge of the arts was far too expansive and manifold for anyone to comprehend.[18] However, proper knowledge did not consist of data as such, but an understanding of its contours and coherence; the structure of knowledge. Lying at the basis of this was mathematical reasoning, which was exemplified by Euclid's *Elements,* and that had begun to be recovered by Peurbach, Copernicus and Regiomontanus. The parts of the mathematical proposition were the proper way to order and distinguish judgements, and to reason. Lucid and unequivocal definitions were crucial for both the finder (*inventor*) and student; dichotomy was the proper way to map knowledge. Equiverbalism (*selfwoordicheyt, anaphora*) was a stylistic tool to enhance lucidity. The last part contained Stevin's renowned exposition on the integration of theory and practice, which distinguished the 'science of arts' from the arts proper.

4 Blending Theory and Praxis

The 'Wijsentijt' was an exposition on the epistemic value of the arts. Like Stevin's programmatic statements in general, it contained a lot of original and inventive ideas and arguments. Stevin did not operate in a vacuum, however, and his views were relatively well grounded in contemporary developments and ideas. The problem is that he rarely explicated the issues and debates he was addressing; his expositions are rather self-contained. One example of this is the final part of 'Wijsentijt' on the integration of theory and practice. 'Vant menghen der spiegheling en daet' is a renowned passage that is often held up as an example of Stevin's visionary ideas, but it tends to be dissociated from the rest of the *Eertclootschrift*. Stevin's views on 'spiegheling' and 'daet' were original, but we should be cautious when taking them out of their sixteenth-century context. After all, it was an era in which practitioners gained prominence in intellectual and cultural affairs. Calls for practice-oriented scholarship and reflections on the integration of theory and praxis were common;

18 Stevin, *Eertclootschrift* (*Wisconstighe ghedachtenissen* volume 1, part 2), 40. 'Ymant onbekent sijnde d'oirden des Wysentijts, soude meughen segghen dat de leering en t'onthoudt van soo veel verscheyden diepsinnighe stoffen als de vrye consten in haer hebben, niet gheschien en can sonder een verdrietich leven, en verswijmnis van ander saken die de schoorsteen doen rooken, en totte huyshoudinge noodich sijn: ….'

Francis Bacon was but one of the last in a long line of writers to address these themes.

Although the meaning of 'reflection' and 'act' seems evident, Stevin's views were not as straightforward as they might appear. 'Spiegeling' and 'daet' – 'reflection' and 'act' – are usually translated as 'theory' and 'practice'. The Latin terms Stevin himself provided already reveal that this is not entirely correct: *theoria* and *effectum*. Furthermore, directly after having translated 'daet' as 'effectum', he linked 'weeghdaet' to 'praxis'. Thus, 'daet' has at least a dual meaning of both 'act' and 'effect'. Likewise, it remains to be seen whether a sixteenth-century 'theoria' can be translated as 'theory', with all its modern epistemic connotations. It is connected to *weegconst,* but in the same passage, *constigh* is also linked to *daden*. There is, in other words, a lot of flexibility – not to say ambiguity – in Stevin's use of the terms 'spiegeling' and 'daet'.

If we consider the content of the 1585 *Weeghconst* and *Weeghdaet*, the distinction 'theory' and 'practice' does not apply directly; at least, not our modern one. Both the *const* and the *daet* concern theoretical expositions of weighing: principles, theorems and deductions. The difference between the two lies in the subject matter: the 'const' is used to address weight *as such* – without natural matter (*sonder natuurlijke stof*) – and the 'daet' addresses tangible weight in concrete objects. In the preface to the principles of *Waterwichts,* Stevin elucidated the difference, drawing a comparison with geometry:

> … as in the principles of geometry the points should be taken without length, the line without width, the plane without thickness, so it is necessary in the principles of the weighing of waters, to state vessels without physical magnitude and without weight, in order to treat them mathematically, …[19]

This refers to the early modern distinction between abstract and material quantity, between pure and mixed mathematics. In mechanics, this regards the material aspects of gravity and friction, as Van Dyck shows in his discussion of the *Weeghconst.*[20] In the 'Wijsentijt', Stevin wrote:

---

19 Stevin, *Waterwichts*, 5. '… gehlijck inde Beghinselen der Meetconst, t'punt ghenomen moet worden sonder lengde, de line sonder breede, t'vlack sonder dicte, alsoo ist inde Beghinselen des Waterwichts noodich om Wisconstlick daer in te handelen, vaten te stellen sonder lichamelicke grootheyt, ende sonder ghewicht, …'.

20 See Van Dyck's chapter in this volume.

> Reflection is an imagining activity without natural matter, like the reflection of the reflector Euclid among others, acting by stating magnitudes and numbers, but each separated from natural matter. Act is an activity that proceeds essentially with natural matter, like measuring land and banks, counting the quantity of rods or feet that are contained, and the like.[21]

The distinction here is between the subject matter of inquiry and the goal. Early modern mathematics had another distinction for the latter: speculative versus practical mathematics. In modern science, both dimensions have been conflated into the distinction between pure and applied mathematics, but in early modern thinking these were two different things. Stevin's characterization of 'spiegheling' and 'daet' applies to the pure/mixed distinction, the subject matter of mathematics. In the 'Anhang' to the *Weeghconst,* he explicitly referred to the category of mixed sciences when he addressed the question of whether the art of weighing is a liberal art.[22] The pure/mixed distinction is no longer very familiar to us, and we tend to understand it in terms of pure and applied. Stevin said little about the goal of inquiry, the relationship between speculative and practical. It appears that he was not interested in speculation at all, believing it not worthy of consideration. In the two treatises, both the *const* and the *daet* of weighing are very practical – geared towards 'application'. At the same time, the treatment of both was formal, aimed at laying down the principles of statics in abstract and material aspects.

In 1600, Stevin got the chance to realize his ideas in an educational setting:

> As it has pleased His Excellency, Count Maurits of Nassau, Stadholder of Holland, and Captain General, that, for the benefit of the state, here in the university should be taught in good Dutch language the art of counting and surveying, principally for the advancement of those who should want to become engineers ...[23]

---

21 Stevin, *Eertclootschrift* (*Wisconstighe ghedachtenissen* volume 1, part 2), 46. 'Spiegheling is een verdochten handel sonder natuerlicke stof, ghelijck onder anderen sijn de Spieghelinghen des Spieghelaers Euclides, handelende deur stelling van grootheden en ghetalen, maer elck ghescheyden van natuerlicke stof. Daet is een handel die wesentlick met natuerlicke stof gheschiets, als lant en wallen meten, de menichte der roen of voeten tellen diederin sijn, en dierghelijcke'.

22 Stevin, *Weeghconst*, 'Anhang' chapter III, 68. He actually asked whether 'weeghconst' is a free 'wisconst' – *ars mathematica.*

23 Molhuysen, *Bronnen*, 122. 'Alsoo Sijne Excellentie, Grave Maurits van Nassau, Stadthouder van Hollant, ende Capiteyn Generael, tot dienst van den lande goetgevonden hadde, dat in de Universiteit alhyer soude worden gedoceert in goeder duytscer tale die telconste

This program of mathematics in Dutch for engineers is usually called 'Duytsche Mathematique'. Formal classes would be offered in winter, and in the summer engineers would be trained in the field. The mathematical curriculum was drawn up by Stevin.[24] It prescribed in detail what mathematics the instructors should teach: those parts of geometry that were relevant to engineering. It also included a practical part on making and designing actual constructions. This practical part was not the training in the field – Stevin did not address this – but consisted of the manufacture and use of architectural *models*. In his architectural writings, Stevin set great store by the use of models for developing and communicating a design. This was the hands-on practice to be learned in the school.

For Stevin, it went without saying that practice would benefit from theory. This worked the other way around as well: both Dutch and mathematics had their proper place in the academy. This was not a common view among academics or practitioners. Why would you train practitioners at the university? What good does formal learning do an engineer? It is a very basic (and still relevant) question: why would practice improve with theory? Does education in formal mathematics help an engineer in the field? By 1600, the engineers of the States' armies had achieved an undisputed reputation as being masters of innovative battle techniques. A Spanish commander once sighed that he had been defeated by shovels instead of canons. By that time, the heartland of the Revolt was basically secured and the Republic was slowly expanding its hold to the east and the south. The engineers in command had not voiced a desire to make the training more formal; nor, even, had the original superintendent Adriaan Anthonisz., who was a practicing mathematician himself. We may assume that men like him considered field training and practical education proper and sufficient. The contention that elemental mathematics was 'required for practical, common engineering' was very much the view of Stevin (and Maurits, in all likelihood). The value of the vernacular for the sciences went further than making learning accessible for the common people. Stevin was convinced that it was suitable for actual learning: not only for teaching, but also for finding, as he put it in the preface to the *Waterwichts*.

---

ende lantmeten principalycken tot bevordering van de geenen die hen souden willen begeven tottet ingenieurscap ...'.

24 Stevin, *Maniere*; Molhuysen, *Bronnen*, 389*-390*. 'Hyer toe sal men leeren die arithmeticque oft het tellen ende het landtmeten maer alleenlyck van elck soe veel, als tottet dadelyck gemeene ingenieurscap nodich is'.

5 Stevin and Ramus

Although he did not explicitly refer to writers and ongoing debates on learning, Stevin evidently responded to them. With respect to the mathematical sciences, one of the loudest voices in Northern Europe (and Holland in particular) was that of Petrus Ramus (1515–1572). Ramus, in turn, was an exponent of the practice-oriented humanism of Agricola and others, which set the wider stage for learning in the Low Countries. Stevin did not list Ramus (or Agricola) among the men of letters, but – as I will argue – everything suggests that he was well-acquainted with Ramist ideas; and he was quite critical of them.[25]

Stevin's 1585 publications respond to Ramist themes in a number of ways. In *L'Arithmetique,* he expressed his conception of number and the basis of reckoning. He adopted Ramus' opinion that the unity was a number, and also tentatively considered the extension to fractions. The *Dialectike* was related to the debate on logic, dialectics and their place in learning. It went straight to the heart of the Ramist program of reforming the liberal arts. Here, dialectics – understood along Agricolan lines in an operative way – acquired prominence over logic, in the formal way of scholastics. The distinction between style and argument that Stevin makes with respect to anaphora is comparable to that made by Ramus. The *Dialectike* adopted the Agricolan/Ramist conviction that logic is a natural capability, and that the science of logic should reflect this natural reasoning by explicating the rules of reasoning. Stevin added that this meant that the science of logic should also speak natural language, rather than the language of the schools.[26]

Stevin's exposition on language was an epistemic position as much as a linguistic one. Following common lines of Renaissance thinking and in accordance with a kind of nominalism propagated by Ramus and others, Stevin explained how language and understanding coincided. This concerned not only the form of the language used, as elaborated in the communicative measures and conceptual dichotomies of Ramus that he adopted as well; according to Stevin, a specific language existed for true knowledge and this was Dutch, as other languages were not able to express the sagacious thoughts of mechanics.[27]

---

25 On Ramus, Ramism and the links to the Dutch Republic, see Hotson, *Commonplace Learning*.

26 Hotson, *Commonplace Learning*, 50–54. On Ramus and mathematics, see Verdonk, 'Einfluß' and *Ramus en de wiskunde*.

27 Stevin, *Weeghconst*, 9. 'Ghemerct daerbeneven, dat wy voorghenomen hebben inde selve te beschrijuen de WEEGHCONST, wiens diepsinnighe ghedaenten {*Qualitates.*} duer slechter spraken ten eersten niet wel bedietlick en sijn, …'.

As regards the blending of theory and praxis, Van Dyck points out that Stevin emphasized the *distinction* between both. The fruitful combination presupposed 'a clear separation of speculation from practice' rather than some integration or identification. This is in clear contrast to the views of Ramus. He also rejected idle speculation, but he saw practice as the source of knowledge and learning and subordinated the sciences to use. Against this background, it makes sense that Stevin would so systematically and carefully elaborate a vision on 'spiegheling' and 'deat'. He contributed his own position to an ongoing debate and had to express himself in a precise way.[28]

As an academic, Ramus published in Latin, but he did propagate the value of the vernacular in learning. He rejected the literary contempt for experience and manual labor and presented himself as a spokesman of the arts. According to Ramus, the arts were the root of natural knowledge, and the sciences throve where artisans and engineers were encouraged. In the mid-sixteenth century, he considered the German states the ideal place in which to cultivate knowledge; in particular Nuremberg, where the court and the citizens had achieved a great flourishing of the arts. Was this mere *topos,* or does this explain in some way Stevin's dedication of the *Weeghdaet* to the burgomasters and governors of the city of Nuremberg? Coincidental or not, the dedication was the very place where Stevin presented his view of the integration of theory and practice.[29]

Neither Stevin nor Ramus were unique in their praise for artisans and their propagation of practical knowledge and reasoning; this was rather commonplace in the sixteenth century. As a string of historians have argued, the scientific Renaissance was rooted in the blossoming of the arts and the increasing status of artisans. Writers gave voice to these developments and articulated their epistemic, institutional and cultural consequences, paving the way for the new philosophies of the seventeenth century. In Northern Europe, Ramus was one of the most audible voices; his influence had been immense, particularly through less radical second-generation authors such as Alsted and Keckermann. In recent historiography on the artisanal roots of modern science, however, Ramus plays relatively little part, and is seemingly obscured by the

28 Hooykaas, *Humanisme*, 24; 78. See also Verdonk, 'Einfluß'.

29 Hooykaas, *Humanisme*, 81–84; 92–95. On Nuremberg: Ramus, *Scholarum*, 62–64. I have always found the dedication of a Dutch book to a far-away city remarkable, even if they spoke the same language. I do not know the extent to which Stevin maintained relations with the municipality – or with the court of Rudolf, to whom he dedicated the *Weeghconst* – or whether he had any aspirations for patronage in these lands.

likes of Paracelsus.[30] With the exception of Van den Heuvel's study of the *Huysbou*, Ramus is largely absent from the historiography on Stevin as well. Verdonk listed some connections between their publications, but he largely focused on philosophical content and did not consider the wider historical context. Most notable is the almost complete absence of Ramus from the *Principal Works*. Besides rearranging and even dividing his works along modern disciplinary lines, and cutting through the historical connections, the *Dialectike* was not included at all. In his 1943 biography, Dijksterhuis made a few obligatory references to Ramus, but never entered into a serious consideration; his presence was further diminished in the English extract of 1970. The *Principal Works* being the main point of access for non-Dutch scholarship, Stevin is completely invisible from the historiography on Ramus as a result.

The mutual historiographical invisibility of Stevin and Ramus has made an orphan of one of the most complicated parts of Stevin's oeuvre, the 'Wijsentijt'. This discourse was a response to current debates, and in this case, the Ramist context is even more evident. As Goulding has shown, in his discussions of the nature of mathematics, Ramus introduced a highly original history of mathematics that proved very influential in late sixteenth-century debates about the mathematical sciences. In the same vein as Stevin, Ramus argued that the roots of wisdom were pre-classical. Goulding omits to mention Stevin, however, and the Dutch twist he gave to the debate on the roots of mathematics.[31]

In typical Renaissance fashion, the debate on the nature and meaning of mathematics invoked history to substantiate positions. Such histories were crucially informed by the positions taken, of course, and Ramus was no exception. At stake were the issues of how to trace the prehistoric origins of mathematics and how to evaluate the Greek written heritage. In other words: what were the original and pure roots of mathematics? According to Ramus, these roots lay in the practical works of ancient and prehistoric measurers and reckoners. The Greeks had collected and systematized this knowledge and added proofs and formalizations. Ramus considered this formalization to have been a backwards step, transforming pure learning into a philosophical exercise. His *bêtes noires* were Plato, Euclid and Theon, whereas Pythagoras and Heron were the personification of true mathematicians. Ramus' praise for the practical mathematics of his day was thus a way of portraying Nuremberg as a regained

30 That is, the most prominent voices: Smith, *Body of the Artisan* and other writings, in particular. Long, *Artisans/Practitioners* primarily focuses on Italy. Hooykaas, *Humanisme*, 'Empirisme' and Berkel, *Beeckman*, link Ramus to practical humanism.

31 Goulding, *Defending Hypathia*.

paradise. He considered Pythagoras to be the archetypical teacher of mathematics and he assumed he had been the first elementarian.

Stevin expressed a much more nuanced position on this point, particularly with respect to Euclid. Although he agreed that original wisdom had been lost by the age of the Greeks, he did not consider classical mathematics to be a corruption of that wisdom. In a reworking of his discourse on *spiegheling* and *daet*, Stevin characterized Euclid as an archetypal 'Spieghelaer' who provided the reason underlying mathematical truths. In his view, (pre-)Euclidean mathematics represented the ultimate way of thinking, with clear and distinct analysis of concepts and inference. Like *spiegheling* and *daet*, *spieghelaers* like Euclid and men of *daet* were complementary rather than exclusive. (He saw the *Geometria* as preparation for the *Meetdaet*). What was lacking in Euclid was the analytical part of *inventio*, but the synthetical part was exemplary. Unlike Ramus, who published new elements in geometry and arithmetic, Stevin did not call for a replacement of the *Elements*. All that was needed was proper access through translation.[32]

The instances discussed above suggest that Stevin was well aware of Ramist ideas about mathematics, learning and wisdom. Moreover, he appears to have actively engaged with them, taking a balanced and critical stance. The historiography has tended to neglect this Ramist context of Stevin's thought, partly because Stevin himself did not make explicit references. I am convinced that our understanding of Stevin is enhanced by allowing for the intellectual context that surrounded it, such as (post-)Ramist conceptions of learning. The 'Wijsentijt' provides an illuminating example: when we place Stevin's reflection upon the origins of wisdom in the context of Ramus' history of mathematics, this rather elusive text turns out to be a rather subtle response to contemporary views. Stevin emphasized and elaborated the prehistoric origins of mathematics, and offered a Dutch interpretation of the Josephan story about the transmission of mathematical knowledge through the Flood. Furthermore, Stevin used his views on pristine knowledge to articulate the epistemic value of the arts. Building upon ideas he had been elaborating since he embarked on his publishing career, he offered a comprehensive view of the nature and order of the *consten*. His vernacular epistemology was empirical, collaborative, civic, popular, and structured by natural logic.

6 Stevin's Persona

As Van de Heuvel and Vermij also argue in this volume, the 'Wijsentijt' forms an integral part of Stevin's thought, and its revival a program for learned pursuits.

32 As Verroten would provide thirty years later.

Although this may be true, we should also note the differences in style and context between the writings of 1585 and 1608. The *Wisconstighe ghedachtnissen* sprang from the courtly setting of Stevin's position as an advisor to Maurits. The Stevin of the 1580s was in an entirely different position and, whatever his concrete ambitions, adopted a rather different tone. The mechanical tracts were aimed at enlightening experts and promoting operative knowledge. In the *Wisconstighe ghedachtnissen,* the same content was presented in what was definitely a more scholarly style, and was included in an encompassing and fairly systematic exposition on the sciences. The extent to which this indicates a change in Stevin's *mind* remains to be seen. It remains difficult to know what to make of Stevin. His ideas and aspirations and his cultural profile are notoriously elusive, mainly because he did not leave any correspondence or ego-documents. We largely have to make do with his published work. To the extent that his oeuvre reflects his views and intentions, an ambiguous picture appears regarding his *persona*.

Stevin is commonly described as an 'engineer' and a 'mathematician'. If we understand 'mathematics' in an early modern sense, the latter covers his interests quite well, but it says little about the role he played and his attitude. Calling him an 'engineer' is rather problematic, not least because that label was largely restricted to the military until the nineteenth century. Moreover, Stevin himself was not a true practitioner. He had started out as an office clerk in Bruges, but upon his arrival in Holland he developed a learned profile: during the 1580s, he studied at Leiden University and started his writing and publishing career. He was involved in various technological projects in water management and harbor design and also in entrepreneurial ventures, but he never held office as an engineer or had a post in civil mathematics. In 1593, Stevin formally entered Maurits' service as a quartermaster, advisor and personal teacher. His principal persona seems to have been that of intermediary, making mathematical learning accessible and useable for the commonweal by writing, advising and planning. He was based at the drawing office, where models and plans were made. The commonweal that Stevin served was the young Republic, which was in revolt. The role that he played for Maurits can best be described as that of scientific advisor, drawing attention to the relevant knowledge on fortification, castrametation, bookkeeping and tactics, and how it could be employed.[33]

Stevin's views on knowledge and learning resonated quite well with those of Maurits and the other members of the House of Nassau. Beginning with the brothers Willem and Johan (VI), the Nassaus had developed an educational strategy for the intellectual and administrative reinforcement of their

33 Van den Heuvel, 'Wisconstighe Ghedachtenissen', 107–110; 117–119.

new Calvinist states. It was continued by their sons Maurits and Willem Lodewijk, the stadholders in Holland and Friesland, respectively. Throughout their domains, the Nassaus established universities and colleges to train new ministers, lawyers and doctors. The educational program was based on Ramist ideas, in particular those of Alsted, advocating a worldly-oriented form of learning that critically re-assessed the teachings of the classics. The emphasis was on context-rich, realistic and student-oriented teaching, which followed the proper logic of its subject matter rather than that of classical texts.[34] Mathematics had a prominent place in these educational campaigns; Ramus himself wrote textbooks on geometry and arithmetic that promoted a new approach to elementary mathematics. In Leiden, an outspoken Ramist had been appointed in 1578 in the person of Rudolph Snellius, and Adriaan Metius, the son of the Dutch superintendent Anthonisz., was recruited to Franeker in 1598. On his appointment as a professor, Metius was also given the right to teach mathematics in the vernacular for the benefit of engineers, surveyors and navigators.[35] The Duytsche Mathematique in Leiden was similar to the arrangement in Franeker, and reflected the ambitions of the Nassau family.[36]

With his publications and programmatic ideas regarding the *consten,* Stevin contributed to the ambitions of the new state. As regards their content, the mathematical sciences clearly took center stage, but they were far from confined to 'wisconst'. Like 'theory and praxis', Stevin's category of 'wisconst' is not self-explanatory. The classical author exemplifying the 'wisconstenaar', Archimedes, was pictured as much as a mathematician as he was a magician. The key feature was his ingenuity with hydraulic works – pumps, fountains, sluices, mills, dikes and polders – which was also prominent in Stevin's oeuvre. In Stevin's conception, the *consten* covered the whole domain of operative knowledge, placing the mathematical sciences alongside alchemy and natural magic. The common denominator was thus 'ingenuity', as reflected by the adage 'A miracle is no miracle', which Stevin had employed since his earliest publications. 'A miracle is no miracle' ties in not only with the mathematical analysis of the *clootcrans* (wrath of spheres) in the *Weeghconst,* but also more generally with the uncovering of the powers of nature in natural magic. In this view, the workings of nature are on the one hand obscure and not directly

---

34 Hotson, *Commonplace Learning.*

35 Dijkstra, *Between Academics and Idiots.*

36 Hotson, *Commonplace Learning,* 38–51. Dijkstra, *Between Academics and Idiots,* 47–88. The extent to which Stevin's reflections were guided towards the Nassau ideology is in need of further consideration.

accessible, as in the Aristotelian observation; but on the other hand, they can be revealed indirectly and brought under control through inventive means. Uncovering the wonders of nature was the true wisdom brought forth by the *consten*, and Stevin explained how to cultivate them. He did so not only by means of programmatic exposition, but primarily by means of 'essays' such as the *Weeghconst*.

Bibliography

Berkel, Klaas van, *Isaac Beeckman on Matter and Motion : Mechanical Philosophy in the Making*, Baltimore: Johns Hopkins University Press, 2013.

Dijksterhuis, Eduard Jan, *Simon Stevin: Science in the Netherlands around 1600*, The Hague: Martinus Nijhoff, 1970.

Dijksterhuis, Fokko Jan, "Duytsche Mathematique and the Building of a New Society: Pursuits of Mathematics in the Seventeenth-Century Dutch Republic", in: eds. Lesley B. Cormack, Steven A. Walton and John A. Schuster, *Mathematical Practitioners and the Transformation of Natural Knowledge in Early Modern Europe*, Dordrecht: Springer, 2017, 167–181.

Dijksterhuis, Fokko Jan, "Stevin, Huygens, and the Dutch Republic", *Nieuw Archief voor Wiskunde* 9 (2010) 100–107.

Dijkstra, Arjen, *Between Academics and Idiots: A Cultural History of Mathematics in the Dutch Province of Friesland (1600–1700)* (Enschede: dissertation Universiteit Twente, 2012).

Euclides, *Zes eerste boekken, vande beginselen der wiskonsten* (Translated into Dutch by J.W. Verroten); Hamburg, Gaubisch, 1633.

Goulding, Robert, *Defending Hypatia: Ramus, Savile, and the Renaissance Rediscovery of Mathematical History*, Dordrecht : Springer, 2010.

Heuvel, Charles van den, "Wisconstighe Ghedachtenissen. Maurits over de kunsten en wetenschappen in het werk van Stevin," in: *Maurits, Prins van Oranje*, ed. K. Zandvliet, Amsterdam/Zwolle: Rijksmuseum/Waanders, 2000, 106–121.

Heuvel, Charles van den, *'De Huysbou' – A Reconstruction of an Unfinished Treatise on Architecture, Town Planning and Civil Engineering by Simon Stevin*, Amsterdam: KNAW, 2005.

Hooykaas, Reyer, "Pierre de la Ramée et l'empirisme scientifique au XVIe siècle", in: *La Science au seizième siècle*, Paris: Hermann, 1957, 297–311.

Hooykaas, Reyer, *Humanisme, science et réforme. Pierre de la Ramée, 1515–1572*, Leiden: Brill, 1958.

Hotson, Howard, *Commonplace Learning. Ramism and its German Ramifications, 1543–1630*, Oxford: Oxford University Press, 2007.

Lipsius, Iustus, *Epistolarum Selectarum Centuria Prima Miscellanea*, Antwerp: Plantijn, 1614.

Long, Pamela, *Artisan/Practitioners and the Rise of the New Sciences, 1400–1600*, Corvallis: Oregon State University Press, 2011.

Molhuysen, P. ed., *Bronnen tot de geschiedenis der Leidsche Universiteit*. Vol. 1 (Rijksgeschiedkundige Publicatiën 20), Den Haag, 1913.

Ramus, Petrus, *Petri Rami Scholarum mathematicarum libri unus et triginta*, Basel: Episcopius, 1569.

Secretan, Catherine, "Simon Stevin's Vita politica. Het burgherlick leven (1590): A Practical Guide for Civic Life in the Netherlands at the End of the Sixteenth Century", *De Zeventiende Eeuw: Cultuur in de Nederlanden in Interdisciplinair Perspectief*,28-1 (2012) 2–20.

Secretan, Catherine, and Pim den Boer eds., *Simon Stevin. De la vie civile 1590*, Lyon: ENS éditions, 2005.

Smith, Pamela H., *The Body of the Artisan : Art and Experience in the Scientific Revolution*, Chicago: University of Chicago Press, 2004.

Stevin, Simon, *De thiende leerende door onghehoorde lichticheyt allen rekeningen onder den menschen noodich vallende, afveerdighen door heele ghetalen sonder ghebrokenen*, Leiden: Plantijn, 1585.

Stevin, Simon, *Dialectike ofte bevvysconst. Leerende van allen saecken recht ende constelick oirdeelen*, Leiden: Plantijn, 1585.

Stevin, Simon, *L'arithmetiqve*, Leiden: Plantijn, 1585 (includes *La pratique d'arithmetique*).

Stevin, Simon, *De Beghinselen der Weeghconst*, Leiden: Plantijn/Raphelinghen, 1586.

Stevin, Simon, *De Weeghdaet*, Leiden: Plantijn/Raphelinghen, 1586.

Stevin, Simon, *De Beghinselen des Waterwichts*, Leiden: Plantijn/Raphelinghen, 1586 (includes 'Anvang der Waterwichtdaet).

Stevin, Simon, *Vita politica. Het burgherlick leven*, Leiden: Ravelenghen, 1590.

Stevin, Simon [attributed to], *Maniere ende ordre die sijn Extie verstaen heeft dat sal worden achtervolght int doen van de Nederduytsche lessen in de universiteyt tot Leyden tot onderrichtinghe der ghene die hun ten dienste van den lande sullen begheven tot oeffeninghe van het ingenieurschap ende andere mathematische consten*, Leiden: Paedts, 1600.

Stevin, Simon, *Wisconstighe ghedachtenissen*, Leiden: J. Bouvvensz., 1608 (title-pages volumes 2–4: 1605) (French and Latin translations were printed by Paedts in the same year: *Memoires mathematiques*; *Hypomnemata mathematica*).

Stevin, Simon, *The Principal Works of Simon Stevin*, Amsterdam: Swets & Zeitlinger, 1955–1966.

Verdonk, J.J., "Vom Einfluss des P. de la Ramée auf Simon Stevin", *Centaurus*, 13-3,4 (1969) 251–262.

Verdonk, J.J., *Petrus Ramus en de wiskunde*, Assen: Van Gorcum, 1966.

Winter, P.J. van, *Hoger beroepsonderwijs avant-la-lettre. Bemoeiingen met de vorming avn landmeters en ingenieurs bij de Nederlandse universiteiten van de 17e en 18e eeuw* (Verhandelingen der Koninklijke Nederlandse Akademie van Wetenschappen, Afd. Letterkunde, Nieuwe Reeks, deel 137) Amsterdam, 1988.

Wal, Marijke J. van der, "Logic, Linguistics, and Simon Stevin in the Context of the Sixteenth and Seventeenth Centuries", in: ed. K.R. Jankowski, *History of Linguistics 1993*, Amsterdam: John Benjamins, 1993, 147–156.

Wal, Marijke J. van der, "Verrotens taalbeschouwing. Een onbekend voorbeeld van Stevinreceptie", *Tijdschrift voor Nederlandsche Taal- en Letterkunde*, 109-4 (1993) 328–347.

Zilsel, Edgar, "The Sociological Roots of Science", *American Journal of Sociology*, 47-4 (1942) 544–562.

CHAPTER 8

Simon Stevin's Age of the Sages

In Search of an Alternative Renaissance: Exploring Scientific Methods Based on Pre-Classical Authorities, Empirical Data and Pure Languages

Charles van den Heuvel

1 Prelude: the Fantastical Theory of the Age of the Sages

'The whole theory forms a typical example of how the most rational and scientific of minds may at the same time foster the most irrational and phantastical ideas on topics lying outside the sphere of his specific competence', wrote Eduard Jan Dijksterhuis in 1955.[1] The 'phantastical' theory that this renowned expert on the life and works of the Flemish-Dutch engineer was referring to was the 'Wijsentijt' ['Age of the Sages'], which Simon Stevin (1548–1620) published as part of the second volume of his *Wisconstighe ghedachtenissen* [Mathematical Memoirs]. The *Wisconstighe ghedachtenissen* was a partially unfinished, revised compilation of the private lessons that Stevin had given to Stadholder Prince Maurits on the mathematical sciences, physics and mechanics. In his 'Age of the Sages', Stevin sets out his dream to restore a golden age of the sciences. It is not clear when this age occurred, but from the evidence that Stevin gathers to prove its existence, it would have preceded the Egyptian, Greek and Roman civilizations. In the period between 6/700 and 1450 AD, which Stevin called an 'Age of the Ignorant' (*Leeckentijd*, in the margin *Barabarum Soeculum*), the knowledge of the liberal arts and sciences inherited by these civilizations was lost in the attempt to erase their texts about the pagan gods they worshipped. This knowledge had been partially restored in the last 150 years, between 1450 and 1600; the period that we would call the Renaissance. Stevin also referred to this period as the Age of the Sages, but called it an Age of the Ignorant in comparison with the lost Age of the Sages.

At first sight, Stevin's theory of the Age of the Sages might indeed seem an odd fantasy, particularly in the context of Stevin's rather rigid works on

---

1 Dijksterhuis, E.J., 'General Introduction. The Life and Works of Simon Stevin', in: Crone et al., *Principal Works of Simon Stevin*, I, 8.

 | DOI:10.1163/9789004432918_010

the natural sciences. This might be the reason why the 'Age of the Sages' is hardly discussed in the historiography on Stevin's life and work. One exception is the doctoral thesis by Kees Schilt, who used the 'Age of the Sages' as a fundamental text to explain Stevin's scientific method from an original, but hardly convincing, hermetist perspective.[2] Charles van den Heuvel drew on the text in an attempt to (re)construct the order of Stevin's *Huysbou* [Architecture], gleaned from tables of contents and posthumous fragments in manuscripts and partial publications.[3] Other than this, the 'Age of the Sages' has only been discussed from a narrow perspective to underpin Stevin's views on the vernacular. The 'Age of the Sages', together with the 'Uytspraeck' van de Weerdicheyt der Duytsche Tael' [Discourse on the Worth of the Dutch Language] that Stevin published as an introduction to his work on mechanics, *Weeghconst,* is indeed crucial for understanding his views on the superiority of the Dutch language. Probably for that reason, much attention was paid to the 'Age of the Sages' in Flanders in the second of half of the nineteenth century, as part of the Flemish emancipation and the linguistic conflict between advocates of the French and Dutch languages in the young Kingdom of Belgium.[4]

Here, we shall advance the claim that the 'Age of the Sages' is not only important for Stevin's views on the Dutch language, but is also illustrative of the impact of the educational turn taken by his career when he started to teach the stadholder, later Prince Maurits. Furthermore, we propose that the development of Stevin's ideas on the Age of the Sages might explain the discontinuity in the history of his most important publication, the *Wisconstighe ghedachtenissen,* and some of the contents of his unpublished works on architecture, music and military affairs. To this end, we will demonstrate that the 'Age of the Sages', which Dijksterhuis dismissed as 'outside the sphere of his [Stevin's] competence', is in fact crucial for understanding developments in Stevin's scientific methodologies; that is, the shift from theory toward the use of empirical data, and the extension of his views on the purity of the Dutch language to other (mathematical, musical, architectural) languages. Finally, the 'Age of the Sages' sheds light on Stevin's views on natural order and his quest for an alternative Renaissance that challenged the reception of classical authorities. Moreover, it reveals that Stevin's interest in the legacy of pre-classical sources was different from that of his contemporaries.

---

2 *Simon Stevin en het hermetisme.*

3 Van den Heuvel, '*De Huysbou*', 18–33, in particular 28–33.

4 See for instance Van Duyse, *Simon Stevin.*

2 The 'Age of the Sages': Signs Based on Philological Fragments

Stevin's 'Age of the Sages' forms a section (the 6th definition) of the introduction to Stevin's *Vant Weereldtschrift* [Geography], which is the first of five parts of the second volume of the *Wisconstighe ghedachtenissen* (1608). Stevin defines this era as follows:

> The Age of the Sages we call that time in which exceptional learning was to be found among men, a fact which we perceive with certainty from certain signs, but without knowing among whom, where or when.[5]

Whether Stevin had always planned to include this section in his work on geography is unclear. Posthumously published sections on architecture and music intended for the *Wisconstighe ghedachtenissen* contain references to the Age of the Sages as well. In a note to the reader, Stevin explains that he had assembled part of his publications for the *Wisconstighe ghedachtenissen* a long time ago, but that he had not been able to finish several sections of this chapter in time for the printer, and therefore planned to publish these at a later stage. That said, however, the first of the arguments – to use his term, 'signs' – that Stevin introduced to underpin his claim for this mythical age is indeed based on his views about geography; that is, 'cosmography' rather than geography, because Stevin's *Weereldschrift* also included astronomical descriptions of the motions of planets and fixed stars in support of Copernicus.[6] Stevin's first 'sign' of the existence of the Age of the Sages is that Ptolemy was not the first to observe the 'second inequality in the motion of heavily bodies', but that this had become known via Arabic texts, with images that showed that the Egyptians marked fixed stars differently on their globes. It was the orientalist scholar Josephus Scaliger who had shown Stevin these figures. This is just one of various examples of textual or visual Arabic sources used by Stevin as evidence that several astronomical doctrines were not transmitted by the Greeks before ending up in Latin, but must have originated from another highly-developed civilizations. His claims are not only based on works by other scientists, but also on his own observations. During a visit to the royal palace in Cracow, Poland, Stevin apparently noticed species of animals on wall paintings with the caption 'Signa Hermetis' (signs of Hermes); these could not be found in Ptolemy's

5 'Wysentyt noemen wy die, waer in by de menschen een seltsaem wetenschap gheweest heeft, t'welck wy deur seker teyckens ghewisselick mercken, doch sonder te weten by wie, waer, of wanneer', *Principal Works of Simon Stevin*, III, 592.

6 See Vermij, *Calvinist Copernicans* and his chapter in this volume.

work, and must therefore have been transmitted by another source. From the astronomical arguments, Stevin moves on to other disciplines that are also related to the general themes mathematics and natural philosophy of his *Wisconstighe ghedachtenissen*.

The second sign that points to the existence of the Age of the Sages can also be found in Arabic texts, this time in writings on algebra. Here Stevin claims that neither the Chaldeans nor the Hebrews, the Greeks or the Romans, 'all of whom were no arithmeticians worth the name', could have had an understanding of the tenth progression of numbers, and that they had completely misunderstood the original use of a 'point' [also named point] for the number '0' that Scaliger had shown him in Arabic books from the Middle East. Stevin used European sources as well. Regarding the question of how the right number of divisions in the ratio between the diameter of a circle and its circumference had been calculated for the first time to represent the radius in the tables of sines, Stevin referred to a treatise on trigonometry by the Austrian mathematician, Georg von Peuerbach. Peuerbach had explained that the 20,000.000 to 62,832 ratio, almost similar to the one he calculated himself, had already been used by the Indians, Egyptians and Arabs. Stevin therefore concluded that such ratios were already in use long before the Greeks and Regiomontanus, probably in the Age of the Sages. Stevin's exposition of the subsequent signs of the Age of the Sages is considerable shorter. Whereas the description of the first two signs took up six pages, the last four fit into one. In his exposition of the third sign, Stevin claims that the way Euclid represented the rules of geometry was consistent with the systematic order of mathematics in the Age of the Sages.

The next argument – the fourth sign – was not based on a complete discipline, but on a discussion of the height of the atmosphere, once again transmitted by Arabic texts which reveal that in former times (i.e., prior to the period of Greek civilization), mathematicians were already concerned with the form of the world and its properties. The next, fifth sign, the 'miraculous' domain of alchemy, likewise reveals a certain disdain on Stevin's part for the knowledge of the Greeks. According to him, alchemy, important for understanding the nature of substances, was unknown to the Greeks.[7] Stevin considered interpretations of nature by the Greeks and their followers, 'who are called Philosophers', praiseworthy, given the fact that according to him, 'they had it all from hearsay, with little real knowledge, with many errors and without knowledge of the

7 This is one of the very few references to alchemy and Hermes in Stevin's oeuvre. These concepts do not return in discussions of his method and therefore provide no real evidence for Schilt's claim (see note 2) that Stevin was a hermetist.

causes'.[8] The real transmitter of the knowledge of alchemy, according to Stevin, was Hermes Trismegistus, whose origins – like those of the Age of the Sages – were unknown, but thought to be very ancient.[9] The sixth and final sign that Stevin recognized, 'magic', is rather vague and somewhat contradictory. Stevin described the practice of magic as 'built on ignorance', but simultaneously stated, without further explanation, that it is based on a profound understanding of mathematics. Furthermore, it might be read as an allusion to his motto, 'Wonder en is gheen wonder' ('A miracle is no miracle'). By this he meant that certain observations at first sight make us wonder, but once we have learned their true nature, revealed by proof (i.e., by mathematical proof), they are no longer miracles to us. Stevin considered this learning from mathematics to be the cause of the thorough knowledge that had existed in the Age of the Sages.

It cannot be denied that as evidence for this mythical age, Stevin's 'signs' are rather incoherent. Sometimes they represent a complete discipline and in other cases just a minor debate or discussion. Stevin's repeated rebuke that the Greeks acquired their knowledge 'by hearsay' is also somewhat surprising in the context of his own references to conversations with Josephus Scaliger. In a similar way, Stevin obtained information from his friend, the jurist and humanist Hugo Grotius (1583–1645), who was a pupil of Scaliger. In 1602, Grotius had published his *Parallelon rerumpublicarum liber tertius,* in which he compared the nature and culture of people in the Republic of Holland with those of Rome and Athens. For this work, Grotius had tried to link features of the ancient Batavian Republic with those of the civilizations that had preceded the Greek and Roman Republics. The information that Grotius gave Stevin was probably based on his notes on these very ancient civilizations that he collected when writing his *Parallelon*. We do not know in which form Grotius provided this information. However, the section that Stevin included in his Geography, under the title *Ghetuychnissen eens seer ouden wysen-tijts, by een vergaert deur den Hoochgeleerden Heer Doctor Huych de Groot* ['Testimonies of a very old Age of the Sages compiled by the very learned doctor Huych de Groot'], was a summary of authors and texts purporting to provide evidence of knowledge of older civilizations predating, and unknown to, the Chaldeans, Egyptians and Greeks.

In short, Stevin seems to have been rather eclectic in providing evidence of the existence of the Age of the Sages, and he is far from systematic. Nevertheless, after a discussion of Stevin's strategies to restore the Age of the Sages, we

8 *Principal Works of Simon Stevin*, III, 605–607.

9 See notes 2 and 8.

will explain how the themes that were addressed for first time in his *Geography* were developed in more detail in other, later texts that were announced, but never published as part of the *Wisconstighe ghedachtenissen*.

3 The Restoration of the Age of the Sages

Stevin is certain that mankind once possessed a period of great learning. As there is no reason, in his view, to believe that human intelligence has deteriorated since then, he concludes that it can be restored if one uses the same methods as in the past. Stevin proposes a four-step strategy for this purpose, which he briefly introduces in four sections or articles:

1. The first problem that Stevin observes concerns the lack of a large body of data obtained by practical experience, which will only be restored if many people set about creating such body of data (similar to what we could call crowd-sourcing today).
2. To get more people involved, the sciences need to be practiced in their native language. To achieve good results, this language should be exceptional. According to Stevin, this has not happened since the Age of the Sages – with the exception of Greek civilization, but then only for the domain of geometry.
3. In order to obtain exceptionally good languages, one needs to know what the excellent qualities of a language are.
4. Finally, it is crucial to choose the best order with which to describe scientific problems and questions.

After this brief overview, Stevin describes each article in more detail, arguing why it is necessary to restore the Age of the Sages. Stevin's argument for the need for empirical data is mainly based on examples of astronomical observations. One brief intermezzo in the domain of geography is of interest, however, because it introduces the concept of probability based on empirical experience. Stevin provides the thought experiment of a man sailing along the coast of Australia, who, seeing the mouth of a large river, might come to the following conclusion: Along large rivers lie fertile lands. People choose to live in fertile lands. Where many people live, good, prosperous cities will arise.

As we shall see later, Stevin developed this example in more detail in his work on architecture. Noteworthy here, however, is Stevin's claim that if the person were to map such land and cities on the shores of Australia, they would be very similar to the large cities on river mouths in other places. Stevin uses this *pars pro toto* description to explain how conclusions can be drawn with a certain degree of certainty, thereby introducing the concept of probability.

From the observations by this single sailor, Stevin moves on to how large numbers of people can gather information about space. Rather like the Internet's most successful citizen science project, Zooniverse, which started off by crowdsourcing observations of space, Stevin uses the example of astronomy to explain how knowledge in the Age of the Sages was obtained by gathering large quantities of empirical data. It is impossible for one person to observe the position of the planets by night and by day. Moreover, one person's data, even if they are precise, are not sufficient to develop theories, because they have not been checked. Stevin criticizes the view that astronomy is too elevated a subject to be studied by ordinary people. Ultimately, when it comes to obtaining greater certainty and knowledge of the causes, the 'big data' produced by ordinary people is more important than the individual expertise of the lauded scholar. According to Stevin, the sciences and liberal arts are usually discussed in Latin and whilst some parents have their children taught in this language so that they may study law, theology or medicine, most people will not be reached. Although these privileged children might gather 'flowers of speech', hardly any of them will dedicate themselves to mathematics, and for that reason it is better to use a native language. Furthermore, not any native language will do; this language also needs to be a good language.

For Stevin, the problem of language is potentially the most serious obstacle to returning to the Age of the Sages. The reason for this is that with the decline of many sciences from the Age of the Sages, men also seem to have lost their understanding of the qualities of a good language. For instance, the French use their native language far more to describe the liberal arts than other peoples. According to Stevin, however, French is less suitable for technical terms, because it makes use of words in other languages, such as Latin, which ultimately results in poor definitions. The use of poor definitions can have a huge impact. Stevin explains this by referring to an example from music theory, where the word 'proportion' instead of the Dutch word 'evenredenheydt' (equiratio) has often been used, which since Pythagoras has led to misunderstandings in tuning.[10]

After this explanation of how certain languages are inadequate for expressing certain concepts, Stevin elaborates on the properties of a good language. Stevin praises the quality of the rich Greek language, which allows for the creation of 'constwoorden' (*vocabula atrium*) or definitions by joining syllables in compound words. According to Stevin, however, the Dutch language is even

10 See Cohen, 'Simon Stevin's equal division of the octave' and the chapter by Cohen and Julia Kursell in this volume.

better, because it consists of more concise, monosyllabic words. To prove his point, Stevin provides a list of 742 monosyllabic words followed by their French and Latin translations with words of two or more syllables. Those who understand the power of the monosyllabic words in the Dutch language, particularly in the purest form spoken in the province of North Holland, can also learn the causes of the signs of the Age of the Sages.

After explaining the need for the development by ordinary people of 'big data' for empirical research, of the importance of using a good native language, and thereby understanding what the qualities of a good language are, Stevin comes to his final point: that of order in the liberal arts and sciences. Without such order, such a high level could not have been attained in the sciences in the Age of the Sages. Stevin describes this scientific order in five articles:

1. Of the Parts of a Proposition (*Membris propositionis mathematicae*)
2. Of Definitions (*Definitionibus*)
3. Of Dichotomy (*Dichotomia*)
4. Of Anaphora (*Anaphora*)
5. Of Theory and Practice (*Theoriae & praxis*)

Regarding the formulation of good propositions, Stevin immediately refers to Euclid's *Elements,* which he believes to have been a remnant from the Age of the Sages. These elements consist of propositions in two forms, theoretical arguments and problems (*Theorematae & Problemata*). These, in turn, consist of various parts: the supposition, the property to be proved, the construction and the actual proof (in the margin: *Datum, Quasitum, Constructionem and Demonstrationem*). These parts cannot be mixed. First, the theoretical proposition has to be understood fully before one can move on to the practical case; as Prince Maurits did, Stevin hastens to say. Once again following Euclid, Stevin states that these propositions should be preceded by good definitions. These definitions are based on the use of correct terms, which are not only always available in books, but also need to be gathered by talking with practitioners from various professions:

> At times I needed to know more about earthworks, mat works, carpentry, bricklaying, smithing (and similar subjects to be discussed in *Huysbou* in greater detail) than I was able to find out from books or mathematicians: here the workmen provided [me with] the best teachers, each in their own field: [but] to make good headway in my investigation, I followed the method used in the Age of the Sages, enquiring among excavators about earthworks, among matters about osier work, among carpenters about carpentry and so forth, concerning the meaning of the particular terms that I required and did not understand, which, having

> collated the definitions and learned these by heart, I was able to use in conversation with them so that they understood me [...] so, just as it is advisable to collate and learn the definitions before approaching subjects in the field of mechanics the same is true of the liberal arts.[11]

Whilst Stevin's correct definitions can be seen as the ingredients of a good proposition, the presentation of the contents of a book in the form of a dichotomy, as he suggested further, can be understood as a kind of menu. The advantage of using a dichotomy, which was also in use in the Age of the Sages, is not only that all parts of the whole can be seen with much certainty, but also that such a dichotomy is based on the logic that each part is always preceded by another that has already been explained or proven. Moreover, not only do the successive parts build up a scientific argument, but the latter can also be reinforced by repetition. Stevin suggests the use of so-called 'anaphora' for this purpose, a rhetorical style figure in which specific terms are repeated to stress a particular argument. The fifth and final section of Stevin's scientific order concerns the combination of theory and practice. Stevin argues that one of the advantages of explaining a theory with practical examples is that many people can be reached in this way. At the same time, he stresses that underpinning measurements in practice with theory is important for our understanding of the common ground that is necessary for the application of the same rules in other circumstances. According to Stevin, however, it was not necessary to be practitioner to contribute to practical knowledge:

> As the theorist Euclid, who we do not find to have been a practitioner, nevertheless described proposals which have clearly been useful to actual architects and land surveyors and other practitioners [...] Thus a

11 'Tis ghebeurt dat ick van eerdewercken, rijswercken, timmering, metsing, smeding (en dierghelijcke daer inden Huysbou breeder af gheseyt sal worden) meer behoufde te weten dan ick deur boucken of Wisconstnaers leeren conde: Hier toe verstreckten my de wercklyuden elck in haer const voor beste meesters: Maer om in mijn leering goede voortganck te crijghen, ick volghde de Wysentijts oirden, vraghende voor al den Eerdewerckers in eerdewercken, Rijswerckers in rijswercken, Timmerlien in timmering &c. na de beteyckening van haer eygen woorden die ick my noodich bevant, en niet en verstont, welcke ick als bepalinghen oirdentlick opgheteyckent hebbende, en die van buyten gheleert, ick sprack terstont met hun soo datse my verstonden [...] Nu ghelijck dit versamen en leeren der bepalinghen eermen ande saeck comt, oirboir is in tuychwerckelicke consten* [mechanicus artibus] alsoo ist oock inde vrije* [Liberalibus Theoriae]', Stevin, *Wisconstighe ghedachtenissen*, 'I Bouck des Eertclootschrifts', 43.

> theorist's theories which also serve practitioners are useful even though he is not a practitioner.[12]

Many authors who have read the work of Stevin in purely rationalist terms have understood his argument to be that a theory was only useful if it could be applied in practice. Despite Stevin's arguments for combining theory and practice and the value of the practical application of theory, however, as we shall see, he was convinced that theories could exist in their own right.

4 Anticipation of the Restoration of the Age of the Sages in Stevin's Early Work

Stevin mentioned the 'Wysentijt' explicitly for the first time in his *Wisconstighe ghedachtenissen*, but the scientific methods that he discussed to restore this period of great wisdom were already anticipated in his very early works. Long before the 'Age of the Sages', Stevin published a work on dialectics: *Dialectike ofte Bewysconst* [Dialectics or the Art of Demonstration] (Leiden 1585).[13] As shown by its subtitle, Stevin intended to explain how one could establish sound scientific proof in order to open up the most profound secrets of nature. The introduction mentioned yet another work that revealed Stevin's obsession with the Dutch language: 'Uytspraeck van de Weerdicheyt der Duytsche Tael'.[14]

Taking a closer look at the five sub-articles mentioned in the discourse on the Age of the Sages reveals that Stevin used these articles in his earlier works as well, albeit inconsistently. The clear presentation of propositions and the introductory use of definitions are present in all of his work, mathematical and otherwise. Anaphora are also present in Stevin's early work, albeit that they are used differently from those in rhetoric. Whereas in rhetoric synonyms are commonly used to reinforce a statement, Stevin repeats exactly the same words. Stevin uses dichotomies to give a clear overview of the contents in several of his earlier books; for instance, in his *Tafelen van*

---

12 'Als den Spieghelaer Euclides, die wy nieten bevinden Doender gheweest te hebben, heeft nochtans voorstellen beschreven den dadelicken Boumeesters en Landtmeters, en ander doenders seer voordelick [...] Daerom een Spieghelaers Spieghelinghen die anders Doenders te sta commen, en sijn niet onnut al en is hy gheen Doender'. Stevin *Wisconstighe ghedachtenissen*, 'I Bouck des Eertclootschrift', 47.

13 See the chapter by Buning in this volume.

14 *Principal Works of Simon Stevin* I, 58.

Interest (1582), *Problemata Geometrica* (1582), *The Thiende* (1585), *L'Arithmetique* (1585) *Sterctenbouwing* (1594), *Deursichtighe* (1605), but, strangely enough, not in the works that appeared after the publication of the 'Age of the Sages' in 1608, *Castrametatio. Dat is Legermeting* and *Nieuwe Maniere van Sterctebou door Spilsluysen* of 1617. Finally, the idea of combining theory with practice that could co-exist in their own right is reflected in Stevin's earlier work, too: his *Arithmetique* was followed by the *Pratique d'arithmetique*, the *Weegconst* [Art of Mechanics] by the *Weegdaet* [Practice of Mechanics], and the *Problemata geometrica* [Geometrical problems] by the *Meetdaet* [The Practice of Measuring].

Although Stevin's views on scientific method in the 'Age of the Sages' were already anticipated in his early work, they seem to form part of a long theoretical (Aristotelian and Euclidean) tradition of the art of reasoning. In what follows, we explain how Stevin's teaching to Prince Maurits, the (partial) publication of the *Wisconstighe ghedachtenissen* and his later role in the States Army would result in changes to the presentation of his scientific arguments.

5 Stevin's Teaching and Its Impact on Scientific Methods

Opinions differ as to when the stadholder, later Prince Maurits, first took lessons from Stevin. It has been suggested that Maurits was already the latter's pupil when they both enrolled at Leiden University in the early 1580s.[15] Others have proposed a date some ten years later. In fact, it seems likely that Stevin became Maurits' tutor around 1593. In December 1603, Maurits urged the States General to provide Stevin with a salary, since he had served ten years as quartermaster in the army without drawing pay. E.J. Dijksterhuis rightly concluded that as Stevin had received no wages for his services in the army, he must have been paid by Maurits.[16] It has generally been assumed that the publication of *Sterctenbouwing* [Fortification] in 1594 was the result of Stevin's contact with Maurits, and that he had had a hand in the military reforms upon which his pupil and Maurits' cousin Willem Lodewijk had embarked around 1590. However, this assumption requires some qualification.[17] It is striking, for

15 Dijksterhuis, *Simon Stevin*, 10 note 1.

16 Ibidem.

17 For the military reforms, see Werner Hahlweg, *Die Heeresform der Oranier* and idem, 'Aspekten'. For the exaggeration of the part played by Maurits and Stevin in military reform and the underestimation of that of Willem Lodewijk, see Postema, *Johan van den Corput*, chapter 6.

example, that the *Sterctenbouwing* contains no reference whatsoever to Maurits or his involvement in military affairs. Moreover, the work was completed in Delft and not in The Hague, where Stevin settled almost immediately to be near the stadholder. And finally, the work is not intended as a detailed study of the construction of Dutch fortifications in practice; it presents existing knowledge rather than a new theory of fortification.[18] Indeed, Stevin may have commenced his *Sterctenbouwing* before he became directly concerned with the military, and Maurits may have only decided to involve his teacher more closely in the reform of the army after he had become acquainted, through his mathematics lessons, with this first essay by Stevin on the construction of fortifications. Although the *Sterctenbouwing* was probably fully underway when Stevin became Prince Maurits' tutor a year before its publication, his interest in teaching is already evident from this work on fortification. For instance, in his description of the design of the hexagonal fortress, Stevin stresses the need for didactical clarity in one's argument, in the statement: 'op dat leering met gheen strijding verduystert en worde [because the teaching should not be complicated by arguments]'.[19]

Of particular interest with regard to Stevin's view on methods is his dedication of the *Sterctenbouwing* to Hendrick van Brienen, deputy of the States General for Gelderland. Stevin is very explicit about his views on the theory and practice of science, and positions himself in the middle of a debate between adherents of Aristotelian theory and its critics, followers of Ramus:

> You held that the Aristotelians disputed violently enough about method, but in practice used little method; that Ramus' opinion on this matter tended to produce good results, but all the same you followed your own way. [...] that your honor called those who only study Euclidean principles, without proceeding to practice, 'sham fighters', who violently fight against their shadow without touching it [...] Now, although I have always been of the opinion that the theories of the theoreticians can serve to further practical work of the practitioners, yet in order to prove with the documents how much your honor's view about the combination of these two is to my liking and because I do not wish you to reckon me among the simple 'sham-fighters', I have sent you this actual example, which, though still they are only (as is said) castles in the air, or, even more properly expressed, bastions of paper, yet come much

18 Van den Heuvel, *'Papiere Bolwercken',* 141–142.

19 Stevin, *Sterctenbouwing*, 8, in: *Principal Works of Simon Stevin*, IV, 67.

nearer to the matter than theoretical ideas of magnitudes separated from matter'.[20]

Authors such as the mathematician Verdonk have quoted this passage to stress the impact of Petrus Ramus on Stevin's work.[21] At the same time, however, this passage reflects Stevin's reservations about Ramus' ideas. Stevin did not share Ramus' views on Aristotle or Euclid. However, the dedication also reveals the extent to which Stevin approved of the combination of theory and practice. Although they could exist in their own right, in Stevin's opinion, the emphasis placed by Ramus and his followers on the practical application and lucid presentation of theories must have appealed to the Dutch scholar's imagination. Ramus, like Stevin, was a teacher at heart, who stressed the importance of the methodical ordering of one's subject matter. He maintained that the truth of a theory should be demonstrated by a certain arrangement of arguments and sub-arguments, which he called the natural sequence. Thus, the correct classification was neither arbitrary nor subject to change; there could be only one correct method, namely the natural method, following a natural hierarchy of arguments and sub-arguments. Ramus and his adherents also emphasized the visual presentation of logic. This middle position occupied by Stevin also becomes apparent in the representation of his methods. Stevin presented the table of content in dichotomies, but did not follow this visual lay-out throughout. Instead, he stuck to Euclid's example of first explaining the terms, and then described his design for a hexagonal fortress according to the Vitruvian division into plan, elevation and perspective.[22]

---

20 'V.E. hieldet daer voor, dat de Aristotelischen bestelick ghenough vande Oirden [methode] streden, maer metter daet weynich Oirdens ghebruyckten: Dat Ramus meining van dies tot een goet einde streckte doch volghdet ghy noch dijn eyghen sin [...] dat V.E. den ghenen die hun niet dan in Euclidische beghinselen en oeffenen, sonder tot de daet te commen, Spieghelvechters noemde, die haer schaduwe heftelick bestrijden sonder gheraken [...] Nu hoewel mijn meyning altijt was, dat der Spieghelaers spieghelingen, tot voordering van der doenders daden connen strecken: Nochtans om mette Sticke te bewijsen, hoe my de vermenghinghe deser twee bevalt, ende dat ick niet en begheer datse my onder de eenvoudighe Spieghelvechters rekent, soo heb ick haer dit werckelic voorbeelt toegheschickt: Het welcke, hoewel dattet noch al maer (ghelijck men segt) Burghen inde locht, ofte noch eyghentlicker, papiere bolwercken en sijn. Doch commet de saecke veel naerder, dat Spieghelische ghedachten der grootheden gescheyden van stof', Stevin, *Sterctenbouwing*, 2–4, in: *Principal Works of Simon Stevin*, IV, 43–47.

21 Verdonk, 'Vom Einfluß', 251–262.

22 Stevin, *Sterctenbouwing*, 8, in: *Principal Works of Simon Stevin*, IV, 67. Stevin emphasizes: 'It is also to be understood that in doing so I follow no method of halving the whole into parts but set down first that which for effective design ought be described first'.

A second work that is directly related to education is Stevin's design of 1600 for a training program on land-surveying and fortification, at the instigation of Prince Maurits: *Maniere ende Ordre die sijn Excie verstaen heeft sal worden achtervolght int doen van de Nederduytsche lessen in the universiteyt tot Leyden tot onderrichtinghe ende bevorderinghe der ghene die hun ten dienste van den lande sullen begheven tot oeffeninghe van het ingenieurschap ende andere mathematische consten* [Manner and order commanded by his Excellency to follow lessons in the Dutch language at the University in Leiden for the training and advancement of those who will serve the country in the exercise of engineering and other mathematical arts].[23] The order of this course, the so-called 'Duytsche Mathematique', was typical of Stevin's approach. First, the students were introduced to basic geometry and trigonometry as applied to the practice of surveying. Then, they were taught to calculate the volumes of the walls and bastions for cost estimates. When the pupils had had sufficient training in drawing and measuring on paper, they started to practice in the field to get used to compasses, rulers and other land-surveying instruments. Furthermore, they made wooden and earthen models of bastions and walls. During this phase, they also practiced reading aloud the correct terminology for the parts of fortresses, to get acquainted with the art of fortification. Finally, in the summer months they assisted engineers and soldiers in the field with practical tasks, and deepened their theoretical knowledge in the classroom in the winter.

In short, the course was very practical, with a significant focus on learning by practical example. This was quite different from the teaching of land-surveying and fortification in Latin in the more historical-philological manner at Leiden University, using the works of Lipisus and Scaliger.[24] The students of Duytsche Mathematique were taught 'only so much as is needed directly for an engineer in general'.[25] Below, it is claimed that the *Wisconstighe Ghedachtenissen* must be read as the culmination of Stevin's experiences as a private tutor to Prince Maurits, a role that caused him to reconsider his views on scientific methods.

23 There is a printed, five-page copy of this work at the Dutch National Maritime Museum in Amsterdam and a manuscript of the work in Leiden's municipal library (Gemeentearchief Bibliotheek).

24 See for instance: Leiden University Library ms. SCA 6 – *Agrimensoprum Romanorum fragmenta of Sacliger* and *Justi Lipsi, De militia romana libri quinque, commentaries ad Polybium* 765 B 16 and 1413 G 21.

25 Molhuysen ed., *Bronnen*, vol 1, bijlage 338, 398*. For the Duytsche Mathematique in general: Taverne, *In 't land van belofte*, 61–69 and 75–81, Van Maanen, *Facets*, Van Winter, *Hoger onderwijs*, 14–36, Van den Heuvel, 'Traité incomplet' and 'Vesting', Krüger, 'Lessons' and 'Frans van Schooten Sr'. and Dijksterhuis, 'Duytsche Mathematique'.

6 *The* Wisconstighe Ghedachtenissen *and the 'Age of the Sages'*

The preface to the *Wisconstighe ghedachtenissen* is quite clear about its educational purpose: Stevin explains how Prince Maurits had studied books and manuscripts during military campaigns and sieges, and, out of fear of losing the exercises and his 'eygen vonden' [own inventions], had decided to publish them. One of these inventions was an adaptation of an instrument used by Albrecht Dürer to make perspective drawings. Stevin relates how Maurits, having learned how to draw plans and elevations, wished to move on to what Vitruvius termed 'scenography' (perspective). He had taken lessons from the eminent artists of his day, but was not satisfied with their explanation by 'sight or guesswork'. Rather, he wished to be able to design exactly the perspective of any figure 'with knowledge of the causes and its mathematical proof'.[26] At first glance, this praising of Prince Maurits' skills seems rhetorical. However, the exercises of Prince Maurits and his conversations with Stevin are a recurrent theme; in fact, it is so ubiquitous that some contemporary scholars even considered the *Wisconstighe ghedachtenissen* to have been a work completely, or at least partly, written by Prince Maurits.[27] The link between Stevin's teaching of Prince Maurits and his intentions with regard to the publication of the *Wisconstighe ghedachtenissen* raise the question of whether the 'Age of the Sages' should be read in similar educational terms as well.

The examples that Stevin gave to support his theory of the Age of the Sages suggest that he had been hesitant about whether to include this text for the *Wisconstighe ghedachtenissen* in the section on geography or on astronomy. It was most likely a last-minute decision. As noted above, Stevin had failed to finish in time the fifth book of the second volume of the *Wisconstighe ghedachtenissen* (1608), dedicated to 'Various Matters'. The end of this book includes Stevin's list of sections that he had not included. This concerned a short text about 'refraction' destined for the published part of the *Deursichtighe* [Perspective], a short text on hydraulics as part of the appendix of the *Weegconst* [Mechanics], and a piece on 'lochtwicht' (upward forces), intended for the same appendix, which was lost. Then Stevin lists a couple of chapters intended for the fifth section on *Various Matters: Telconstighe anteyckeninghen* [arithmetical notes], *Spiegheling der Singconst* [Music Theory], *Huysbou* [Architecture], *Chrijchshandel* [Military Affairs] and, finally, *Verscheyden anteyckeninghen* [Various Notes].

26 'Eerste bouck der deursichtighe, Vande verschaeuwing', [4], in: *Principal Works of Simon Stevin*, II b, 800. On Maurits' interest in drawing, see Van den Heuvel, 'Wisconstighe Ghedachtenissen', 2000.

27 See for this co-autorship Van den Heuvel, 'As the Author Intended', 2017.

These listed texts were only very partially reused in later publications on different topics during his lifetime; the majority of them appeared posthumously in print or were not published at all.

From the tables of contents of manuscripts compiled by the scholar Isaac Beeckman, excerpts of manuscripts and posthumous publications by Simon's son Hendrik Stevin, Constantijn Huygens and others, based on these manuscripts, we can deduce that these chapters on arithmetic, architecture, music theory and military affairs were already at an advanced stage.[28] Of course Stevin could have written substantial parts between the publication of the second volume of the *Wisconstighe ghedachtenissen* in 1608 and his death in 1620. However, the only two works that Stevin still published during his lifetime, *Castrametatio. Dat is Legermeting* and *Nieuwe Maniere van Sterctebou door Spilsluysen,* both from 1617, were directly related to his career as a quartermaster of the army camps of the States Army and his advisory role to Prince Maurits in fortification and civil engineering. For that reason, it seems more likely that after the publication of the *Wisconstige ghedachtenissen*, Stevin did little additional work on the manuscripts that had been intended in the first place to document his tutoring of Prince Maurits for a wider public.

This is of importance for two reasons. First of all, if the hypothesis is correct that the contents of these texts on architecture, music theory, military affairs, and so forth, changed little after the publication of the *Wisconstighe ghedachtenissen,* they can be linked to Stevin's intentions in this work with regard to mathematical matters. Second, because the text of the 'Age of the Sages' includes examples from architecture and music theory, this fragment can be linked to Stevin's unpublished texts on these topics intended for the *Wisconstighe ghedachtenissen*.

7 The 'Age of the Sages' and Stevin's Unfinished Works

It is important to realize that the unfinished works by Stevin that he had announced in his *Wisconstighe ghedachtenissen* are only indirectly known to us via tables of contents and fragments in manuscripts, or via partial publications by others. There are no surviving autograph manuscripts by Stevin of his incomplete chapter of this work on mathematical affairs. We also know that later editors of his texts frequently reordered them. For instance, when Hendrik

28 The unpublished *Telconstighe anteyckeninghen* might have been calculations by Stevin that he was unsure whether to include in the publication on arithmetic as part of the fifth book of the *Wisconstighe ghedachtenissen*: 'Various Matters'.

Stevin decided to publish some of his father's unfinished manuscripts, he did not envisage an unchanged print of the original, but a representation 'as the author intended'.[29] In a preface, he set out his reasons for this decision:

> Given that it was not the intention to reprint the original edition of *Wisconstighe ghedachtenissen* which is arranged according to the writer's own wishes, [...] I have not considered myself to be bound to the order of some of the pieces or parts as described in the summary of the fifth section of the aforementioned first edition, but have arranged these as I deemed necessary in the circumstances and according to their internal logic.[30]

Such editorial decisions by later authors demonstrate how carefully we need to compare the order of the texts in these posthumous fragments and the scientific order set out by Stevin in his 'Age of the Sages'. It is hard, perhaps even impossible to assess how Stevin's five articles for restoring the Age of the Sages can be retrieved exactly in these later fragments. In order to get a better understanding of Stevin's intentions with the 'Age of the Sages', we shall therefore need to content ourselves with a far more abstract exploration of the scientific order in these unfinished works.

We shall now explore the intentions in these unfinished works with reference to three themes: Stevin's desire to establish a pre-classical authority, his emphasis on empirical 'big data' and, finally, his obsession with pure languages.

7.1 *Pre-classical Authority*

From the early days of his career, Stevin had been interested in scientific argumentation to disclose the secrets of nature. It seems that Stevin initially tried to position his argumentation in a framework of contemporary studies, often with references to classical authors. When Stevin started to teach Prince Maurits, however, it seems that educational interest gradually drove him towards the more empirical approach of Petrus Ramus and his followers. Stevin would

29 Van den Heuvel, 'As the Author Intended', 2017.

30 'Ghemerckt het voornemen niet en was de eerstgedruckte Wisconstighe ghedachtenissen, welcke na des Schrijvers eyntlicke meyning gheschickt sijn, te herdrucken [...] soo heb ick my an d'oirden die sommighe der stucken of deelen van die in't Cortbegrijp des vijfden Stucks derselve eerstgedruckten ghegeven was, niet ghenoodsaeckt gevonden te verbinden, maer die na 't ghene my die de omstandighen en eyghen beweghinghen schenen te vereysschen, geschickt', Hendrik Stevin, 'An den leser' [To the reader], in *Materiae Politicae* [not numbered].

remain faithful to his Aristotelian and Euclidean roots, but he was searching for a pre-classical authority, as evidenced by the inclusion of the philological references to the conversations with Scaliger and Grotius. This shift from the classical to the pre-classical, from Antiquity to the Age of the Sages, is probably the reason why the publication of the *Wisconstighe ghedachtenissen* was delayed. There are a number of manuscript excerpts of Stevin's *Singconst* that represent various stages towards its publication in print. In the introduction to the final manuscript version, Stevin states that he intends to include in an appendix his critical remarks on the masters of the *Singconst* (music) of his time, who blindly follow the Greeks. Stevin's questions regarding the authority of classical authors such as Vitruvius and his followers in the Renaissance might have been the reason why he started a completely new treatise on architecture, which he failed to finish in time. In the preface to 'Vande Deursichtige' of 1605, Stevin stated that he had already written a 'Huysbau' some years previously, for private use. It is not clear which excerpts published by Isaac Beeckman and Hendrick Stevin belonged to this private copy or to the one that was intended for publication in the *Wisconstighe ghedachtenissen*.[31] However, it seems that Stevin referred to this earlier work on architecture in his treatise on perspective because it originally followed in a more Vitruvian tradition, whose third representation of architecture was the 'scenographia'. The fact that Stevin no longer followed the classical authority of Vitruvius and his exponents becomes apparent from his statement on his unfinished work on architecture:

> Architects fail to agree [...] that one should follow the rules written down by Vitruvius in their entirety [...] myself considering the pros and cons of this discourse, have preferred the latter option, because no other rules can bind everyone in general, except those that come from nature.[32]

For Stevin, the scientific order transmitted by the anonymous authors of the Age of the Sages was closer to the rules of nature than the one promulgated by

31 Van den Heuvel, *'De Huysbou'*, Chapter 1, in particular, 5–8.

32 'De Boumeesters en commen niet over een, aengaende de maten der pilaren met datter toebehoort tot cieraet streckende, waer af ettelicke achten, datmen de beschreven regelen van Vitruvius volcommelick moet volgen, sonder daer af te weycken; Ander en willen hun daer an niet verbinden. Ick overleggende me en tegen [In the margin: 'Pro & contra'] van dit geschil, heb my de meyning der laeste, best laten bevallen, uyt oirsaeck dat geen ander regelen, alleman int ghemeen en connen verbinden, dan die uyter nature'. Simon Stevin, 'Aenteijckeninghen Op Serlijns En Anderen/ Inhoudende Sommige Geschillen [1] Geschil op maten der pilaren met haer aencleving Op maten der pilaren met haer ancleving', in: Hendrik Stevin, *Materiae Politicae, Byvough der Stedenoirdening*, 107.

the authors and artists of Antiquity and the Renaissance. As we already noted, in his 'Age of the Sages', Stevin declared that it was not the observations of such authorities, but rather the empirical data of ordinary people that was more important for explaining the causes of phenomena in nature.

7.2 *Empirical Data*

Stevin's brief reference to the likelihood of big cities being built at the mouths of rivers in his 'Age of the Sages' is developed in more detail in his *Huysbou* [Architecture], in the part on *Vande oirdening der steden* [Town planning], published posthumously by his son Hendrik in his *Materiae Politicae* of 1649. In the chapter entitled 'On the selection of sites if one wishes to lay out towns', Stevin remarks upon the exceptional fortune of the Dutch, given the position of Holland on the coast at the mouths of the great navigable rivers of the Rhine and Meuse, coming from distant populated countries. In other countries, according to Stevin, it would have taken far more time to obtain more than a hundred thousand valiant inhabitants. In the chapter entitled 'On the means with which great and powerful cities can be created', Stevin argued against those scholars, such as Aristotle, who preferred medium-sized cities. According to Stevin, one of the arguments in favor of large, highly populated cities concerned knowledge production:

> In such cities one hears what happens everywhere in the world, yes the foreign things of the world can be found there. [...] Those who want to exercise themselves in letters, have the university close at hand, that is also convenient for citizens who want to bring up their children there, without having to send them out of the country at great expense, this being also the cause of the many scholars that one finds in such cities: further he who wants something artfully made, or wants to learn himself to make it or that his children learn it has many artists ready, since according to the proverb where there is carrion the birds gather.[33]

33 'Men hoort in sulcke Steden watter over al inde werelt ommegaet, ja de vreemdicheden des Werelts vergaren daer. [...] Die hun inde bouckstaven [in margin Letteren] willen oeffenen, hebben het hooch School bijder hant, dat oock bequaem is voor Burgers die hun kinderen daer in willen optrecken, sonder die tot groote cost buijtens slants te moeten senden, oock oorsaeck wesende van veel geleerden diemen in sulcken Stadt crijcht: Voort die ijet constelick wil gemaeckt hebben, of selff leeren maecken of zijn kinderen doen leeren, heeft menichte van constenaers al gereet, want na het spreeckwoort daer het aes is vergaren de vogelen', Simon Stevin, 'Van de middelen daermen groote machtige steden door maken can', Eenighe Stucken der Chrychconst, KB 128-A-9-II, fol. 3r. published in Hendrik Stevin, *Materiae Politicae, 1 Onderscheyt van de Oirdeningh der Steden, 8.*

7.3 *Pure Languages*

Empiricism as the key to gaining a better understanding of the rules of nature is also an essential aspect of Stevin's music theory. Stevin does not use the Dutch word for 'musica' (*muziek*), but the word *Singconst* ('Art of Singing'). Although for Stevin, music in principle encompassed all musical expressions, his views on musical theory were in the first place empirical, based on 'natuerlicke sanck' ('natural singing'). Similar to what we observed in Stevin's work on perspective, however, it is not sufficient to follow our senses. In order to create rules from empirical observations by sight or sound, they need to be compared and underpinned with mathematical rules.

> Again, just as all the ratios of two given rectilinear plane figures or solids cannot be recognized by sight, but obey geometrical rules teaching us how to find them, so all the ratios of two given sounds cannot be judged by hearing, but they are revealed by means of the musical rules governing them.[34]

As we already noted, Stevin used an example from music theory in this passage on the restoration of the Age of the Sages to explain that the use of the Dutch language could prevent misunderstandings. This example, based on the term 'equirationality', is explained in more detail in his posthumously published musical theory.

> The very thorough understanding of equirationality was not found among the Greeks and their ancestors [...] The cause of these errors is that Greek could not interpret this term properly together with all mathematical terms as properly as this one [the one in the Dutch language] [...]
>
> But DUTCH serves to teach one to fathom the hidden secrets of nature, and to prove that miracle is no miracle.[35]

34 'Wederom ghelyckmen alle Redens van twee voorghestelde platen of lichamen duer tghesicht niet bekennen en can, maer hun meetconstighe regels hebben, leerende hoemen die vinden sal, also en sijn alle Redens van twee voorghestelde gheluyden uyt het gehoor niet te oirdeelen, maer sij worden openbaer duer haer Singconstighe reghels'. Simon Stevin, 'Van de Spiegheling der Singconst', in: *Principal Works of Simon Stevin*, V, 428.

35 'Dattet recht grontlick verstandt der Everedenheyt byde Grieken ende hun navolghers niet gheweest en heeft. [...] Doirsaeck dier dwalinghen is dat hun spraeck dit woort medtsgaders al d'ander Wisconstighe namen niet soo eyghentlick beteeckenen en conden als dese [...] Maer het DUYTSCH om de vrie consten daer in te leeren, om de natuerens verborghentheden daer in te duergonden ende te bewysen dat wonder gheen wonder en is', Ibidem.

We find the same argument in the *Huysbou* [architecture], when Stevin claimed that if the ancients had used the term *lycksijdigicheyt* (like-sidedness) – his neologism in Dutch for mirror symmetry – instead of the Greek *symmetria* (*saematicheyt*) in the sense of proportionality, misunderstandings in architecture could have been prevented.[36] Stevin recognizes his interpretation of symmetry in nature, and it is this natural law, not classical authority, that is binding.

Although Stevin had already explored the qualities of an 'excellent' language in his 'Age of the Sages', in his works on musical theory and architecture he seems to extend the superior qualities of the Dutch language to reveal the secrets of nature – based on concise and pure expressions – to other non-textual, musical and visual languages. These languages are quite distinct from the vocabularies used in the Renaissance. Whereas the architects of his time debated the correct proportions of columns, Stevin considered this to be a superficial discussion. In his work on architecture, Stevin presents plain facades without classical features, and explains that embellishments and ornaments can lead to distractions and misconceptions of the essential order in architecture:

> Indeed, inexperienced owners of buildings may be more interested in artistically-fashioned decoration than in the best layout, making the worst choices.[37]

Although Stevin, in his *Dialectike ofte Bewysconst* and 'Uytspraeck van de Weerdicheyt der Duytsche Tael', already had underlined the role played by language in revealing the secrets of nature, it seems that since the 'Age of the Sages', he wanted to go beyond proving the superiority of the Dutch language by including other audio and visual languages in his quest for purity. The cross-references in his works on music and architecture, to each other and to underlying principles in mathematics, reveal that Stevin tried to compare the qualities of various languages. Apart from this more inclusive approach, it seems that the quest for pure languages not only needed to support the representation of the best scientific or natural order, but that purification also became an active design instrument. The plain architectural style based on mirror symmetry had to support the commissioner in making the right choices with regard to the design and lay-out of the building. In music, he recommended

36 Simon Stevin published in Hendrik Stevin, *Materiae Politicae, 1 Onderscheyt van de Oirdeningh der Steden*, 13.

37 '[w]aer benevens noch gebeurt, dat onervaren eygenaers des gebous, meer siende na 't cieraet door de constige hant gewrocht, dan na de beste oirden het quaetste verkiesen', Simon Stevin published in Hendrik Stevin, *Materiae Politicae, Byvough der Stedenoirdeningh*, 103.

that all major and minor (two in major) steps be equal in the scale of natural singing. In short, Stevin was not just trying to analyze, but also to design pure languages.

8 Epilogue: Restoring the Age of the Sages and an Alternative Renaissance

Stevin's attempt scientifically to describe and restore the Age of the Sages was not uncommon in the Early Modern Period. As Dimitri Levitin recently pointed out, without engaging with histories of historical scholarship – that is, histories of ancient, pre-classical wisdom before the Egyptian, Greek and Roman civilizations – one is sure to misunderstand the great intellectual enterprise of the seventeenth century.[38] Levitin merely focused on early modern England. In the Dutch Republic, apart from Scaliger and Grotius, who provided pre-classical input for Stevin's 'Art of the Sages', scholars such as Gerard Vossius and Jean Leclerc questioned historical argumentation purely based on Judeo-Christian traditions, and tried to trace their pagan origins. As Eric Jorink stressed in his review of Levitin's study, this interest in pre-classical wisdom was not only pursued in ancient texts, but also in images and objects.[39] It is in this context of visual and material culture that we must situate Stevin's quest in a pre-classical period for fundamental rules in the sciences and liberal arts such as painting, music, poetry or architecture, which reflected the secrets of nature and revealed God's orderly creation. In the previous chapter, Fokko Jan Dijksterhuis argued that Stevin's interest in the Age of the Sages could be understood in the context of the quest of Petrus Ramus and his followers for the origins of mathematics. Although Stevin explicitly refers to Grotius and Scaliger as sources of information for his views on ancient wisdom, he makes no reference to Ramus. Nevertheless, as Dijksterhuis points out, one can recognize some implicit similarities. Both Ramus and Stevin were in search of the origins of mathematics, both adhered to empiricism and both stressed the importance of the vernacular. However, there are also fundamental differences. Whereas Ramus used this pre-classical survey of the origins of mathematics to explain why Greek mathematics became corrupt, Stevin made an exception of geometry, where the scientific order of the Age of the Sages had been preserved in the 'beschrijving' (description) of mathematics, in particular in Euclid's *Elements*. This emphasis

38 Levitin, *Ancient Wisdom*, 1–4.

39 Jorink (in press). See further Jorink, *Reading* and *De Ark, de Tempel, het Museum*.

on the description of mathematics is not trivial. Whilst Ramus, in a similar way to Stevin, expressed the need for the vernacular to reach out to more people, which was crucial if empirical experiments were to be based on multiple observations, Stevin went a step further in his conception of the instrumental role of language. Above, we observed that Stevin was convinced that the Greeks and successively the Romans had misunderstood fundamental characteristics of the laws of nature, expressed in scientific order, and in music and architecture, due to the inferior qualities of Greek, Latin and languages derived from these. As Stevin stated in his third article on the revival of the Age of the Sages, good languages are necessary, but they were unfortunately lost, together with other sciences. In order to restore them, Stevin deemed it necessary to investigate the good qualities of languages; qualities that he recognized in the Dutch language.

Neither Stevin's attempt in his 'Age of the Sages' to restore the pre-classical authority of science nor his claim to explain the secrets of nature by mathematical proof were new. Although this was consistent throughout his career, his tutoring of Prince Maurits resulted in a gradual shift from scientific argumentation based on a comparative study of contemporary literature with references to classical studies toward a more empirical one. This shift might explain why Stevin wrote a new work on scientific method after his *Dialecticke of the Bewysconst*. We have also established similarities between themes touched upon in the 'Age of the Sages' and more extensive elaborations of these in fragments of his unfinished chapter on 'Various Matters' for this mathematical work. His obsessional belief in the superior ability of the Dutch language to unravel and present the secrets of nature, as expressed in his *Dialecticke of te Bewysconst* and 'Uytspraeck van de Weerdicheyt der Duytsche Tael', remained unchanged. However, it seems that Stevin wanted to extend the concise and pure qualities that he had extrapolated from the Dutch language to designing audio and visual languages as well. These languages, explained in his *Singconst* and *Huysbou,* were quite different from the vocabularies used by his contemporaries. Stevin considered the latter impure misreadings of the sciences and liberal arts that preceded classical Antiquity. Stevin's approach to pre-classical languages was therefore quite different from that of his contemporaries. Whilst the latter primarily studied ur-languages (expressed both in phonograms and logograms) to trace their origins and establish their authority in relation to biblical languages, Stevin's focus was on the re-use of the pure, elementary qualities of these languages to express fundamental principles of scientific argumentation and to reveal the essential characteristics needed to understand and even to design the world around him.[40]

40 Compare Levitin, Ancient Wisdom, 83, 152, 155 and 222–224.

As we already noted, Stevin considered the Renaissance an Age of the Ignorant in comparison with the lost Age of the Sages. Rather than accepting the classical authorities and then fine-tuning, within that framework, how these texts should be interpreted and their principles applied to the arts and sciences, as most of his contemporaries did, Stevin often questioned these authorities altogether. For him, how to follow Vitruvius' rules was not the real question; instead, he was interested in whether this authority had himself understood the underlying principles of nature. Stevin thereby distanced himself from his contemporaries and searched for his own, alternative Renaissance, by (re)constructing his imaginary Age of the Sages. Stevin's Renaissance of the Age of the Sages was not philological in the sense that Scaliger and Grotius had tried to gain a better historical understanding of the relations between pagan wisdom, Jewish and Christian knowledge. Bearing in mind the dichotomy, as set out in the introduction to this volume, that distinguishes between polyhistorians and polymaths, Scaliger and Grotius were better examples of the first, Stevin of the second definition.

Although Stevin referred to Scaliger and Grotius for a historical legitimization of the Age of the Sages, his interest in this imaginary period seems to have been far more active and instrumental. His Renaissance was directed at a (re-) creation of a scientific order based on natural principles, which could be analyzed empirically, underpinned by mathematical proof, and clearly expressed in pure spoken, written, audio and visual languages. In his view, the Renaissance of his contemporaries was guided by a blind belief in classical authorities, who had themselves often misunderstood and wrongly described nature, and had therefore also misguided the authors of his time. In short, Stevin's 'Age of the Sages' can be read as quest for and construction of an alternative Renaissance in two ways: in its different use of pre-classical sources and in its different reception of classical authorities.

Stevin's attempt to reconstruct the Age of the Sages was only partially successful. Josephus Scaliger and Hugo Grotius had provided him with the philological signs of such a period. Stevin was able to connect these ideas with some of the texts that he published in his *Wisconstighe ghedachtenissen,* such as the section on astronomy.[41] However, an apparently pushy printer and the demands of his practical work as a military advisor stood in the way of a further exploration and implementation of the scientific order of the Age of the Sages in a more coherent presentation of the 'Various Matters' in his *Wisconstighe ghedachtenissen*. Although Stevin had needed philological advice from

41 See the chapter by Vermij in this volume.

Scaliger and Grotius, the 'Age of the Sages' by no means lay 'outside his specific competence'; instead, it is crucial as a means of understanding his unfinished work, *Wisconstighe ghedachtenissen*. Although these works have often been read in practical terms, the role played by the 'Age of the Sages' in this publication and the directly related unfinished works cannot simply be understood in 'polypragmaric' terms, based on pretentious claims to omniscience.[42] Although Stevin did not succeed in translating the scientific order that he recognized in the Age of the Sages and sometimes made matters even more confusing – for instance, when he tried to introduce precise mathematical ratios to music intervals – his attempts to find a common (natural) order in the arts and sciences were the genuine attempts of a polymath.

Stevin's 'Age of the Sages' clearly shows that the dichotomies between polymathy and *polyhistorie* on the one hand, and between polymathy and polypragmatism on the other, are too limited for those who wish to understand the complex relationship that early modern scholars had with the period of ancient wisdom. With that in mind, Stevin's 'Age of the Sages' is of interest to the historiography on the history of science. Eduard Jan Dijksterhuis' reading of Stevin as a rational scientist with a strong interest in the application of theoretical thought in daily practice fits perfectly with the common reading of the history of science of the early modern period, based on the Scientific Revolution and the emergence of empiricism. Stevin's search for an alternative Renaissance allows for a more comprehensive reading of the history of science in the early modern period that combines the history of 'the Scientific Revolution' with philological historicism. Stevin's references to the Age of the Sages in the *Wisconstighe ghedachtenissen* and in his unfinished publications – brief, scattered and incoherent as they may be – can play a role in bringing the histories of science and of knowledge closer together.

Bibliography

Archival sources

Beeckman, Isaac, *Loci communes* [*etc.*], Middelburg, Zeeuwse Bibliotheek, manuscript HS 6471.

Huygens, Constantijn, *Adversaria*, The Hague, National Library, KA XVLII and XVLIII.

Stevin, Hendrik, *Eenighe Stucken der Crychconst. Beschreve deur Simon Stevin*, The Hague, National Library in the Netherlands 128 A 9–II.

42 See for this Burton's dichotomy between polymathy and polypragmatism in the introduction.

Stevin, Hendrik, *Vijfde deel der gemengder stoffen van de Krijgkonst*, The Hague, National Library in the Netherlands 128 A 10.

Stevin, Simon, *Maniere ende Ordre die sijn Excie verstaen heeft sal worden achtervolght int doen van de Nederduytsche lessen in the universiteyt tot Leyden tot onderrichtinghe ende bevorderinghe der ghene die hun ten dienste van den lande sullen begheven tot oeffeninghe van het ingenieurschap ende andere mathematische consten* Gemeentearchief Bibliotheek Leiden LB 36949 portefeuille.

Printed sources before 1800 and later editions/ reprints

Beeckman, Isaac, *Journal tenu par Isaac Beeckman de 1604 à 1634 publié avec une introduction et des notes par Cornelis de Waard*, 4 vols., The Hague: Martinus Nijhoff, 1939–1953.

Grotius, Hugo, *Batavi, Parallelon rerumpublicarum liber tertius*, edited by Johan Meerman. Haarlem: Loosjes, 1801–1803.

Molhuysen, P.C., *Bronnen tot de geschiedenis der Leidsche Universiteit 1574–1811* (*Rijks Geschiedkundige Publicatiën* 20), The Hague: Martinus Nijhoff, (7 vols.) 1913–1924.

Stevin, Hendrik, *Materiae Politicae Burgherlicke Stoffen Vervanghende Ghedachtenissen der Oeffeninghen des Doorluchtichsten Prince Maurits van Orangie, &c. Gouverneur, Opperste Velt-Heer, Admirael-Generael vande Verenichde Nederlanden, &c. Beschreven deur Simon Stevin van Brugge, Superintendent van zijne Finance, Quartiermeester Generael van 't Leger,&c. En uyt sijn naegelate Hantschriften by een gestelt door Sijn Soon Hendrick Stevin, Heere van Alphen, van Schrevelsrecht,&c. Gedruckt tot Leyden*, Leiden: Adryanen Rosenboom, 1649.

Stevin, Simon, *Dialektike ofte Bewysconst*, Leiden: Christoffel Plantijn, 1585.

Stevin, Simon, *De Beghinselen der Weeghconst*, Leiden: Christoffel Plantijn by François van Raphelingen, 1586.

Stevin, Simon, *De Sterctenbouwing, beschreven door Simon Stevin van Brugge*, Leiden: Françoys van Ravelenghien, 1594.

Stevin, Simon, *Maniere ende Ordre die sijn Excie verstaen heeft sal worden achtervolght int doen van de Nederduytsche lessen in the universiteyt tot Leyden tot onderrichtinghe ende bevorderinghe der ghene die hun ten dienste van den lande sullen begheven tot oeffeninghe van het ingenieurschap ende andere mathematische consten*, Leiden: Jan. Paedts Jacobsz., 1600.

Stevin, Simon, *Wisconstighe ghedachtenissen, inhoudende t'ghene daer hem in gheoeffent heeft den Doorluchtichsten Hooghgheboren Vorst ende Heeren, Maurits, Prince van Oraengien*, Leiden: Jan Bouwensz., 1605–1608.

Stevin, Simon, *Hypomnemata Mathematica, hoc est eruditus ille pulvis, in quo se exercuit* [..] *Mauritius Princeps Auraicus*, Leiden: Ex Officina Ionnis Patii, Academiae Typographi, 1608.

Stevin, Simon, *The Principal Works of Simon Stevin*, eds. E. Crone, E.J. Dijksterhuis, R.J. Forbes, M.G.J. Minnaert and A. Pannekoek, 5 vols., Amsterdam: C.V. Swets & Zeitlinger, 1955–1966.

Literature

Cohen, H. Floris, "Simon Stevin's Equal Division of the Octave", *Annals of Science* 44, no. 4 (1987) 471–488.

Dijksterhuis, Eduard Jan, *Simon Stevin*, The Hague: Martinus Nijhoff, 1943.

Dijksterhuis, Fokko Jan. "Duytsche Mathematique and the Building of a New Society: Pursuits of Mathematics in the Seventeenth-Century Dutch Republic", in: eds. Lesley B. Cormack, John A. Schuster and Steven A. Walton, *Mathematical Practitioners and the Transformation of Natural Knowledge in Early Modern Europe. Studies in History and Philosophy of Science*; Vol. 45. Berlin: Springer, 2017, 167–181.

Hahlweg, Werner, *Die Heeresform der Oranier und die Antike*, Berlin: Juncker und Dünnhaupt Verlag, 1941.

Hahlweg, Werner, "Aspekten und Problemen der Reforms des niederländischen Kriegswesens unter Prinz Moritz von Oranien", in: *Bijdragen en Mededelingen betreffende de Geschiedenis der Nederlanden*, vol. LXXXVI (1971) 161–177.

Heuvel, Charles van den, *'Papiere Bolwercken': De introductie van de Italiaanse stede- en vestingbouw in de Nederlanden (1540–1609) en het gebruik van tekeningen*, Alphen aan den Rijn: Canaletto, 1991.

Heuvel, Charles van den. " 'Wisconstighe Ghedachtenissen'. Maurits over de kunsten en de wetenschappen in het werk van Stevin", in: ed. Kees Zandvliet, *Maurits Prins van Oranje*, 245–257. Amsterdam/Zwolle: Rijksmuseum/Waanders, 2000, 106–121.

Heuvel, Charles van den, "Le traité incomplet de l'Art Militaire et l'instruction pour une école des ingénieurs de Simon Stevin", in: *Simon Stevin (1548–1620) L'emergence de la nouvelle science*, Exhibition catalogue, Catalogue Bibliothèque Royale Albert I, Bruxelles, (17-09-2004/30-10-2004), Turnhout: Brepols, 2004, 101–111.

Heuvel, Charles van den, *'De Huysbou': A reconstruction of an unfinished treatise on architecture, town planning and civil engineering by Simon Stevin*, Amsterdam: KNAW Edita, 2005.

Heuvel, Charles van den, "De vesting als mathematisch en culureel kennissysteem. Het onderwijs in de vestingbouw aan hovelingen, kooplui en ambachtslieden", in: eds. Geert Vanpaemel and Ad Meskens, *Vóór en na de scheiding. De wiskundige cultuur in de Lage landen 1550–1650.* [*themanummer*] *Scientiarum Historia* 32 no. (2006) 99–117.

Heuvel, Charles van den, " 'As the Author Intended': Transformation of the Unpublished Writings and Drawings of Simon Stevin (1548–1620)", in: eds. Sietske Fransen, Niall Hodson and Karl A.E. Enenkel, *Translating Early Modern Science*, Intersections, vol. 15, Leiden: Brill Publishers, 2017, 119–153.

Jorink, Eric, *Reading the Book of Nature in the Dutch Golden Age, 1575–1715*, Leiden: Brill, 2010.

Jorink, Eric, *De Ark, de Tempel, het Museum. Veranderende modellen van kennis in de eeuw van de Verlichting: Inaugurele rede uitgesproken bij de aanvaarding van het ambt van buitengewoon hoogleraar 'Verlichting en religie' vanwege Teylers Stichting*, Haarlem: Gottmer Uitgevers, 2015.

Jorink, Eric, "Review-essay. Dmitri Levitin, Ancient Wisdom in the Age of the New Science. Histories of Philosophy in England, c. 1640–1700. Ideas in Context 113. Cambridge: Cambridge University Press, 2015," *History of Universities*, [in press].

Krüger, Jenneke, "Lessons from the Early Seventeenth Century for Mathematics Curriculum Design", *BSHM Bulletin Journal of the British Society for the History of Mathematics*, 25 no.3 (2010) 144–161.

Krüger, Jenneke, "Frans van Schooten Sr. (1581–1645) Lecture Notes for the First Dutch Course for Engineers, Leiden, 1600–1681", in: *Researching the History of Mathematics Education*, edited by Fulvia Furinghetti and Alexander Karp, *ICME-13 Hamburg 2016. Monographs,* Cham Switzerland: Springer International Publishing AG, 2018, 143–166.

Levitin, Dmitri, *Ancient Wisdom in the Age of the New Science. Histories of Philosophy in England, c. 1640–1700*, Cambridge: Cambridge University Press, 2015.

Maanen, Jan A. van, *Facets of Seventeenth Century Mathematics in the Netherlands*, Utrecht: Elinkwijk, 1987.

Postema, Jan P.J., *Johan van den Corput (1542–1611)*, Kampen: IJsselakademie, 1993.

Schilt, Kees, *Simon Stevin en het hermetisme: Een cultuurhistorische studie naar zijn methodologie*, Unpublished Ph.D. thesis, Vrije Universiteit Amsterdam, 2013.

Taverne, Ed R.M., *In 't land van belofte: in de nieue stadt. Ideaal en werkelijkheid van de stadsuitleg in de Republiek 1580–1680*, Maarssen: Gary Schwartz 1978.

Van Duyse, Prudens, *Simon Stevin naer Voorduin's Bekroond Werk, met enige dichtstukken*, Brussel: Van Daele, 1846.

Verdonk, Johannes J., "Vom Einfluß des P. de la Ramée auf Simon Stevin," *Centaurus*, 13, no 3/4 (1969) 251–262.

Vermij, Rienk, *The Calvinist Copernicans. The Reception of the New Astronomy in the Dutch Republic 1575–1750*, Amsterdam: Edita KNAW, 2002.

Winter, P.J. van, *Hoger beroepsonderwijs avant-la-lettre. Bemoeiingen met de vorming van landmeters en ingenieurs bij de Nederlandse universiteiten van de 17e en 18e eeuw*, Amsterdam/Oxford/New York: Noord-Hollandsche Uitgevers Maatschappij, 1988.

CHAPTER 9

Politics in the Vernacular

The *Vita politica. Het burgherlick leven* (1590) as a Practical Guide to Civic Life

Catherine Secretan

In the past two decades there has been a renewal of interest in Simon Stevin, and the various aspects of his work have given rise to enlightening studies and interpretations.[1] One of his texts in particular has attracted new attention: the *Vita politica. Het burgherlick leven*.[2] Until recently this text was rather neglected, with the exception of the 1939 edition by Annie Romein.[3] The reason for this prevailing lack of concern was probably the fact that politics seemed to have played a very marginal part in Stevin's preoccupations. Moreover, the *Vita politica* was a short text that apparently lacked true originality and, generally speaking, stood apart from the humanist political trend. This paper aims to challenge this opinion and show that the *Vita politica* (1590) should not be considered in isolation, either from Stevin's other works or from his time. It also suggests that being a practical manual, written in Dutch instead of Latin, the *Vita politica* may be interpreted, to a certain degree, as 'politics from below'.

1 In addition to the well-known study on Stevin's life and achievements by Dijksterhuis, *Simon Stevin*, an excellent synthesis on Stevin's life, work and activities has been recently published by Devreese and Vanden Berghe, *Wonder en is gheen wonder.* One should also mention the richly documented and illustrated book published on the occasion of an exhibition devoted to Stevin in 2004, at the Belgian Royal Library in Brussels: *Simon Stevin 1548–1620. L'émergence de la nouvelle science.*

2 *Vita politica. Het burgherlick leven, Beschreven deur Simon Stevin.* Tot Leyden, By Franchoys van Ravelenghien, 1590.

3 Romein-Verschoor and Overdiep, *Het burgherlick leven.* For my part, several years ago I highlighted the importance of the *Vita politica* in two papers: 'L'état de fait comme légitimité ... ' (1986), and 'L'émergence de l'idée d'individu ...' (1990). I recently published a French translation, with an introduction, of the *Vita politica* in Secretan and Den Boer, eds., *Simon Stevin. De la vie civile, 1590.* A new edition in Dutch has also been published, edited by P. den Boer and A.C.G. Fleurkens: *Simon Stevin, Het burgherlick leven.* All quotations from the *Vita politica* in this paper, unless otherwise specified, are drawn from the English version published in: *The Principal Works of Simon Stevin,* vol. v.

 | DOI:10.1163/9789004432918_011

1 *The Publication and Context of the* Vita politica. Het burgherlick leven

The *Vita politica* (*VP*) should be taken into consideration if only because it was reprinted nine times after its first publication in 1590.[4] Time and again, contemporaries evidently felt the topicality of this discourse. When it was printed for a second time in 1611, a note was added to the title page, underlining the usefulness of the *VP* in 'desen beroerlicken tijdt [in these troubled times]'; a reference to the harsh political crisis triggered by the signing of the Twelve Years' Truce (1609–1621). Likewise, by the end of the century, the title page of the 1684 edition bore mention of the relevance of this text 'for any *Hoekse en Kabeljauwse* time, and particularly for our internal discords in Holland'.[5] With this comparison between the current antagonism over relations with France between the stadholder Willem III and the Amsterdam regents, and the *Hoeks* and *Cabeljauwen* conflict that originated in 1350 and lasted more than a century, the political value of the *VP* was again confirmed.[6]

Nevertheless, the text that Franchoys van Ravelenghien of Leiden published for the first time in 1590 under the title *Vita politica. Het burgherlick leven* may have surprised the contemporary reader, both because of its style and its content. Neither a pamphlet nor a real treatise, the *VP* was published as a manual on politics as one of the various *consten* that Stevin was simultaneously describing and ordering, replacing Latin and conventional words with vernacular neologisms. In fact, Stevin warned the reader in the 'Argument' (*Cort begrip*) that he would not proceed 'step by step from genus to species, for which we lack time and inclination', but that he would 'only touch upon the principal and most essential doubtful points, on which there are daily the most violent differences of opinion among people'.[7] It has been suggested that the *VP* could

4 1611, by Jan Andriesz.; 1646, by Nicolaes van Ravesteyn; 164[9], by Thomas Fonteyn; [1649], by Justus Livius (to this posthumous edition was added an 'Appendix' or *Anhangh*, published for the first time, in which Stevin denounced Machiavel's amorality); 1658, by Jaques Fierens; [1660], by Adryaen Rosenboom; 1668, by Hero Galama; 1684, by Abraham Olofsz.; 1686, by Abraham Troyel and Cornelis de Graaf. See the bibliography in P. den Boer and A.C.G. Fleurkens, eds., *Simon Stevin, Het burgherlick leven*, 213–219.

5 '[o]m in alle Houkse ofte Cabeljaawse tijden, ende bysonderlik gedurende onse binnenlandse verschillen in Holland, geleesen te werden'.

6 Israel, *The Dutch Republic*, 19–21.

7 'Alsoo ons voornemen niet en is, dese stof des Burgherlicx levens, int langhe ende breede te beschrijven, noch de sake voorvoets op van gheslacht tot afcomsten te vervolgen, waer toe ons tijdt en wille ghebreect, sullen alleenlick roeren de voornaemste ende nootelickste twijselachtighe punten, van welcke tusschen de menschen daghelicx de heftichste verschillen vallen', *VP*, 482.

be considered as a 'réduction en art du politique'. By this is meant that it was written on the model of an engineering manual, whose aim was to gather together all of the parts of a scattered body of knowledge, put them in order and give them precise definitions, so that the person who needed this specific knowledge – be he a merchant, jurist, sailor or statesman – could use it properly and efficiently.[8] This is exactly what Stevin was doing in the *VP* when he declared that he would not proceed 'van gheslacht tot afcomsten', as a political writer usually does, but would only deal with some given facts.

Focusing on political and religious strife was a common thread in many political discourses at that time, and a frequent motive for political writing. Indeed, although the *VP* appears unusual in terms of its format and style, on the other hand, its starting point connected it with an ongoing debate. Be it approached from the question of religious wars and true sovereignty (as in Jean Bodin's *Six books of the Republic*, 1586), from that of political prudence and conservation of the state (as in Giovanni Botero's *Della ragion di stato*, 1589), or with the aim of giving advice to a prince (as in Justus Lipsius' *Politica*, 1589), the claim was always to shed light on the foundation of civil concord; and this is what such authors had in common with Simon Stevin. But what distinguished the content of the *VP* was the fact that Stevin did not pretend to investigate the causes of civil and religious conflicts, nor did he suggest any general classification or general rule. Rather, he claimed that he would only stick to the facts (the conflicts) and, on this basis, help the citizen to find a practical solution. Moreover, Stevin explicitly declared that he had drawn his inspiration from reflections on contemporaneous history and would deal with citizenship as a real-world experience, instead of quoting 'endless examples and different cases'.[9]

Another bewildering aspect of the *VP* was also the fact that it was written by an author who, until then, was mainly known for completely different publications that had nothing to do with politics. Unfortunately, Stevin's biography does not offer any details that might help to solve the puzzling status of this text. The reader is reminded that we know almost nothing about the context in which Stevin wrote the *VP*. We know only that he had settled in Leiden in 1581 and registered at the city's university, probably to follow the mathematics lectures given by Rudolphus Snellius.[10] But we have neither letters nor any kind of personal documents that can tell us why he was taking part in political debates. Thus, at first sight, the fact that Stevin was

8 Hélène Vérin, 'La pensée politique d'un ingénieur: l'exemple de la *Vita politica*', 109–130.

9 '....oneindelicke voorbeelden, ende verscheyden ghevallen', *VP*, 478.

10 *Simon Stevin 1548–1620. L'émergence de la nouvelle science*, 23.

not a writer on political theory, but an engineer and a mathematician, might have appeared unexpected. With the exception of the *Materiae politicae. Burgherlicke Stoffen* (1649) – which mainly deals with urban administrative issues, such as the recruitment of civilian officers – the rest of his works were devoted to technological achievements and theoretical mathematics. Chronologically speaking, the VP was published between the *Beghinselen der Weeghconst* (1586) and the *Appendice algébraïque* (1594). Therefore, the question about architecture raised by Charles van der Heuvel ('Why did Stevin, essentially a mathematician and engineer, write a treatise on architecture?') seems to apply to his writing on politics, too.[11] Remarkably, such a question had been anticipated by Stevin himself, as shown by the *captatio benevolentiae* he pleads for in the dedication of his discourse to Govert Brasser, burgomaster of Delft:

> As to the shortness of this work, which might be reproved with some reason in view of the endless examples and different cases with which it might have been amplified, the excuse for this must on the one hand be my love of briefness and on the other hand my present mathematical work.[12]

Stevin's awareness of the unusual character of the VP compared to other contemporaneous political treatises provides us with a clue to understanding the meaning and the scope of his discourse, a clue that belongs to the political context of the time. The second half of the sixteenth century was dominated by civil and religious wars. Social unrest was everywhere. The 1580s in particular were the period of the 'Ten years' (1588–1598); that is to say, the birth of the new state after the repudiation of Philip II, by the Act of Abjuration of 1581. The main issues at stake were the right of civil power to impose a single religion on the population and the question of the best form of government for the seven provinces of the Union of Utrecht (1579). During the English interlude (1585–7), harsh debates had taken place between François Vranck, the town pensionary of Gouda, who wrote a defense of the authority of the States, and Thomas Wilkes, one of the English members of the Council of State, who argued for popular sovereignty under the administration of the Governor-General, who, at that time, was the Earl of Leicester.[13]

11 Van den Heuvel, '*De Huysbou*,' 8.

12 VP, 479–481.

13 Israel, *The Dutch Republic*, 220–228.

It comes as no surprise, then, that Stevin perceived the contemporary situation as one that urgently required reflection and advice. A single remark in the dedication of the *VP*, mentioned above, testifies to Stevin's concern about the situation:

> Because in many different places great changes take place in the government, with one adhering to this and another to that side and each considering his own party best, this gives rise to great difference of opinion among people.[14]

From this it is clear that the pressure of contemporary circumstances spurred Stevin's decision to write on politics, and it explains why he decided to set aside his current scientific work in order to involve himself in political thinking. It also explains why he made many more references to recent history and cases belonging to 'most of the present European governments' than to the remote past.[15]

2 *The Content of the* Vita politica *and the Two Options for a 'Civic Life'*

The goal of the *VP* was to be a practical guide to overcoming various kinds of conflicts as though they were technical challenges. This aim is reflected in the approach followed by Stevin's reasoning:

> First defining civic life, we shall subsequently say what is the rightful authority governing a citizen; next, how one is to behave as a good citizen in civil wars: in government; in the laws which we regard as not binding us, those which are dubious, and those which contradict each other; in religion; and finally of civic life in general, the whole to be contained in eight chapters.[16]

The *VP* does not pretend to do more than give useful advice, easily accessible to all, on how to remain simultaneously in harmony with the government of one's country and with the dictates of one's conscience. From this point of view, what Stevin considered as 'burgherlikheyt' mainly required the citizen's consent to the political regime of the city or the state where he lived:

14 *VP*, 475.
15 *VP*, 503.
16 *VP*, 483, 485 (slightly modified).

> A man who so behaves himself in it that the greatest stability and welfare of the community results from it in this life is called a civic person [*burgherlick persoon/politicus*]. And such proper practice is called a civic life [*Burgherlick leven/Vita politica*].[17]

The word 'politicus' or 'politica' is one of the most ambiguous and difficult words to translate from Stevin's time, due to its variety of meanings. Deriving from the Greek *polis*, the word could mean either the city or the State, or its constitution, or the statesman as an individual. As noted by Nicolai Rubinstein, the term entered medieval political language after the mid-thirteenth century as the result of the translation of Aristotle's *Politics* and, to a lesser extent, of his *Nicomachean Ethics*.[18] One of its most current senses – besides qualifying man as a political animal – was that of 'governing according to the law' from the side of the statesman.[19] As a result of Florentine history and political uncertainty in the city, the word tended to be used more narrowly to define every government that followed the law ('regulated by laws', in Machiavelli's words) in contrast to 'absolutism' (governments that evade the law).[20] Moreover, in this precise linguistic context, the 'regimen politicum' generally referred to a republican form of government, to the exclusion of any other (monarchy or democracy).

In the VP, this normative value is applied to citizenship and should be understood as the meaning of 'burgherlick' in the quotation given above, i.e., the citizen who promotes civic values and obeys the law. The frequent recurrence of the word 'eerlick' (honorable) related to the best political choice or to the most appropriate behavior obviously suggests that the moral aspect of politics was at stake.[21] Admittedly, in the VP the word 'politicus' is used both in 'the wider sense as related to the state in general' and the narrow one, implying 'virtuous' as being in conformity with the law.[22] But the Latin equivalent given in the margin of the VP each time the Dutch term recurs gives rise to some confusion, due to the range of possible meanings suggested by Stevin: 'political', 'civil' or 'urban', according to the context. This ambiguousness bears some

17 VP, 489.

18 Rubinstein, 'The History of the Word *politicus*,' 41–56.

19 Ibidem, 44–45.

20 Atkinson and Sices, *The Sweetness of Power* ..., I, 58, 148.

21 ' ... de eerlickste syde te anveerden ', VP, 500; ' ... so ist openbaer onburgherlick ende oneerlick te wesen ', 524; ' ... sy achtent al voor eerlick voor Burgherlick haer eedt wel ghequeten te sijne', 532.

22 Rubinstein, 'The History of the Word *politicus*', 43, 46.

similarity to that of the Italian 'civile' of the fourteenth century. As is well known, the frequently used expression 'vivere civile' in Florentine humanist politics referred to a form of virtuous participation or commitment on the part of the individual citizen to the welfare of the city, which, in turn, would provide the citizen with true liberty. In this sense, Stevin's approach recalls works such as those by Leon Battista Alberti (*I Libri della Famiglia*, 1437) or Matteo Palmieri (*Vita Civile*, 1438). This equation between virtue and political support for the city state brings us very close to the sense of 'burgherlichheyt', understood as civic conduct beneficial to the whole community. Such a positive value given to the word 'politicus' as the Latin equivalent of 'burgherlick persooon', and 'politicè' as the equivalent for 'burgherlick' behavior, has induced me to translate 'Vita politica' as 'Vie civile' instead of 'Vie politique', which might have implied something of a calculated or hidden strategy. That said, the relevance of the comparison between Stevin's argument and the Italian ones of the Renaissance goes no further than this general meaning conveyed by the use of a same vocabulary, for two reasons. The first is the lack of a theory of the State in the *VP*, and the second, the fact that exile is given as the sole alternative in the case of disagreement about the current political authority.

Stevin focuses on two points only: how to gain a true understanding of the actual political constitution, and the correct way to behave in cases of civil conflicts. What seemed more important to him was the way in which a given form of government conformed to its theoretical definition, because this alone granted legitimacy to the civil authority. For example, the Doge of Venice would have become illegitimate if he had replaced the Senate with an Inquisition.[23] Therefore, the first duty of a true citizen is to obtain a fair understanding of the political constitution of the country in which he lives, since it is only with this knowledge that he will be able to respond personally to situations of conflict. But the precondition for doing this successfully is to make correct use of words and language in general.[24] For the author of the *VP*, political understanding, like any other field of knowledge, was highly dependent upon linguistic ability. Like a leitmotiv, verging on an obsession, Stevin repeatedly warned against using any improper word, metaphor or term that did not make sense.[25] In his view, 'metaphor' (*lyckspreuck*) and 'words without reason' (*woorden sonder reden*) had often been the origin of civil wars, because

23 *VP*, 409.

24 *VP*, 503, 535.

25 On the use of metaphor as a cause of misunderstanding, see Verbeek, 'La *Dialectike* de Simon Stevin', 181–190.

they introduce erroneous definitions and these generally lie at the root of conflicts.[26]

Once this has been established, Stevin lists, according to the classical rule, the different forms of government, of which he finds four: 'absolute monarchy, democracy, constitutional monarchy, and aristocracy'.[27] Whilst monarchy is by its very nature absolutist and equated by Stevin with dictatorship – such as, 'at the present time the Turk, the Grand Duke of Moscovia, and formerly Moses, Tamerlane, and the like' – democracy is the opposite form, consisting of a government 'which is subject to the community, according to whose ordinances different persons, who are called parliament, have come to the government [...] as formerly the democracy of Athens, of Rome; at the present time of Switzerland'.[28] Aristocracy exists 'when the government consists of the most suitable, most honorable, or most distinguished persons of the country'.[29] To the conventional tripartition of different types of government, Stevin adds a fourth, for which he coined a neologism – 'staetvorstheyt' – and to which he devotes a longer description. With this word, Stevin refers to the 'Princes who do not govern alone as absolute monarchs, but have a parliament at their side', as in Brabant ('a Duke with a parliament'), France ('a King with a parliament'), Venice ('a Duke with a Senate') and Spain ('a King with an Inquisition'). He emphasizes the fact that this type of government 'stands for that which we in Dutch generally call parliament'.[30] From this, the reader may infer that Stevin was both distancing himself from those who favored the emerging absolutist monarchies and promoting what the rebels defended against the Spanish centralized government.

The second duty of a true citizen is to decide whether or not he will adhere to a given government. Here lies the other option suggested by Stevin, which situated him in opposition to a strong political trend of his time: the theory of political resistance as lawful opposition to a persecuting government, as advanced by Theodore Beza in his *Du droit des Magistrats* (1574) and by the anonymous *Vindiciae contra tyrannos* (1581). Stevin openly challenged any contest or revolt against the actual ruling party:

> Even if some of their rules were unreasonable [...] this does not entitle anyone to remain with them against their will in order to correct those bad rules as he thinks suitable.[31]

---

26 *VP*, 503, 535.

27 *VP*, 501.

28 *VP*, 503.

29 *VP*, 505.

30 *VP*, 503.

31 *VP*, 575, 577.

In that sense, the rule of 'burgherlijkheid' implied submission or exile. On this point, Stevin's opinion was totally at variance with those of the Calvinists authors who inspired the rebels during the Revolt.[32] While political resistance was seen as a legitimate choice by Calvinist theoreticians, and a reason for stoic resignation or 'constantia' by Lipsius, for Stevin, on the contrary, it implied either acceptance or going into exile:

> Therefore, when it says to you […]: 'Look here, these are the general rules in accordance with which we all live; if you wish to live with us in our country, you must conform to them; if not, look for another place', one is obliged either to conform thereto or to leave the country.[33]

A third option was outlined in the last chapter of the *VP* ('Of civic life in general'), where Stevin explains how to turn political resistance into what he calls 'honest force' ('eerlijcke gewelt').[34] Once outside the country, a citizen may openly attack his government without being guilty of bad citizenship. This was an unusual alternative for the time, rightly interpreted by Mario Turchetti as a new vision of lawful political resistance.[35] Thus, the advice that Stevin gave to those citizens who could not endure the situation in which they found themselves and felt trapped by conflicts, whether about the legitimacy of the present government (chaps. II-IV), incompatible laws (chap. V) or religious freedom (chap. VIII), was to choose another country instead of opposing the authorities of their own. Stevin was rethinking the *topoi* of his time in a practical manner, starting on one side from individual feeling as a kind of empirical data, and on the other side, from the political demands of a burgeoning new state (the Dutch Republic in the 1590s) in need of citizens' obedience.

---

32 Secretan, *Les privilèges, berceau de la liberté*, 41–49; Van Gelderen, *The Political Thought*, 110–120.

33 *VP*, 575.

34 *VP*, 579.

35 Turchetti, 'Notes sur la *Vita politica*. Contribution à la terminologie politique du XVIe siècle ', 229–230. A fine essay was recently published by Daniel P. Maher (who was kind enough to send it to me), 'Simon Stevin's *Vita politica*: Pre-provisional Morality?' In this essay, Daniel P. Maher examines the possible influence of Stevin's political thought on Descartes' provisional morality. The argument is based on the fact that Descartes was directly familiar with some of Stevin's mathematical work and that the two may have met in Breda in 1618, when Descartes was there to join the army of Prince Maurits. Moreover, one can infer that Isaac Beeckman, who knew of the unpublished *Anhangh* (see his Journal, in Cornelis de Waard ed., *Journal tenu par Isaac Beeckman*, vol. 2: 1619–1627, 420), was also familiar with Stevin's *Vita politica* and may have informed Descartes about this text.

The *VP*'s central concern with peace and stability reflects the historical situation in which it was written, as suggested above. Intended for the individual caught up in the religious and political conflicts of the time, Stevin's aim was to express the standard that established all political legitimacy for the government's existence, that is to say, the *de facto* principle.[36] Remarkably, this *de facto* idea has enjoyed certain renown as a political principle, at least if one notes that it recurred in the famous treatise by Pieter de La Court, the *Politike Weegschaal*, in the 1660s, together with the quotation of a quatrain by the Seigneur de Pibrac. Guy du Faur de Pibrac (1529–1584) was a French jurist and poet, and a close counsellor of King Charles IX of France. He gained enduring fame for his *Quatrains* of 1598, which had been translated by Dirck Graswinckel, and, particularly, for his championing of political obedience on the part of citizens.[37] One of these quatrains was often quoted for its lauding of the *de facto* principle, as we can see at the end of the *Politike Weegschaal,* when Pieter de La Court concludes that 'De wesende Regeering is de beste':

> If [a man] revolts against the government which exists (de facto), he introduces the state of nature, that is to say the war of all inhabitants against each other, which is the most miserable situation one can ever think of for a human community.[38]

Following this assessment, a few pages further on we encounter the famous quatrain by Pibrac:

> Aime l'estat tel que tu le vois estre,
> S'il est Royal, aime la Royauté,
> S'il est de Peu, ou bien Communauté,
> Aime l'aussy, car Dieu t'y a fait naistre.[39]

This is the meaning of the first 'general rule' of the *VP*, which appears as a sort of leitmotiv throughout the text:

36 On the *de facto* principle, see my essay '*L'état de fait* comme légitimité', 418–434.

37 *De Vierlingen van den Heer van Pybrac.*

38 'Want men, rebelleerende teegen de Regieeringe die in wesen is, (de facto) dadelik invoert de natureliken stand ofte Oorlog aller Ingezeetenen tegens malkander, 't welk de ellendigste stand is die men onder eenige vergaaderinge van menschen zoude konnen bedenken', V.H., *Consideratien van Staat; ofte Politike Weeg-Schaal*, 666.

39 Ibidem, 669.

> Everyone must always consider as his rightful authority those who at present are actually governing the place where he has chosen his dwelling.[40]

Such a rule can be interpreted as support for the newly independent state of the United Provinces, which in 1581 had solemnly renounced its sovereign, Philip II of Spain, and was setting forth as a self-governed 'republic'. Although Stevin did not express a preference for a particular form of government, the examples chosen in the VP (Holland, Venice, Brabant), together with the importance attributed to the 'staetvorstheyt', suggest that his discourse endorsed the new Dutch power.

3 *The* Vita politica*: Not a 'Mirror for a Prince' But a Practical Guide for the Citizen*

Unlike Justus Lipsius' *Politica*, a text almost contemporaneous with Stevin's discourse, the VP did not address the Prince but the citizen. Whether Stevin was deliberately opposing Lipsius is impossible to establish; but he could not have ignored the *Politica* altogether, which had been published just one year before his own text, by the same publisher (Franchoys van Ravelenghien). Moreover, both Stevin and Lipsius attended Leiden University at the same time (Lipsius arrived in 1578 as a successor to Scaliger, and Stevin enrolled in 1583). Last but not least, even if Stevin had been unable to read Latin, the *Politica* was translated and published in Dutch in the same year of its publication in Latin, thanks to Marten Everaert, a mathematician and physician from Bruges, who was in Leiden around 1590.[41] Thus, a body of historical and intellectual circumstances, added to the format and style given to the VP, suggest that the VP could not have been conceived in 1590 in the absence of Lipsius' *Politica*.

This may explain why Stevin's work cannot be classified under the heading of the 'mirrors for princes', and why it is dedicated to a regent of Delft (Govert Brasser) instead of the Prince Maurits of Nassau (whose service he would enter only in 1593). Fundamentally, his initial motive for writing a political discourse was to act as a guide for any individual who found himself caught up in social conflicts due to 'misunderstanding'. By contrast, the political 'mirror' tradition, which goes back to Seneca's *De Clementia*, was conceived as an ideal model to be presented to the prince in order to encourage his moral improvement.

40 *VP*, 493.
41 See Serrure, *Vaderlandsch museum voor Nederduitsche letterkunde*, 453–462.

Although he departed from this model, Stevin appears to have been well acquainted with this literary tradition and referred to it when quoting an example drawn from Spanish history:

> Although the parliament of Spain has consented to it, it is now greatly regretted by the imprudent deposed body, which in the beginning did not expect such an end, and thus they hold such a mirror to all rulers that no parliament of any country consents to such a change.[42]

Such a category is better suited to Lipsius' *Politica*, a work intended to give advice to a prince and about which its author explicitly declared:

> The work which I now present to you is a Politics. In which it is my aim, just as in *De Constantia* I equipped citizens for endurance and obedience, now to equip those who rule for governing.[43]

Instead of a 'mirror for princes', Stevin provides a practical manual for the citizen, written in Dutch so that it might be read by anyone living in a political state, be they a prince or a common citizen. This specific purpose of the *VP* is confirmed by the style of the text itself, namely the translation into the vernacular of all the usual scholarly Latin vocabulary. It is also confirmed by the fact that Stevin never refers to any classical examples or quotations from commonplace books, but only quotes examples borrowed from recent, or even contemporary, history:

> Let us take as an example the United Netherlands, whose inhabitants as a community have agreed upon certain laws and general rules, to which they all conform and which they engage to follow. In addition, each separate province, such as Holland, Zeeland, Friesland, etc., has its own rules.[44]

Such a reference to a contemporary political situation is typical of Stevin's style and quite the opposite of Lipsius', who, on the contrary, advocated the humanist style of quoting classical examples:

---

42 *VP*, 511, 513.

43 Lipsius, *Politica. Six Books of Politics or Political Instruction*, 261. The original edition was published in 1590, by the same Leiden publisher who published the *Vita politica*: Franchoys van Ravelenghien; see Mout, 'Justus Lipsius between War and Peace', 149, 156.

44 *VP*, 537.

> To teach you how you can set out on the right course in Civil Life, and pursue it correctly, is my project; and to do this not by means of my own counsel, but by that of the ancients, and even in their very words.[45]

Moreover, the relationship between *spiegeling* (theory) and *daet* (practice) – a hallmark of Stevin's methodological approach – forms a guiding principle in the VP, as it did in his other works.[46] He is known for having insisted on the importance of practice and striven to give it a status equal to that of theory. Among the various accounts that he gave of the link between theory and practice, the clearest one is to be found in the *Wysentyt* ['The Age of the Sages']. But at the beginning of the *Anhangh,* published for the first time in the 1649 edition of the VP, we also encounter a long passage in which Stevin describes the case of the architect (*boumeester*) who, like Vitruvius, needs to possess both theoretical knowledge and the technical know-how of his profession.[47] In Stevin's view, all knowledge, including political knowledge, was a 'combination of theory and practice', since theory (*spiegheling*) was as essential as practice (*daet*), and vice versa. The significance he gave to practice is captured by the formula used in the sixth chapter of the VP, 'Reason confirmed by practice' ('de reden bevesticht met de daet').[48] This largely corresponded to one of the period's widespread preoccupations. The urgent need for the revolted provinces to find an immediate solution to their military and political conflict with Spain was a decisive impulse for the development of practical knowledge of matters such as military fortification, navigation, astronomy and cartography.[49] But one should also think of the discussions about architecture and architects' skills that were launched by Pieter Coecke van Aelst (1502–1550) in Antwerp. A Flemish painter, sculptor and architect, Pieter Coecke was the official painter to Charles V and librarian of Mary of Hungary, and he had travelled in Italy with the emperor. He was close to the learned scholars of Antwerp. This explains why he published, among other things, a Dutch translation of Book IV of Serlio in 1539 (*Generale regelen der architecturen of de vijf manieren …*), and

45 Lipsius, *Politica,* 261.

46 He was scornfully mentioned on this point by Justus Lipsius, who described Stevin in a letter of 1584 as '… a pure mathematician, someone who has no knowledge of any other science, and knows almost nothing about language; besides, he is much more an engineer than a theoretician'. Quoted by Jozef T. Devreese, 'Le nouveau départ de la mécanique après 1800 ans de léthargie', 143.

47 *Principal Works of Simon Stevin,* vol. III, 619; Den Boer and Fleurkens eds., *Simon Stevin. Het burgherlick leven,* 134.

48 VP, 557.

49 See Van Berkel, *Citaten uit het boek der natuur,* 38–48; Van Bunge, 'Before Philosophy', 2–7.

of Book III (1546), the rest being posthumously published by his widow.[50] Also important is the fact Pieter Coecke was the author of the first book on architecture to be published in the Netherlands, a synthesis of some passages taken from Vitruvius' *De Architectura*, entitled *Die inventie der colommen met Haren coronementen ende Maten* (Antwerp, 1539). In this book, Pieter Coecke makes a clear distinction between theory and practice, but he stresses, as Stevin would do later, that the two should not be separated, and that an architect should be both learned in 'timmering' (the framework) and skilled in 'overlegginge' (speculative reflection on his art).[51]

Turning back to the *VP*, this view entailed that civic virtue is a mix of experience and knowledge and, furthermore, it justified Stevin's decision to write on politics:

> As one may ask me if this matter belongs to my profession, I will answer to this point in the following terms [...]: the first question is about civic subjection, the other, about government. Concerning subjection, since I have been living under it from my childhood till now, without ever playing a part in the government, I wrote about a matter I did practice myself, or at least practiced as long as I could. There is no better knowledge of subjection than the one subjection itself can teach, this is what the most experienced rulers themselves declare.[52]

Thus, Stevin's confidence that he was entitled to write on politics seems to have been grounded in his practice of citizenship, and consequently in opposition to both the 'mirrors for princes' literary genre and the humanists' claim of a privileged relationship with the supreme political power. Arguing from a very different point of view, he asserted the ability of any citizen to make a statement on politics.

4 Conclusion

Although Stevin did not express a preference for a particular form of government, the 'indubitable general rule' of 'burgherlijkheid', repeated throughout the text ('Everyone must always consider as his rightful authority those who at present are actually governing the place where he has chosen his dwelling'),

50 See Offerhaus, 'Pierre Coecke', 443–452.

51 Miedema, *Theorie en praktijk*, 165.

52 Den Boer and Fleurkens, eds., *Simon Stevin. Het burgherlick leven*, 130, 132.

can be interpreted as support for the newly independent state of the United Provinces; all the more so as a barely disguised allusion (reflecting the wording of the rebels) to the political conflict between two 'legitimate rulers' and the defense of privileges may be interpreted as support for the Dutch revolt against Philip II:

> But if both contesting parties are headed by legitimate rulers, one should consider the form of the government with the limited powers of each of them, and besides this the matter in dispute, and if one knows this well, one should oppose the one who *against his oath wants to change the good laws, privileges, offices* [...] *or abolish them altogether*, and support those rulers who, according to their oath, protect them.[53]

Thus, although it recommends that a citizen should accept the existing government or choose exile, the *VP* should not be considered a 'conservative' treatise. Two other features point in this direction.

First, Stevin's reasoning is based on a new vision of politics, which emerges from the way in which he approaches the matter of the 'civil life'. In the quotation from the 'Argument' given above, the allusion to the mathematical work with which Stevin was engaged at the time when he was writing the *VP* should not be only interpreted as an excuse for the 'brevity' of this text, but also as a justification for his manner of reasoning in this political discourse. A new vision of politics, morality and religion was taking shape, one that later in the seventeenth century would be transformed by mechanistic natural philosophy.[54] This new approach entailed a new assessment of personal opinion or sentiment. One of the original aspects of the *VP* is the crucial role assigned to personal judgment in the citizen's political decision.[55] Even if Stevin urged people to gain a true understanding of the current political situation, he nevertheless left the final choice to personal feeling. Unexpectedly, Stevin's discourse frequently makes use of such notions as 'personal feelings', 'individual feelings' (*yemants ghevoelen, eyghen ghemoet*), 'innermost feeling' (*innerlijk ghemoet*) and inclination (*gheneghentheyt*).[56] Although religion is the main issue at stake here, the 'good citizen' is therefore he who, above all, agrees with the laws of his country; and in this case, 'inclination' and 'personal opinion' play a decisive role. Political choice should not be a judgement on the 'truth' of

53 *VP*, 507, 509 (emphasis added).
54 Kossmann, *Political thought in the Dutch Republic*, 58–83.
55 See my essay 'L'émergence de l'idée d'individu dans la *Vita politica* de Simon Stevin', 45–60.
56 See my essay 'Vérité et politique dans la *Vita politica*', 201–214.

a political system, but a question of the correct understanding of politics and of harmony between individual feelings and the state.

Second, the *VP* is original because of the democratization of politics implied by the choice of the vernacular. Stevin's interest in linguistics should, of course, be connected to the vernacular movement that started in the early Renaissance culture, with Dante (*Of Eloquence in the Vulgar Tongue*, 1303) and the first translations into French of Aristotle's' *Politics* by Nicolas Oresme, around 1371.[57] Stevin's contribution to the vernacular, which is well known, should be seen in this context, if only to underline the fact that his choice of the Dutch language did not stem from hostility to Latin as the language of the intellectual elite, but from his interest in politics and the linguistic identity of the new state. But the use of the vernacular in political discourse was not yet common practice. Jean Bodin, for example, deemed it necessary to justify his decision to write the *Six Livres de la République* in French:

> J'ay entrepris le discours de la Republique et en langue populaire, tant pour ce que les sources de la langue latine sont presque taries, et qui seicheront du tout si la barbarie causee par les guerres civiles continue, que pour estre mieux entendu de tous Français naturels.[58]

Likewise, the use of the Dutch language in the *VP* can be interpreted as a way to address the 'simple people without civic experience'.[59] In this sense, Stevin's use of Dutch had more to do with a learned vulgarization. In addition, the fact that he addressed the citizen first, instead of the Prince, indicated that politics was ceasing to be solely dependent upon a Prince's 'prudentia' and was becoming a shared concern.

Bibliography

Atkinson, James and Sices, David, eds. and transl., *The Sweetness of Power. Machiavelli's "Discourses" and Guicciardini's "Considerations"*, Illinois: Northern Illinois University Press, 2002.

Berkel, Klaas van, *Citaten uit het boek der natuur. Opstellen over Nederlandse wetenschapsgeschiedenis,* Amsterdam: Bert Bakker, 1998.

57 Briguglia and Ricklin, eds., *Thinking politics in the vernacular,* 11–12; Marmursztejn, 'Nicole Oresme', ibidem, 102–127.

58 *Les six livres de la République*, Préface, 10.

59 *VP*, 491.

Bodin, Jean, *Les six livres de la République*, Paris: Corpus des œuvres de philosophie en langue française, 1986.

Boer, Pim den, and Anneke C.G. Fleurkens eds., *Simon Stevin. Het burgherlick leven. Vita Politica*, Utrecht: Erven Bijleveld, 2001.

Briguglia, Gianluca and Ricklin, Thomas, eds., *Thinking Politics in the Vernacular from the Middle Ages to the Renaissance*, Fribourg: Academic Press Fribourg, 2011.

Bunge, Wiep van, "Before Philosophy. Theory and Practice in the emerging Dutch Republic, 1580–1620", *The European Legacy*, 4, n°5, 1999, 2–7.

V.H. [Johan and Pieter de la Court], *Consideratien van Staat, ofte Politike Weeg-Schaal*, 3rd rev. ed. (Amsterdam, 1662).

Devreese, Jozef T., and Vanden Berghe, Guido, *Wonder en is gheen wonder. De geniale wereld van Simon Stevin. 1548–1620*, Leuven: Davidsfonds, 2003 (English transl.: *'Magic is No Magic'. The Wonderful World of Simon Stevin*, Southampton, Boston: WIT Press, 2008).

Devreese, Jozef T., "Le nouveau départ de la mécanique après 1800 ans de léthargie", in *Simon Stevin 1548–1620. L'émergence de la nouvelle science*, Turnhout: Brepols Publishers, 2004, 143–150.

Dijksterhuis, Eduard Jan, *Simon Stevin*, The Hague: Martinus Nijhoff, 1943.

[Du Faur, Guy, Seigneur de Pibrac], *De Vierlingen van den Heer van Pybrac, uit het Frans in het Nederlands en in het Latijn vertaald door Mr. Dirck Graswinckel*, Ingeleid en verzorgd door Geerten Gossaert, Stichting De Roos: Utrecht, 1955.

Gelderen, Martin van, *The Political Thought of the Dutch Revolt. 1555–1590*, Cambridge : Cambridge University Press, 1992.

Heuvel, Charles van den, *'De Huysbou'. A reconstruction of an unfinished treatise on architecture, town planning and civil engineering by Simon Stevin*, Amsterdam: Edita KNAW 2005.

Israel, Jonathan, *The Dutch Republic. Its Rise, Greatness, and Fall.1477–1806*, Oxford: Clarendon University Press, 1995.

Kossmann, Ernst H., *Political Thought in the Dutch Republic*, Amsterdam : KNAW, 2000.

Lipsius, Justus, *Politica. Six Books of Politics or Political Instruction*, edited with translation and introduction, by Jan Waszink, Assen: Royal Van Gorcum, 2004.

Maher, Daniel P., "Simon Stevin's *Vita Politica*: Pre-provisional Morality?" in *Interpretation. A Journal of Political Philosophy*, 43/2, 2017, 215–232.

Marmursztejn, Elsa, "Nicole Oresme et la vulgarisation de la *Politique* d'Aristote au XIVe siècle", in Gianluca Briguglia, Thomas Ricklin, eds., *Thinking politics in the vernacular from the Middles Ages to Renaissance*, Fribourg: Academic Press Fribourg, 2011, 102–127.

Miedema, Hessel, *Theorie en praktijk. Teksten over schilderkunst in de Gouden Eeuw van de Noordelijke Nederlanden*, Hilversum: Verloren, 2017.

Mout, Nicolette, "Justus Lipsius between War and Peace. His Public Letter on Spanish Foreign Policy and the respective Merits of War, Peace or Truth (1595)", in Judith Pollmann and Andrew Spicer, eds., *Public Opinion and Changing Identities in the Early Modern Netherlands*, Leiden: Brill, 2006, 142–162.

Offerhaus, Johannes, "Pierre Coecke et l'introduction des traités d'architecture aux Pays-Bas (1595)", in *Les Traités d'Architecture de la Renaissance*. Etudes réunies par Jean Guillaume, Paris: Picard, 1988, 443–452.

Romein-Verschoor, Annie, and G.S. Overdiep eds., *Het burgherlick leven. Herdruk naar de eerste uitgave van 1590*, Amsterdam: Wereldbibliotheek, 1939.

Rubinstein, Nicolai, "The History of the Word *politicus* in Early-Modern Europe", in Antony Pagden, ed., *The Languages of Political Theory in Early-Modern Europe*, Cambridge: Cambridge University Press, 1987 (paperback edition 1990), 41–56.

Secretan, Catherine, "L'état de fait comme légitimité : le pragmatisme politique de Simon Stevin (1548–1620) dans sa *Vita politica*", in *Revue des Etudes Germaniques*, 1986/4, 418–434.

Secretan, Catherine, *Les privilèges, berceau de la liberté*, Paris: Vrin, 1990.

Secretan, Catherine, "L'émergence de l'idée d'individu dans la *Vita politica* de Simon Stevin", in *Nouvelles de la République des Lettres*, Naples, Prismi, 1990/II, 45–60.

Secretan, Catherine, and Pim den Boer, eds., *Simon Stevin. De la vie civile 1590*, Lyon: ENS éditions, 2005.

Secretan, Catherine, " Vérité et politique dans la *Vita politica* : la norme individuelle", in Catherine Secretan and Pim den Boer, eds., *Simon Stevin. De la vie civile 1590*, Lyon: ENS éditions, 2005, 201–214.

Serrure, C.P., *Vaderlandsch museum voor Nederduitsche letterkunde, oudheid en geschiedenis*. Deel 2 (1858), Gent: H. Hoste, 1863.

Simon Stevin 1548–1620. L'émergence de la nouvelle science, Turnhout: Brepols Publishers, 2004.

Stevin, Simon, *Vita politica. Het burgherlick leven, Beschreven deur Simon Stevin*. Tot Leyden, By Franchoys van Ravelenghien, 1590.

Stevin, Simon, *The Principal Works of Simon Stevin*. 5 vols., Amsterdam: C.V. Swets & Zeitlinger, 1955–1966.

Verbeek, Theo, "La *Dialectike* de Simon Stevin", in Catherine Secretan and Pim den Boer eds, *Simon Stevin. De la vie civile 1590*, Lyon: ENS éditions, 2005, 181–190.

Vérin, Hélène, "La pensée politique d'un ingénieur: l'exemple de la *Vita politica*", in Catherine Secretan and Pim den Boer, eds., *Simon Stevin. De la vie civile 1590*, Lyon: ENS éditions, 2005, 109–130.

Turchetti, Mario, "Notes sur la *Vita politica*. Contribution à la terminologie politique du XVIe siècle", in Catherine Secretan and Pim den Boer, eds., *Simon Stevin. De la vie civile 1590*, Lyon: ENS éditions, 2005, 229–230.

CHAPTER 10

Simon Stevin's Music Theory Revisited

A Dialogue

H. Floris Cohen and Julia Kursell

This chapter is a dialogue in four parts. In the first, Cohen briefly restates the interpretation of Stevin's treatise on music theory at which he arrived in the 1980s – an interpretation that has remained unchallenged to date.[1] In the second part, Kursell nevertheless challenges this interpretation in various respects. In the third, Cohen gratefully accepts some of her objections, and uses others for further clarification. The outcome for the reader, as we suggest in the fourth part, is a fresh view – in part, at least – of what Stevin really intended in his treatise on music theory.

Part 1, by H. Floris Cohen

A private tutoring course in mathematical science, such as the one that Simon Stevin arranged for Prince Maurits, could hardly fail to address the science of music. Music theory was part and parcel of the *quadrivium*, on a par with geometry, arithmetic and astronomy. To be a polymath and *not* to address music theory in addition to these other three parts of the *quadrivium* – now *that* would require a special explanation. Numerous contemporaneous but also slightly later writers, such as Kepler, Galileo, Beeckman, Descartes and Mersenne, wrote about music theory and contributed now and then novel insights to it. So did Stevin – if only to include the subject in the curriculum to which he subjected his supreme commander. As he announced in 1605, in the first volume of *Wisconstighe ghedachtenissen* (*Hypomnemata mathematica*), he intended the second volume to contain a treatise on music theory – he even provided a brief abstract. Yet no such treatise appeared in that second volume three years later, though not for lack of writing one. He even wrote two versions, and he sent the second to an organist of his acquaintance. Alas, Abraham Verheyen

1 H.F. Cohen, *Quantifying Music. The Science of Music at the First Stage of the Scientific Revolution, 1580–1650* (Dordrecht: Reidel, 1984), 45–74; idem, 'Simon Stevin's Equal Division of the Octave'. In: *Annals of Science* 44, September 1987, 471–488. Every statement made in the account that follows goes back to sources and literature listed and extensively referenced in *Quantifying Music*.

 DOI:10.1163/9789004432918_012

responded with a critique as cogent as it was devastating. We may suppose, but cannot be certain, that Verheyen's stern objections were what caused Stevin to withhold his treatise 'Vande spiegheling der singconst' ['On the Theory of Music'] from the public. We do know that the manuscript was read in subsequent decades by a few knowledgeable commentators, notably Isaac Beeckman and Christiaan Huygens. For a few more, such as René Descartes, Stevin's brief abstract in the *Wisconstighe ghedachtenissen* sufficed to provoke the same verdict that would be reached by Beeckman and Huygens, namely, that Stevin had it profoundly wrong. And then the manuscript disappeared, only to resurface in the 1880s with the preparation of Christiaan Huygens' *Oeuvres Complètes.* The treatise was speedily published in a very competent edition.[2] Much had changed in the meantime in the practice and theory of music, however, so that Stevin's core message now appeared in a radically different light. In a naïvely progressivist reading, Stevin came to be hailed as a man who, with remarkable foresight, had not only produced a refreshingly sober-minded plea for *equal temperament,* but had also calculated this now-standard way of tuning modern pianos with far greater accuracy than anyone had ever managed to do before. It is this caricature that, notably in volume 5 of Stevin's *Principal Works,* has made its way to the English-speaking world. E.J. Dijksterhuis, in his masterly book on Stevin of 1943, did not allow himself to be fooled, yet the posthumous, poorly abbreviated and poorly translated version of his book was not a good vehicle for setting things straight in this regard.

The point is – and Dijksterhuis knew this well – that 'Vande spiegheling der singconst' is not about equal temperament, even though a twentieth-century reader could recognize equal temperament in the figures that Stevin calculated. But the conceptual framework in which he made his calculations is quite different. The treatise is not even about temperament at all; rather, throughout his treatise Stevin insisted that he was computing *the one and only natural tuning system.* And it was precisely this outrageous thesis that none of the contemporaries who read it or heard about it could take seriously.

At the heart of Stevin's treatise is his assertion, supported by many ingenious arguments, that the ancient Greeks had been utterly wrong (and so had everybody else since) in maintaining that consonant musical intervals are given by ratios of the first integral numbers. So, what concept of a consonant interval was Stevin arguing against?

2 Simon Stevin, 'Vande spiegeling der singconst' (this 1884 edition by Bierens de Haan is vastly superior to the 1966 one by Fokker in vol. 5 of Stevin's *Principal Works,* on the grounds set forth in notes 46 and 48 of *Quantifying Music*).

The expression 'consonance' stands for two different musical notes that sound well together. The idea is this: take a string, sound it, then take half the string and sound it again – what you hear is two notes that seem to sound the same, only the pitch of the latter note is higher than that of the former. This particular – very particular – musical interval is known as the octave. In other words: the consonant interval known as the octave is given by string lengths in a ratio of 1:2. In similar fashion, you can produce other intervals that count as consonant, though with not quite so striking an effect of seeming coalescence. The next-best-sounding consonance is produced by using strings whose lengths are as 2:3 – the resulting interval is called the fifth. In ancient Greek musical theorizing, the list of consonant intervals ends with the fourth (3:4). In European music-making from the High Middle Ages onward, and in European music theorizing from the mid-sixteenth century onward, further intervals came to be acknowledged as consonant – the major and minor thirds and sixths, with ratios of 3:5, 4:5, 5:6 and 5:8.

The big problem, in musical practice as well as theory, was (and in a sense still is) that it is mathematically impossible in any tuning system to have all consonant intervals pure at the same time. By way of exception, at least in European tuning systems, the octave, so as to maintain the effect of coalescence, is always tuned to the ratio of 1:2. But *within* the compass of one octave, it is simply impossible to have all of your fifths, fourths, thirds and sixths pure at the same time.[3] How, then, to 'divide the octave'?

About half a century prior to Stevin's treatise, two highly consequential developments had turned this into a burning issue. One was the rise of polyphony; that is, a way of making music that arose in the twelfth century, in which several relatively independent melodies are sung simultaneously by different voices (tenor, alto, etc.). As a consequence, the distinction between consonance and dissonance became a good deal more acute than in the monophonic musical practice of the ancients. The other, much more recent development was the 'emancipation' of the musical instrument, which was being employed more and more often, not only to accompany vocal music, but even for purely instrumental pieces. And whereas both singers and players of many an instrument can adapt to some extent how they produce their notes to what sounds best, players of instruments with fixed pitches, such as the lute and the

3 The shortest answer to the question of why this is so is that we happen to hear logarithmically, and that powers of figures not divisible by each other can never be equal. For a slightly more expansive, yet still concise explanation of this and several other issues concerning the arithmetic of music and the division of the octave, see Prins and Cohen, 'Knowing Heaven and Earth Through Music', 18, and several more passages.

organ and, a little later, other keyboard instruments such as the virginal and the harpsichord, lack such flexibility. That is why the problem of how to divide the octave became most urgent for lutenists and keyboard players. From the mid-sixteenth century onward, dozens and dozens of divisions of the octave were being proposed and fiercely debated, until the pianoforte came into fashion in the early nineteenth century and the debate was settled in favor of equal temperament.

What concessions to tonal purity are you making, then, when you adopt equal temperament? The concessions are huge – indeed, the only consonant interval that remains pure in equal temperament is the octave. All of the others – all fifths, fourths, major and minor thirds and sixths – are deliberately mistuned to a smaller or larger but always noticeable extent. 'Noticeable? What do you mean? Is it not the case that few people nowadays are even aware of the out-of-tune-ness of these intervals?' Yes, that is so, and the reason why is that we have become thoroughly accustomed to equal temperament, if only because hardly any alternatives exist anymore. But you need only hear a musical composition performed in some other temperament where the inevitable impurities are distributed differently, in a manner that spares them more, albeit in fewer keys, to realize at once how crude a solution equal temperament has been and how much tonal purity has been lost in the process.

Only, Stevin would not have agreed. In his quite deliberately dissenting view, apart from the octave, the consonant intervals are not given by the ratios of the first few integral numbers 1 to 6. Instead, so he maintains over the length and breadth of his treatise, they are given by what he calls the 'natural', that is, the equal division of the octave. Thus, the consonant interval that we call the fifth is not given by 2:3, but by the twelfth root of one half raised to the power of seven. And similarly so for all other musical intervals in the octave.

Stevin argued his case in two principal ways. He sought to prove it, and he went on to support his proof by means of three distinct arguments.

Let us consider his proofs first. These are, inevitably, based on a *petitio principii*, that is, they can be seen in effect to presuppose what they are meant to be proving. However, the proofs are so ingenious that when I investigated the matter some forty years ago, I had real trouble pinpointing precisely where in the deep layers of his proof he had (surely unwittingly) smuggled in the theorem that he had set out to prove. Among these layers is his appeal to one specific, empirical trial, of a kind that has led Dijksterhuis to suspect that Stevin was unable to hear a tone difference of almost half a semitone. The implication (to which I return) is that Stevin would have been tone-deaf to an astonishingly high degree.

Meanwhile, Stevin's typical approach and style of reasoning are captured in a delightful manner in his three supporting arguments. I have called one of these the argument from rationality, another the argument from proportionality, and the final one the argument from nature. To set forth the argument from nature would send us deeply into matters a little too arcane perhaps for the musically uninitiated, so I confine myself here to the first and the second argument.

The argument from rationality states that it is a bad mistake (an ancient Greek mistake, to be sure) to call integral numbers 'rational' and thereby to privilege them over all other, allegedly irrational numbers. To link the consonant intervals up with the first integral numbers is therefore nothing but superstitious number-juggling. If you compare, for instance, the number two with the square root of eight, there is simply no good reason to attribute their incommensurability solely to the square root of eight; hence, it makes no sense to call the square root of eight rather than the number two irrational or irregular or any such locution. Stevin had already made the general point in his early work *L'Arithmétique*; in 'Vande spieghelingh der singconst' he refers to this work and then bluntly states that he is not going 'an donuytsprekelicke onredelicheyt ende ongheschicktheyt van sulcken misverstant hier te leeren duytsprekelicheyt redelicheyt gheschicktheijt ende natuerlicke constighe volmaecktheyt deser ghetalen' ('from the inexplicable irrationality and absurdity of such a misconception to teach here the explicability, rationality, commodity and natural ingenious perfection of these numbers').

In the present case, the English translation provides the non-Dutch reader with a satisfactory rendition of Stevin's nicely concise, late sixteenth-century Dutch. But this is not true of Stevin's next argument, the argument from proportionality, which only a person knowledgeable in Dutch can truly grasp. That is so because it brings out Stevin the linguistic purist. His argument from proportionality is a perfect exemplar of his conviction that Dutch lends itself more readily to scientific objectives than any other language.[4] In Dutch more than in any other language, Greek definitely included and Latin even more so, you can with great facility make composite words, which is an important feat, 'overmidts der dinghen namen daer duer oock hare corte bepalinghen sijn' ('since the names of things are also thereby their brief definitions'). 'Evenwijdig' for 'parallel' already indicates by way of the expression itself that the width ('wijdte') is equal ('even'). 'Stelkunde' for algebra already indicates that

4 The remainder of this paragraph and the entire paragraph that follows is identical to a summary description in my 'From *Philosophia Naturalis* to Science'.

this is the discipline that provides expert knowledge ('kunde') of what we 'stellen' ('suppose'). And so on, to the very point where we see Stevin arguing that very long ago, in the 'Wijsentijt' ('Age of the Sages') when all the knowledge was already available that the present age is busily seeking to recover, the one and only language spoken was Dutch.

The question that Stevin raises in 'Vande spiegheling der singconst' is how the Greeks, who were clever enough to hit upon the true, equal division of the octave, came up instead with their childish division according to these ratios of the first few integral numbers, such as 2:3 or 3:4 or 4:5. Why did it require a Dutchman to find out that all consonant intervals flow from the equal division of the octave into twelve semitones? Language is the reason why. The Greeks should have derived the division of the octave from the only proper kind of proportionality, *geometric* proportionality, which unlike so-called arithmetic or harmonic proportionality leads at once to the equal division. But they failed to recognize this. Neither the Greek nor the Latin words express the relationship between the ratios of the terms in question. Greek has λογος for 'ratio', and ἀναλογια for their relationship. Latin, along with all the languages that stem from it, is even worse: it provides no connection at all between *ratio* and *proportio*. But luckily there is one language that can express the connection quite clearly and concisely. This language is Dutch, and the proper word for it has been coined by Stevin himself – 'everedenheyt' (or, in modern Dutch, 'evenredigheid'). This, Stevin says, is a word 'of infinite power', for it is the 'bepaling sijns gronds' ('the definition of its own essence'). Indeed, 'te seggen dat 6, 4, 3 en dierghelycke [,] Singconstighe everedenheyt maecken [,] daer oneindelycke ydelheden uijt volghen ende besloten worden; Men antwoort duer beweghing van tvoornoemde gheluijt, hier van syn gheen even Rhedens, daerom oock gheen Everedenheyt'. Or, translated into a language that is quite unfit to make this particular point, since all terms in question derive straightforwardly from the very Latin Stevin deemed unfit for scientific discourse: 'from the statement that 6, 4, 3 and the like constitute a harmonic proportion ['everedenheijt'], an infinite series of vanities follows and is made to follow. One responds by pronouncing the aforesaid word: here are no even Ratios, therefore, there is no *Everedenheyt* ['proportionality']'.[5]

One interesting aspect of this wonderful consideration is that it provides Stevin with the perfect justification for his deliberate deviation from every single account of utterly basic musical theory ever produced from Pythagoras onward, up to and including his own time; only a Dutchman is in a position

5 On *evenredenheyt,* see also the chapter by Van Dyck in this volume.

to grasp what true geometric proportionality is like and that, therefore, the division of the octave into twelve equal semitones is the only properly proportional division!

Let us now return to Stevin's proofs. Leaving out all of the technicalities, as I am doing throughout this brief account to the largest possible extent, the outcome of the empirical trial that forms part of one of his proofs is entirely dependent on how the two keyboards that he needs for it have been tuned. If in equal temperament, the trial confirms the property in need of proof by definition; if in any other temperament, the trial cannot fail to disprove it. The interval in question is the sum of three whole tones, the augmented fourth ('tritone') F-B. If you pile two such tritones upon one another, the outcome is, in equal temperament, a pure octave, but, if tuned such that its constituent intervals are in what anyone other than Stevin calls pure intonation, then the outcome is an interval that deviates from a pure octave to quite a large extent. The size of the difference is close to half a semitone; connoisseurs know it as the 'lesser diesis', which is characterized by the ratio 125:128, and comprises 41.1 cents. This 'cent' unit is a measure for interval size that was invented in the late nineteenth century. Certainly with respect to the octave an ordinary person's hearing is capable of distinguishing between pure and impure from a difference of about 2 cents onward. Therefore I find it very hard to share Dijksterhuis' belief that Stevin would not have been able to hear the difference, which after all is no less than twenty times as large as is needed for any ordinary person. And therefore I assume that if Stevin carried out the trial the way he summarily described it, the two keyboards that he used had (no doubt inadvertently) been tuned in equal temperament, thus yielding a significant portion of the *petitio principii* that (as I noted earlier) Stevin can be seen to have committed in this proof. Indeed, there are reasons for supposing that Stevin failed to see how having the two keyboards in equal temperament was bound to invalidate his own proof, the main reason being that at various places in his treatise he makes it clear that he does not understand what temperament as such is about in the first place.

So what is going on here? Whether Stevin was tone-deaf or not must remain anyone's guess and is not of much inherent interest anyway, but that he lacked even the tiniest smattering of musical sensitivity is a readily observable fact that jumps from every single page of 'Vande spiegheling der singconst'. Except for the sheer interest evoked by the mathematical problems that quadrivial *musica* carries in its wake, music apparently had no personal meaning for him. Stevin's argument can be enjoyed for its freshness; for his sheer boldness in declaring every single predecessor just plain wrong; for the clear-cut mathematical principles he appealed to; for the attractiveness and (up to a point) persuasiveness of many a supporting argument. But what was missing (and this can so easily happen when one is wholly out of touch with one's subject, particularly if it is about human conduct and accomplishment) was at least a residual sense of the absurd. This precious sense can only be acquired from the kind of personal familiarity with your subject that keeps you in check when you arrive at the borderline that separates refreshing unorthodoxy from sheer irreparable madness. The relentlessly mathematical train of thought that pervades Stevin's treatise is, in the present case, a thoroughly perverted train of thought – perverted by a lack of sensitivity for what, at root, is the human subject it addresses. In the final analysis, I regard 'Vande spiegheling der singconst' as a mirror for all scholars: beware that borderline behind which madness awaits, and make sure that you retain a genuine feeling for what you are doing and why you are doing it.

Part II, by Julia Kursell

Dear Floris,

Thank you for sharing these stimulating thoughts and observations with me, even though I am not a scholar of the early modern period. Following the gist of your last remark, and not being an expert on equal footing with you in the relationship between science and music in that period, I will try to reflect on your remarks from the perspective of a scholar of the history of music. I do so, because I have faith in your support for those who aim to break through the confines of established discourse, from which I and others have benefited so much.

I will touch upon three questions. The first concerns what you call the proof of the augmented fourth. Can there be a mathematical proof that includes applicability to a physical referent and – to bring this question into my own domain – can this referent be music, i.e., a culturally defined phenomenon? I would like to propose that language, and more specifically Stevin's peculiar approach to language, shows why the discrepancy between mathematical proof and the practice of tuning did not hinder him from suggesting such an application.

My second question elaborates on the question of the referent. Even if the music that is generally associated with the problem of tuning – and that is to say, polyphonic music for instruments with keyboards or frets – does not lend itself to an application of Stevin's ideas, this does not necessarily mean that this is true of music in general. There might be music for which Stevin's invoking of a natural quality is appropriate. I will suggest that his elaborations on a natural quality of music do hold for singing, and thus for the part of musical culture in the Low Countries that was most widespread, most accessible and also most highly developed. Your expertise will be all the more helpful here, since you so convincingly argue that Stevin's treatise is not about temperament. The question is, then, what Stevin means by natural *tuning*, rather than just natural *singing*.

Referring to your 1987 article to which you refer in note 1, my third question challenges the claim that Stevin instead participates in the debate about *consonance*. As you argued in that paper, one looks in vain in Stevin's treatise for the arguments that are needed to discuss musical consonance. Instead, you proposed that he sought to find a natural *scale*. While this is in line with the assumption that singing – and more specifically, monophonic singing – formed the basis of his argument, this does not fully explain the remarks that Stevin did make about consonance and dissonance. For this, we will have to consider his elaboration on the interval of the fourth. This also allows us to tackle the question you raise about musical sensitivity. The appendix on the fourth points to a specific type of listening that in today's terms would be called spectral as opposed to tonal listening. While I fully agree that we need not diagnose historical figures with any kind of pathological deficiency, I hope that this distinction will shed some light on the potential that Stevin's treatise has for musical discourse.

1

To begin with, your argument that Stevin only proves what is already contained in his notion of an equidistant scale is convincing. The problem, then, arises from his intermingling of two modes of arguing. On the one hand, he resorts to the demonstration that, when adding two intervals that have been defined according to his own calculus, one obtains another interval that is defined within that same framework. This is a plain syllogism. For a mathematical proof, nothing else is needed. But one should add that if Stevin's claim is true by definition, so is its contestation. When we resort to intervals as they are defined by music theory rather than by Stevin's calculation, we already know from the outset that adding up two of the intervals you indicate will never result in an octave. Two augmented fourths piled upon each other result in a new note: an augmented fourth from C is an F sharp, an augmented fourth from an F sharp results in a B sharp, not in a C. That is the very problem of temperament. Compromises had to be found due to the multitude of note names, and the multitude of intervals that fine ears wish to hear in the music – but which would have resulted in a proliferation of additional keys. Such instruments, as you well know, have actually been built.[6] With a keyboard, even with two keyboards, one can thus neither prove nor disprove Stevin's claim, if one requires the mathematical and phenomenal values to be congruent. When carrying out a division of the monochord that is correct in the terms of Stevin's argument, one will obtain a sound that will not make sense in the terms of such a strict logic of tonality. The problem of tuning can only be mitigated, not solved.

If Stevin understood this, he was not up to comparing a D double flat to a C. Rather, he must have been aware that only his system of steps in the octave produced two notes with the same name when adding three whole tone steps to three whole steps. For this, hearing is not necessary – although this is in fact what Stevin's contemporaries got to hear when they played the most popular instrument of the time, the lute. The alleged predominance of the keyboard instrument is to a certain extent an artifact of the predominance of the piano up to Dijksterhuis' and Fokker's time.

Thus, something else must be at stake in Stevin's argument, and this seems to be an argument about language. As you explain so beautifully, his argument is that Dutch is not better at denoting objects, but at relating terms. In addition, Dutch is not more natural than Greek or Latin because it is a vernacular and thus comes to mind more easily, but because it allows the speaker to relate what needs to be related. This, in turn, enables Stevin to create similar formalisms in language to those he seeks to find in mathematics. For the moment, we can thus say that even if we do not yet have a proper referent for either the language or the mathematical formalism, that is, a referent that is in need of such description, we can at least state that the linguistic and mathematical formalisms are in congruence.

2

Stevin wrote not about music, but about *singconst*, literally translated in the twentieth-century edition as 'the art of singing'. Although his use of this term and its derivatives is congruent to 'music', the fact remains that the term refers to singing in the first instance; and so does the first definition, which concerns *de nateurlicke sanck*. The Dutch word 'sanck' (present-day 'zang'; singing) highlights the most widespread musical practice in his time, and it also demonstrates once more a play with derivative morphological forms. To sing need not always be an art, and what is natural about it is perhaps to be found in the activity rather than in the art of singing. Singing, which is mentioned in many

6 See for instance Prins and Cohen, 'Knowing Heaven and Earth Through Music', 28–29.

instances in the text, undermines the idea that this is a treatise about tuning, since the voice cannot be tuned. Yet, Stevin's treatise does not adhere to the tradition of theorizing vocal music, either, which continued up until the sixteenth century. Natural singing can just as well be understood to refer to some sing-along tune that comes to mind without much thinking and without much musical education. It is not ennobled through application to the heavenly bodies, nor does Stevin discuss any rules of composition. Instead, he picks out the definition of intervals to begin his treatise, and applies this to a mundane understanding of singing.

For many centuries, the very fact that music used intervals and that these could be described as proportions of small integers justified the inclusion of music, or rather *harmonia*, in the *quadrivium* at all. The use of musical instruments began to challenge this basic definition of the musical interval in terms of harmonic proportions only much later, after polyphony was established. Authors such as Gioseffo Zarlino and Vincenzo Galilei became quite specific about what a musician had to do to achieve an equidistant tuning for the frets of a lute. Galilei contested Zarlino's beautiful considerations on how to construct the proper geometrical division of the string, pointing to the fact that such a division does not result in pitches that sound right, because it disregards the reality of the frets, which add tension to the strings. And Zarlino replied that even with the frets being fixed at some particular distance, good musicians would always follow their ear and tweak the pitch so that it would be more harmonious. This is much more concrete than Stevin's account. He neither cared about the sound when he described the division of the *sanglijn*, nor does his elaboration on how to distribute intervals over a keyboard indicate that any actual practice of tuning is involved. I would be curious to hear from you whether Stevin's term 'stellen' for tuning was the regular historical word for the action of tuning. You mention that it was used for algebra, because it resembles the word meaning 'to claim' in contemporary Dutch. Stevin's musical advisor Verheijen duly used it to explain to Stevin why he was wrong.

I would thus claim that Stevin is referring to mundane singing, even in those moments when he mentions more concrete musical and especially instrumental practice. The keyboard, for one, seems to function primarily as a neutral display for the distribution of pitches in his argument. Stevin did not venture to participate in the discussion about how to refer to those notes that are not part of the diatonic system, which he calls 'slit keys', taking the design of contemporary organ keyboards as his model for this designation. When Stevin does use note names beyond the diatonic scale, that is, the notes with names from A to H, he simply continues the alphabet beyond the established letters. Any other choice would have forced him to refer to the note names with the sharps and flats I mentioned earlier, i.e., those dictated by 'pure' intervals.[7]

Concerning the monochord, it is remarkable that Stevin speaks of a *sanglijn*, although he could also have spoken of a 'single string' to refer to this instrument. By his time, the monochord had become a standard item in musical treatises for the visualization of pitches. Nevertheless, it had retained the function from previous treatises of providing support for referring to the elusive quality of the sound. The monochord served to find the correct distances of pitches at a time when there was no other way of referring to them. Stevin's *sanglijn*, while still providing a visualization, refers to a practice that exists without such support.

The most interesting case of a discussion of practice in Stevin's treatise is, in my view, the chapter 'waer in doirsaeck verclaert wort vande on-volmaecktheyt dieder int stellen der orghels ende clavesimbels ghebuert' (179); that is, the chapter that explains the imperfection that occurs when organs and harpsichords are tuned. Stevin not only argued for approximation in music, but he grounded this argument in the inability of the human subject to detect the difference. Singers know intuitively whether they are singing correctly. But as research in experimental psychology, carried out 300 years after Stevin, has shown, what is intuited by a singer as correct intonation can deviate from the (mathematically) correct value by an interval of a third. This is about seven times as much as the two notes that Stevin was unable to distinguish, according to Dijksterhuis. Similarly, a listener can judge a singer to be performing correctly when such deviations are measured. The phenomenon occurs in singers and listeners with decent musical ability and, to a lesser extent, even in individuals with absolute pitch (Abraham 1921).[8] Felt and measured musicality in singing can thus indeed produce deviations that astonished Stevin's modern readers.

3

Consonance is the topic that made his twentieth-century editors wonder whether Stevin ever noticed the phenomena of distortion that can be observed in intervals that deviate from harmonious proportions. You argue that Stevin's

7 One may add to this that in Stevin's time, letters beyond *a* to *h* were also used for lute notation, where they indicate finger positions that have no relation to the note names we associate today with the diatonic scale. On this, see the classic description in W. Apel, *The Notation of Polyphonic Music*.

8 Cf. Abraham, 'Tonometrische Untersuchungen'; on his research into musical intonation, cf. J. Kursell, 'Experimental Cylinders'.

treatise must be seen against the background of the discussion of consonance, rather than of tuning systems. Indeed, musical intervals are defined in terms of consonance and dissonance. A consonance would thus be an interval that is appropriate for music and a dissonance an interval that must be dismissed. This was introduced in Greek music theory and it continued until the point of transition that you highlight. Up to the sixteenth century, music was held to be the art of proportions, consonance in music being one example of this. During the Middle Ages, consonance also became a phenomenon in its own right: the relationship between two simultaneously sounding voices. The treatises accordingly discuss which of these tone relations are appropriate for music and how they should be handled in singing and later in composition. Zarlino famously adjusts this discussion to actual practice, acknowledging the prevalence of triads in the music of his time. And Johannes Lippius, in his *Synopsis musicae novae* of 1612, introduced the term triad, admitting that for the description of such an entity it would be more appropriate to consider the note a 'monad' that enters into a relation of three, rather than to continue speaking of proportions, which require three relations of two elements each for describing just one triad.[9] At this point the concepts of tone and note merge, marking the end of the age of proportions.

The physical explanation of consonance emerged in parallel, as you have explained on many occasions. The distinction between consonance and dissonance in musical practice was by then well established. Physicists could refer to it as a phenomenon: what is it that makes listeners feel that some sounds merge into harmony whereas others do not? Stevin's mathematical model of an equidistant tonal system does not add anything to this. In monophonic singing, the problem is irrelevant. Yet, Stevin does comment on how voices and instruments actually sound together in an appendix on the interval of the fourth.

There, Stevin comments upon a rule in counterpoint that is a corollary of the music that is organized in triads. If the triad is the basis of harmony and in that same capacity a moment of rest, then the role of the fourth changes. Music theory understands a triad to be a combination of tones that piles two thirds upon each other. In modern terms, one could also say that the triad combines a choice of pitches that fit into a Fourier Series, with the lowest pitch being identical to the fundamental of the series. A concept of harmonic rest can thus be said to occur when the lowest pitch is identical to the fundamental.

9 Cf. Lippius, *Synopsis Musicae Novae*; on Lippius cf. Rivera, *German Music Theory in the Early 17th Century*.

This is not the case when the interval between the lowest pitch and the next pitch is a fourth. Then the third partial, or in other words the fifth to the fundamental, is the lowest pitch. Harmonic theory associates this with tension and thus dissonance.

Even though these relations between rest and tension were only made explicit in late nineteenth-century music theory, they nevertheless help us here to explain what Stevin encountered when discussing the problem of the fourth. In short: music theory declared the fourth to be a dissonance when occurring between the lowest voice and the voice above it. I quote from the English translation: 'The fourth is considered a discord by present-day composers, so that in singing with three or more voices it must not be heard against the lowest part; nay, below two voices it is not suffered at all' (483). What Stevin seems to ignore is that this dissonance is a regular feature in a musical setting. That an interval is dissonant does not mean that one may not use it, but that a combination of tones, which is defined as consonant, must then follow. Stevin nevertheless endeavors to prove that the fourth in this position should not be considered a dissonance at all.

At this point, his argument can be called a proof in the sense you invoke. Stevin himself refers to it as 'daetlich bewijs' (175), which is translated as 'empirical proof'. He now does refer to actual sound, and he even introduces controlled variation of the same set-up to make his point. Starting from the observation that singing in different registers may appear to the singers themselves as unison, he argues that it may not always be clear which voice is the lowest and thus with which the fourth must be seen as discordant. Stevin now invites his reader to think of three voices, two of which sing in an octave, whereas the third joins in a fourth to the lower. Since the first two may consider their sound to be in unison, not noticing the difference by an octave, it becomes unclear to which of them the third is in the relation of a fourth. They may not even hear it as such, but as a fifth. He concedes that this may be the effect of combining human voices and suggests repeating the experiment with a flute as a third element. He writes: 'om deser dynghen oirsake wat breeder te verclaren soo is te weten dat als sulcke twee gheluyden tsamen een eerste ofte dobbel eerste maken dalderscherpste ghehoiren en connen niet sekerlick oirdeelen welck van tween het is' (176). In other words, even the most educated musical ear may be unable to distinguish pitches that join in an octave, so that, consequenly, it becomes impossible to say what the interval between the third tone and this joint sound is.

Stevin's hearing, again, is accurate; although music theorists will argue that the function of the fourth, and that is to say, its embedding in a sequence of sounds, is what matters here. That the fourth will be heard as dissonant means that it will be followed by a consonance. Even if one does not detect any unpleasant effect from this discordant sound, it will reveal itself through the next sound as having been in need of resolution into a consonance. Yet, this is not what Stevin's argument is about. He only states in his peculiar terminology that the octave, for which he uses the term 'double first', is not distinguished from unison. The octave is a duplication of a tone that results from half its string length or, in more modern terms, twice its frequency. There are situations in which such a duplication occurs throughout a musically correct musical setting. Organ stops came to make use of such duplication on a systematic basis, and listeners failed to note any contradictions to the rules of musical setting.[10]

This brings me to your last remark – that Stevin lacked any smattering of musical sensitivity, and that his deliberations about music are thus refreshing for having introduced a surprising topic into his mathematical thinking, but have no value for music at all. I think it is fair to say that he did observe music carefully. He does not prioritize the music of the church or court, but I do not see that this deserves our reproach, since it is in line with the values of Dutch society as it emerged in the wake of Stevin's time. He even seems to have had a fine ear. However, he made a curious mistake: he did not obey the rules of musical discourse. The comparison to Descartes may be fruitful in this respect. Descartes, in my opinion, seems to explain that discourse to himself with a sharp mind and little hint of knowing what this discourse is all about. Stevin, by contrast, trusts his hearing, but he underestimates the power of a discourse to shape our hearing. At times when there is no other reference to sound than language and practice, hearing depends on learning what one is supposed to hear, to an even greater extent than this is the case now that exposure to recorded sound has become the rule. Stevin rigorously denies the discourse, introducing instead his own Dutch terms. By the same token, he introduces a new system of reference. This system includes four components: the musical practice of his surroundings, his brilliant ability to create mathematical formalisms, an awareness of these formalisms being never exactly in agreement with empirical reality, and, last but not least, his sharp ear.

Perhaps what kept Stevin from publishing his treatise was something other than being ashamed of his ignorance; he may have understood that the discourse on music in his own time forbade it.

With my warmest wishes and all my gratitude,

Julia

10 One might add, though, that Stevin's example is not very telling, as it does not concern triads, in which case the fourth between the two lowest voices only becomes relevant.

PS: After having re-read Stevin, I noticed that his 'proof' contains another rather clever element. Comparing the effect of piling two tritones upon each other, he adds dissonance to dissonance, each of which is composed of three steps that cannot count as consonances, either. Which ratio should each tone have? 9:8? 10:9? There is no way to define these steps by their relative 'purity' or pure sound, as they are dissonances in any case. The best way to sing several steps with the same name correctly seems to be to sing each of them in the same way. Perhaps he was interested not so much in a division of the octave in the first place, but in a system that could explain uniform steps that made sense when singing a scale – and still did not contradict the octave. In this regard, he was perhaps merely trying to bring some formalism to this obvious empirical problem, as he did in other fields. He just picked the wrong problem to solve, a problem no one noticed but him, with his self-taught ear. Music theorists, by contrast, always took the purity of consonances as the gauge of their tonal systems.

Part III, by H. Floris Cohen

Dear Julia,

Your very welcome comments have given me occasion to clarify some points and to change my views on certain others.

Clarification first. My point about Stevin's *petitio principii* is that (as may happen to every mathematician intending to prove some theorem or other) at some hidden layer in each of his proofs he has inadvertently used what looks like just a link in the chain of the proof as if it were a given; in other words, that the theorem that he aims to prove turns up in the proof itself. You are surely right that music, even such math-infused aspects of music as the theory of consonance and the division of the octave, does not lend itself to mathematical proof of any kind. But that means only that you and I agree that Stevin was trying to do something that cannot be done. His *petitio principii* is just one symptom of the ineluctable fact that anyone who wanted to prove a certain theory of consonance would be bound to run into trouble of this sort. With respect to this, I will now add something that I left out of my original summing-up: Stevin is not even consistent in his efforts to prove his theory of the consonance of the notes of the equally divided octave, for in one of his definitions of 'natuerlicke sangh' ('natural singing' or 'natural music'), he states that these notes are consonant *by nature*. But if a certain quality is such and such by nature, how could it make sense to seek mathematical *proof* for it?

What I like so much about Stevin's efforts to come to grips with a domain where he believes every predecessor to have been utterly mistaken (they didn't speak Dutch so they couldn't really help it) is the deadly rigor with which he keeps pursuing his reasoning. Take his argument about the so-called 'wolf fifth'. This interval arises from the incompatibility of pure octaves with pure fifths (an incompatibility of the kind that I briefly addressed in the first section). To rise from C by seven *octaves* gets you, obviously, to the C seven octaves higher. But to rise from C by twelve *fifths* gets you, *not* to that same C, but to B sharp.[11] Consequently, the interval formed between that B sharp and the F underneath is not a pure fifth. It howls, which is why it has earned the name of 'wolf fifth'. It may actually be situated at any point, wherever you prefer to leave the circle of fifths open, but so much is certain that if you make music both with pure *octaves* (no exceptions here) and with pure *fifths*, you inevitably get into trouble somewhere. Obviously, it is easier for singers to circumvent the problem in practice than, in particular, for keyboard players with their pre-fixed key notes. The main solution that began to be explored in Stevin's time was some form of temperament. It was not until the decades around 1800, which saw the invention of the *pianoforte*, that this was taken one great step further, thus giving rise to our familiar *equal* temperament – a theoretically possible solution that an eighteenth-century organist by the name of Quirijn van Blankenburg found utterly ludicrous, arguing that if you distribute the inevitable falseness equally you find that, having got rid of the wolf, you have now bred a whole nest of wolf cubs.[12] Some hundred and twenty years earlier, Stevin had not been sensitive to arguments of this kind at all, in that for him no wolf fifth existed in the first place. And indeed, in his own perverse system of *naturally* equal semitones, no wolf fifth can occur by definition. With that deadly consistency of his, he rather attributes all notions of the existence of a wolf fifth to a fundamental misunderstanding, namely, to the deeply ingrained notion that the pure consonances are *not* given by the equal division of the octave but by ratios of the first few integral numbers. So, yes, in Stevin's view there is no wolf fifth, and if you believe you hear one it is due to nothing but your own inability to infer from the gap between theoretical preconception and empirical reality that your preconception is just wrong. 'Your honor: case closed'.

11 See, and hear, https://www.youtube.com/watch?annotation_id=annotation_2604736211&feature =iv&src_vid=Au2QFQ2ppyg&v=1DUZsQ2by2s.

12 Van Blankenburg, *Elementa musica*, 114–5: '... maar helaas: Elke wolf heeft op zyn vertrek drie jongen gekregen die hij verstrooit heeft over 't gansche clavier; zo dat al wat men speelt, is een concert van jonge wolven'.

Case closed so much the more because, luckily, there was a witness for the prosecution with impeccable credentials – the Nijmegen organist Abraham Verheyen. His letter to Stevin, faithfully transcribed and printed by Bierens de Haan together with Stevin's two versions of his treatise (the original manuscripts rest in one and the same manuscript box at the Royal Library in The Hague) states in its own terms what is so perversely (and therefore so informatively) wrong with Stevin's argument.

Up to this point, then, I stick to my picture of Stevin as a relentlessly consistent mathematician who, in seeking to drive home a point that by its very nature does not lend itself to mathematical treatment, inevitably, and for most intriguing and revealing reasons, overshoots his mark. So far, so good. But also, no farther than that. You have convinced me that there is a broader way of viewing what Stevin was doing. Quite rightly you remind me that the title 'Vande spiegheling der singconst' may be translated not only as 'On musical theory', but also, and in the first place, as 'On the theory of singing'. Even more importantly, you point out that to consider Stevin's two versions of his treatise from that angle reveals some aspects to which I have been, well, perhaps not deaf but definitely blind.

First of all, I shall consider what you wrote about the problem of the fourth. I pointed out in my earlier work on Stevin that, in the short appendix that he dedicates to the subject, he is explaining all over again that everybody else has been mistaken. But you have made me realize that the grounds for his disagreement really betray a remarkable sensitivity to the kind of simple folk-music he was exposed to on an everyday basis. And of course your point about singers' intuition about their own singing possibly being amiss by no less than a full third is well taken, too. All in all, then, Stevin's concern with music was not quite so exclusively confined to an account of consonance and its application on keyboard instruments, but extended to lute practice and to singing as well.

More importantly still, you keenly observe that there is a quite remarkable novelty in Stevin's treatise – a novelty defined not by a presence, but by an absence; namely, the absence of any reference at all to the music of the spheres. That is definitely not a minor issue; to the contrary, as you rightly point out, just about every learned treatise on music prior to Stevin's took its point of departure in the threefold distinction first made by Boethius in or around the year 510: *musica coelestis* (the harmony, represented by the consonances, between the celestial bodies), *musica humana* (how that harmony is reflected in our own bodies and souls), and *musica mundana* (music actually sung, played, and listened to, all of which thus serves as a representation, an imperfect one of course, of that heavenly harmony). It all goes back, ultimately, to Pythagoras, and even though Stevin is quite explicit in his rejection of a Pythagoras-like approach to numbers, I had not fully realized the consequences until now. Almost contemporaneously with Stevin's career, fierce battles were being fought out in Italy regarding the true nature of consonance, and it was precisely the literally cosmic implications of this that lent the controversy its fierceness. One century later, listening to music had turned forever from a cosmic or at least cosmically-tinged event into what was by and large a purely aesthetic experience. This 'untuning of the sky' was a most incisive process,[13] and I can now see, as I failed to do earlier, that Stevin is, at the very least, a remarkably early representative of that secular shift in sensibility.

I return, finally, to the greater or lesser acuteness of Stevin's hearing and, more generally and more interestingly so, to his musical sensitivity as such. True, I have never gone along with Dijksterhuis' supposition that Stevin must have been tone-deaf to the point of being unable to hear an interval difference of over 40 cents. Still, my mental image of Stevin used to be that of a person to whom music meant little else than one more terrain (belonging as it did to the *quadrivium*) that appeared ready-made for the application of his very considerable mathematical skills. By means of subtle yet effective observations, such as 'Stevin's hearing, again, is accurate', you have managed to shatter that image, and to replace it in my mind with a partially new one. It runs thus. Yes, Stevin did care about music and was to some extent sensitive to it. Yes, he did listen to musical performances (everyday Dutch song in particular). In short, in writing 'Vande spiegheling der singconst', he did know, to some extent, what he was talking about. When the part-musical, part-mathematical problem of consonance came up, however, the mathematician in him ran away with it and arrived at an almost solipsistically wrong-headed yet delightfully rigorous solution, in such a manner that he unwittingly turned whatever aural checks he undertook into hiding places for his own relentlessly mathematical train of thought.

With my grateful thanks for your most helpful comments,
Floris

Part IV, by H. Floris Cohen and Julia Kursell

What does this dialogue add to rethinking Simon Stevin? We propose that a dialogue is a privileged form for doing exactly that: rethinking this difficult

13 The allusion is to the title of John Hollander's book *The Untuning of the Sky*. A brief summary of the entire process can be found in Prins and Cohen, 'Knowing Heaven and Earth Through Music'.

author with respect to musical matters. Yet, a problem with Stevin's work on music is that taking it seriously also means reacting to the way in which it fell outside of the musical discourse of his time. Even though other contemporary writers on music and science were unable to escape idiosyncrasy, to some extent, given the enormous changes in music theory around 1600, Stevin stands out in this respect. The most daring writers – from the Neoplatonic Marsilio Ficino and the famous maestro di cappella of San Marco, Gioseffo Zarlino, to Marin Mersenne and Athanasius Kircher – all came up with radically different ideas, terms and explanations of how music's elements formed a coherent system.[14] Yet none of them renounced the basics that Stevin refused to accept. Since then, most authors have chosen to explain this as driven by mathematical inventiveness, rather than musical listening. We thus propose that Stevin should be considered as a listener in the context of the music of his time. After all, is it really so convincing to project the musical reality of the nineteenth century onto an author from 1600, as so many writers on Stevin did? Of course, following the outstanding mathematics he displayed to calculate this alleged purpose can engage one to the extent that one forgets that it is perhaps not. At least this might be closer to a music theorist's than a historian's standpoint; that is to say, Stevin has always been appealing to writers who wished to disclose his potential for saying something on music as a phenomenon whose inner coherence is more important than its audibility. While music theory has taken the license of treating music in this way ever since Stevin's time, this did not satisfy us when integrating Stevin's interest in music into the more general context of rethinking his interest in solving practical problems through mathematics more generally. We concluded that in order to draw attention to the gap between an accepted discourse and a position that falls beyond its scope, we also needed a new form for our own discourse.

A dialogue not only offers two positions rather than one, it also escapes the traditional role of the author as the strong individual who independently draws conclusions from the state of the art. Luckily for us, with some notable exceptions, research on Stevin has been rather silent since Cohen's first writings on him.[15] Given the multitude of positions, the very rich field of early modern music and science has shed light on many of the peculiarities of individual authors, including the question of whether certain shared concerns

14 See, e.g., Gouk on Fludd, Pesic on Kepler, Buehler on Leibniz, Prins on Ficino.

15 Philippe Vendrix (ed.), *Music and Mathematics*, is particularly rich in this respect. Among several interesting contributions, the fine paper by Brigitte Van Wymeersch stands out. Although she focuses on Descartes, the concept of the 'nombre sourd' is also interesting in comparison to Stevin. Rudolf Rasch's paper again explores the hypothesis, mentioned above, that Stevin proposes equal temperament.

drove their idiosyncrasies. It is this lack of such shared concerns among Stevin and his peers that we propose to the present reader should be rethought; and we do so in harmony.

Bibliography

Abraham, Otto, "Tonometrische Untersuchungen an einem deutschen Volkslied," *Psychologische Forschung: Zeitschrift für Psychologie und ihre Grenzwissenschaften*, 4 (1923) 1–22.

Apel, Willi, *The Notation of Polyphonic Music, 900–1600*, Cambridge, MA: The Medieval Academy of America, 1949.

Blankenburg, Quirinus van, *Elementa musica*, Den Haag: Laurens Berkoske, 1737.

Cohen, H. Floris, *Quantifying Music. The Science of Music at the First Stage of the Scientific Revolution, 1580–1650*, Dordrecht: Reidel, 1984.

Cohen, H. Floris, "Simon Stevin's Equal Division of the Octave," *Annals of Science*, 44 (1987) 471–488.

Cohen, H. Floris, "From *Philosophia Naturalis* to Science, From Latin to the Vernacular," in: J. Bloemendal ed., *Bilingual Europe. Latin and Vernacular Cultures, Examples of Bilingualism and Multilingualism c. 1300–1800*, Leiden: Brill, 2015, 144–160.

Dijksterhuis, Eduard Jan, *Simon Stevin*, The Hague: Martinus Nijhoff, 1943.

Hollander, John, *The Untuning of the Sky*, Princeton, PA: Princeton University Press/ Oxford: Oxford University Press, 1961.

Kursell, Julia, "Experimental Cylinders: Experiments in Music Psychology around 1900," *Journal of Sonic Studies*, 13 (2017), https://www.researchcatalogue.net/view/324247/324248 (April 20, 2018).

Lippius, Johannes, *Synopsis Musicae Novae Omnino Verae atque Methodicae Universae. In Omnis Sophiae Praegustum Parergos Inventae Disputatae & Propositae Omnibus Philomusis*, Strasbourg: Paul Ledertz, 1612.

Prins, Jacomien & H. Floris Cohen, "Knowing Heaven And Earth Through Music" (2018), https://www.ris.uu.nl/ws/files/58621475/Jacomien_Prins_and_H._Floris_Cohen_Knowing_Heaven_and_Earth_Through_Music.pdf.

Rivera, Benito, *German Music Theory in the Early 17th Century. The Treatises of Johannes Lippius*, Ann Arbor, MI: UMI Research Press, 1974.

Stevin, Simon, "Vande spiegeling der singconst," in: Simon Stevin, *'Vande spiegeling der singconst' et 'Vande molens'*, Deux traités inédits, ed. David Bierens de Haan, Amsterdam, 1884.

Index

www.ingramcontent.com/pod-product-compliance
Lightning Source LLC
Chambersburg PA
CBHW071324230926
51849CB00002B/308
* 9 7 8 9 0 0 4 4 3 2 9 0 1 *